Lecture Notes in Mathematics 1939

FONDAZIONE
CIME
ROBERTO CONTI
CENTRO INTERNAZIONALE MATEMATICO ESTIVO
INTERNATIONAL MATHEMATICAL SUMMER CENTER

C.I.M.E. means Centro Internazionale Matematico Estivo, that is, International Mathematical Summer Center. Conceived in the early fifties, it was born in 1954 and made welcome by the world mathematical community where it remains in good health and spirit. Many mathematicians from all over the world have been involved in a way or another in C.I.M.E.'s activities during the past years.

So they already know what the C.I.M.E. is all about. For the benefit of future potential users and co-operators the main purposes and the functioning of the Centre may be summarized as follows: every year, during the summer, Sessions (three or four as a rule) on different themes from pure and applied mathematics are offered by application to mathematicians from all countries. Each session is generally based on three or four main courses (24–30 hours over a period of 6-8 working days) held from specialists of international renown, plus a certain number of seminars.

A C.I.M.E. Session, therefore, is neither a Symposium, nor just a School, but maybe a blend of both. The aim is that of bringing to the attention of younger researchers the origins, later developments, and perspectives of some branch of live mathematics.

The topics of the courses are generally of international resonance and the participation of the courses cover the expertise of different countries and continents. Such combination, gave an excellent opportunity to young participants to be acquainted with the most advance research in the topics of the courses and the possibility of an interchange with the world famous specialists. The full immersion atmosphere of the courses and the daily exchange among participants are a first building brick in the edifice of international collaboration in mathematical research.

C.I.M.E. Director
Pietro ZECCA
Dipartimento di Energetica "S. Stecco"
Università di Firenze
Via S. Marta, 3
50139 Florence
Italy
e-mail: zecca@unifi.it

C.I.M.E. Secretary
Elvira MASCOLO
Dipartimento di Matematica
Università di Firenze
viale G.B. Morgagni 67/A
50134 Florence
Italy
e-mail: mascolo@math.unifi.it

For more information see CIME's homepage: http://www.cime.unifi.it

CIME's activity is supported by:

– Istituto Nationale di Alta Mathematica "F. Severi"
– Ministero dell'Istruzione, dell'Università e delle Ricerca

Daniele Boffi · Franco Brezzi
Leszek F. Demkowicz · Ricardo G. Durán
Richard S. Falk · Michel Fortin

Mixed Finite Elements, Compatibility Conditions, and Applications

Lectures given at the
C.I.M.E. Summer School
held in Cetraro, Italy
June 26–July 1, 2006

Editors:
Daniele Boffi
Lucia Gastaldi

Springer

Fondazione
CIME
Roberto Conti

Daniele Boffi
Dipartimento di Matematica
Università degli studi di Pavia
Via Ferrata 1, 27100 Pavia, Italy
daniele.boffi@unipv.it
http://www-dimat.unipv.it/boffi

Franco Brezzi
Istituto Universitario di Studi Superiori (IUSS)
and Istituto di Matematica Applicata e
Tecnologie Informatiche del C.N.R.
Via Ferrata 3, 27100 Pavia, Italy
brezzi@imati.cnr.it
http://www.imati.cnr.it/~brezzi

Leszek F. Demkowicz
Institute for Computational Engineering
and Sciences
The University of Texas at Austin
ACES 6.332, 105
Austin, TX 78712, USA
leszek@ices.utexas.edu
http://users.ices.utexas.edu/~leszek

Ricardo G. Durán
Departamento de Matemática
Facultad de Ciencias Exactas y Naturales
Universidad de Buenos Aires
Ciudad Universitaria. Pabellón I
1428 Buenos Aires, Argentina
rduran@dm.uba.ar
http://mate.dm.uba.ar/~rduran

Richard S. Falk
Department of Mathematics - Hill Center
Rutgers, The State University of New Jersey
110 Frelinghuysen Rd.
Piscataway, NJ 08854-8019, USA
falk@math.rutgers.edu
http://www.math.rutgers.edu/~falk

Michel Fortin
Département de mathématiques
et de statistique
Pavillon Alexandre-Vachon
Université Laval
1045, avenue de la Médecine
Québec (Québec)
G1V 0A6, Canada
mfortin@giref.ulaval.ca
http://www.mat.ulaval.ca

Lucia Gastaldi
Dipartimento di Matematica
Università degli Studi di Brescia
Via Valotti 9
25133 Brescia, Italy
gastaldi@ing.unibs.it
http://dm.ing.unibs.it/gastaldi

ISBN: 978-3-540-78314-5 e-ISBN: 978-3-540-78319-0
DOI: 10.1007/978-3-540-78319-0

Lecture Notes in Mathematics ISSN print edition: 0075-8434
ISSN electronic edition: 1617-9692

Library of Congress Control Number: 2008921921

Mathematics Subject Classification (2000): 65-02, 65N30, 65N12, 35M10, 74S05, 76M10, 78M10, 58A10, 58A12

Cover design: WMXDesign GmbH

Printed on acid-free paper

9 8 7 6 5 4 3 2 1

springer.com

Preface

This volume is a collection of the notes of the C.I.M.E. course "Mixed finite elements, compatibility conditions, and applications" held in Cetraro (CS), Italy, from June 26 to July 1, 2006.

Since the early 1970s, mixed finite elements have been the object of wide and deep study by the mathematical and engineering communities. The fundamental role of mixed methods for many application fields has been recognized worldwide and their use has been introduced in several commercial codes. An important feature of mixed finite elements is the interplay between theory and application: on the one hand, many schemes used for real life simulations have been cast in a rigorous framework, and on the other, the theoretical analysis makes it possible to design new schemes or to improve existing ones, based on their mathematical properties. Indeed, due to the compatibility conditions required by the discretization spaces to provide stable schemes, simple minded approximations generally do not work and the design of suitable stabilizations gives rise to challenging mathematical problems.

The course had two main goals. The first one was to review the rigorous setting of mixed finite elements and to revisit it after more than 30 years of practice; this resulted in developing a detailed a priori and a posteriori analysis. The second one was to show some examples of possible applications of the method.

We are confident this book will serve as a basic reference for people exploring the field of mixed finite elements. This "Lecture Notes" cover the theory of mixed finite elements and applications to Stokes problem, elasticity, and electromagnetism.

Ricardo G. Durán had the responsibility of reviewing the general theory. He started with the description of the mixed approximation of second-order elliptic problems (a priori and a posteriori estimates) and then extended the theory to general mixed problems, thus leading to the famous inf–sup conditions.

The second course on Stokes problem has been given by Daniele Boffi, Franco Brezzi, and Michel Fortin. From the basic application of the inf–sup theory to the linear Stokes system, stable Stokes finite elements have been analyzed, and general stabilization techniques have been described. Finally, some results on visco-elasticity have been presented.

Richard S. Falk has dealt with the mixed finite element approximation of the elasticity problem and, more particularly, of the Reissner–Mindlin plate problem. The corresponding notes are split into two parts: in the first one, recent results linking the de Rham complex to finite element schemes have been reviewed; in the second, classical Reissner–Mindlin plate elements have been presented, together with some discussion on quadrilateral meshes.

Leszek Demkowicz has given a general introduction to the exact sequence (de Rham complex) topic, which turns out to be a fundamental tool for the construction and analysis of mixed finite elements, and for the approximation of problems arising from electromagnetism. The results presented here use special characterization of traces for vector-valued functions in Sobolev spaces.

We thank all the lecturers and, in particular, Franco Brezzi, who laid the foundation for the analysis of mixed finite elements, for his active participation in this C.I.M.E. course.

Daniele Boffi, Pavia
Lucia Gastaldi, Brescia

Contents

Mixed Finite Element Methods

Ricardo G. Durán

Departamento de Matemática, Facultad de Ciencias Exactas y Naturales, Universidad de Buenos Aires, Ciudad Universitaria Pabellón I, 1428 Buenos Aires, Argentina
rduran@dm.uba.ar

1 Introduction

Finite element methods in which two spaces are used to approximate two different variables receive the general denomination of mixed methods. In some cases, the second variable is introduced in the formulation of the problem because of its physical interest and it is usually related with some derivatives of the original variable. This is the case, for example, in the elasticity equations, where the stress can be introduced to be approximated at the same time as the displacement. In other cases there are two natural independent variables and so, the mixed formulation is the natural one. This is the case of the Stokes equations, where the two variables are the velocity and the pressure.

The mathematical analysis and applications of mixed finite element methods have been widely developed since the seventies. A general analysis for this kind of methods was first developed by Brezzi [13]. We also have to mention the papers by Babuška [9] and by Crouzeix and Raviart [22] which, although for particular problems, introduced some of the fundamental ideas for the analysis of mixed methods. We also refer the reader to [32, 31], where general results were obtained, and to the books [17, 45, 37].

The rest of this work is organized as follows: in Sect. 2 we review some basic tools for the analysis of finite element methods. Section 3 deals with the mixed formulation of second order elliptic problems and their finite element approximation. We introduce the Raviart–Thomas spaces [44, 49, 41] and their generalization to higher dimensions, prove some of their basic properties, and construct the Raviart–Thomas interpolation operator which is a basic tool for the analysis of mixed methods. Then, we prove optimal order error estimates and a superconvergence result for the scalar variable. We follow the ideas developed in several papers (see for example [24, 16]). Although for simplicity we consider the Raviart–Thomas spaces, the error analysis depends only on some basic properties of the spaces and the interpolation operator, and therefore, analogous results hold for approximations obtained with other finite element spaces. We end the section recalling other known families of spaces and giving some references. In Sect. 4 we introduce an a posteriori error estimator and prove its equivalence with an appropriate norm of the error up to higher order terms. For simplicity, we present the a posteriori error analysis only in the 2-d case. Finally, in Sect. 5, we introduce the general abstract setting for mixed formulations and prove general existence and approximation results.

2 Preliminary Results

In this section we recall some basic results for the analysis of finite element approximations.

We will use the standard notation for Sobolev spaces and their norms, namely, given a domain $\Omega \subset \mathbb{R}^n$ and any positive integer k

$$H^k(\Omega) = \{\phi \in L^2(\Omega) : D^\alpha \phi \in L^2(\Omega) \ \forall \ |\alpha| \le k\},$$

where

$$\alpha = (\alpha_1, \cdots, \alpha_n), \quad |\alpha| = \alpha_1 + \cdots + \alpha_n \quad \text{and} \quad D^\alpha \phi = \frac{\partial^{|\alpha|}\phi}{\partial x_1^{\alpha_1} \cdots \partial x_n^{\alpha_n}}$$

and the derivatives are taken in the distributional or weak sense.

$H^k(\Omega)$ is a Hilbert space with the norm given by

$$\|\phi\|_{H^k(\Omega)}^2 = \sum_{|\alpha| \le k} \|D^\alpha \phi\|_{L^2(\Omega)}^2.$$

Given $\phi \in H^k(\Omega)$ and $j \in \mathbb{N}$ such $1 \le j \le k$ we define $\nabla^j \phi$ by

$$|\nabla^j \phi|^2 = \sum_{|\alpha| = j} |D^\alpha \phi|^2.$$

Analogous notations will be used for vector fields, i.e., if $\mathbf{v} = (v_1, \cdots, v_n)$ then $D^\alpha \mathbf{v} = (D^\alpha v_1, \cdots, D^\alpha v_n)$ and

$$\|\mathbf{v}\|_{H^k(\Omega)}^2 = \sum_{i=1}^n \|v_i\|_{H^k(\Omega)}^2 \quad \text{and} \quad |\nabla^j \mathbf{v}|^2 = \sum_{i=1}^n |\nabla^j v_i|^2.$$

We will also work with the following subspaces of $H^1(\Omega)$:

$$H_0^1(\Omega) = \{\phi \in H^1(\Omega) : \phi|_{\partial\Omega} = 0\},$$

$$\widehat{H}^1(\Omega) = \{\phi \in H^1(\Omega) : \int_\Omega \phi \, dx = 0\}.$$

Also, we will use the standard notation $\mathcal{P}_k$ for the space of polynomials of degree less than or equal to k and, if $x \in \mathbb{R}^n$ and α is a multi-index, we will set $x^\alpha = x_1^{\alpha_1} \cdots x_n^{\alpha_n}$.

The letter C will denote a generic constant not necessarily the same at each occurrence.

Given a function in a Sobolev space of a domain Ω it is important to know whether it can be restricted to $\partial\Omega$, and conversely, when can a function defined on $\partial\Omega$ be extended to Ω in such a way that it belongs to the original Sobolev space. We will use the following trace theorem. We refer the reader for example to [38, 33] for the proof of this theorem and for the definition of the fractional-order Sobolev space $H^{\frac{1}{2}}(\partial\Omega)$.

Theorem 2.1. *Given $\phi \in H^1(\Omega)$, where $\Omega \subset \mathbb{R}^n$ is a Lipschitz domain, there exists a constant C depending only on Ω such that*

$$\|\phi\|_{H^{\frac{1}{2}}(\partial\Omega)} \leq C\|\phi\|_{H^1(\Omega)}.$$

In particular,

$$\|\phi\|_{L^2(\partial\Omega)} \leq C\|\phi\|_{H^1(\Omega)}. \tag{1}$$

Moreover, if $g \in H^{\frac{1}{2}}(\partial\Omega)$, there exists $\phi \in H^1(\Omega)$ such that $\phi|_{\partial\Omega} = g$ and

$$\|\phi\|_{H^1(\Omega)} \leq C\|g\|_{H^{\frac{1}{2}}(\partial\Omega)}.$$

One of the most important results in the analysis of variational methods for elliptic problems is the Friedrichs–Poincaré inequality for functions with vanishing mean average, that we state below (see for example [36] for the case of Lipschitz domains and [43] for another proof in the case of convex domains). Assume that Ω is a Lipschitz domain. Then, there exists a constant C depending only on the domain Ω such that for any $f \in \widehat{H}^1(\Omega)$,

$$\|f\|_{L^2(\Omega)} \leq C\|\nabla f\|_{L^2(\Omega)}. \tag{2}$$

The Friedrichs–Poincaré inequality can be seen as a particular case of the next result on polynomial approximation which is basic in the analysis of finite element methods.

Several different arguments have been given for the proof of the next lemma. See for example [12, 25, 26, 51]. Here we give a nice argument which, to our knowledge, is due to M. Dobrowolski for the lowest order case on convex domains (and as far as we know has not been published). The proof given here for the case of domains which are star-shaped with respect to a subset of positive measure and any degree of approximation is an immediate extension of Dobrowolski's argument. For simplicity we present the proof for the L^2-case (which is the case that we will use), but the reader can check that an analogous argument applies for L^p based Sobolev spaces $(1 \leq p < \infty)$.

Assume that Ω is star-shaped with respect to a set $B \subseteq \Omega$ of positive measure. Given an integer $k \geq 0$ and $f \in H^{k+1}(\Omega)$ we introduce the averaged Taylor polynomial approximation of f, $Q_{k,B}f \in \mathcal{P}_k$ defined by

$$Q_{k,B}f(x) = \frac{1}{|B|}\int_B T_k f(y,x)\,dy$$

where $T_k f(y,x)$ is the Taylor expansion of f centered at y, namely,

$$T_k f(y,x) = \sum_{|\alpha|\le k} D^\alpha f(y) \frac{(x-y)^\alpha}{\alpha!}.$$

Lemma 2.1. *Let $\Omega \subset \mathbb{R}^n$ be a domain with diameter d which is star-shaped with respect to a set of positive measure $B \subset \Omega$. Given an integer $k \ge 0$ and $f \in H^{k+1}(\Omega)$, there exists a constant $C = C(k,n)$ such that, for $0 \le |\beta| \le k+1$,*

$$\|D^\beta (f - Q_{k,B} f)\|_{L^2(\Omega)} \le C \frac{|\Omega|^{1/2}}{|B|^{1/2}} d^{k+1-|\beta|} \|\nabla^{k+1} f\|_{L^2(\Omega)}. \tag{3}$$

In particular, if Ω is convex,

$$\|D^\beta (f - Q_{k,\Omega} f)\|_{L^2(\Omega)} \le C\, d^{k+1-|\beta|} \|\nabla^{k+1} f\|_{L^2(\Omega)}. \tag{4}$$

Proof. By density we can assume that $f \in C^\infty(\Omega)$. Then we can write

$$f(x) - T_k f(y,x) = (k+1) \sum_{|\alpha|=k+1} \frac{(x-y)^\alpha}{\alpha!} \int_0^1 D^\alpha f(ty + (1-t)x)\, t^k\, dt.$$

Integrating this inequality over B (in the variable y) and dividing by $|B|$ we have

$$f(x) - Q_{k,B} f(x) = \frac{k+1}{|B|} \sum_{|\alpha|=k+1} \int_B \int_0^1 \frac{(x-y)^\alpha}{\alpha!} D^\alpha f(ty + (1-t)x)\, t^k\, dt\, dy$$

and so,

$$\int_\Omega |f(x) - Q_{k,B} f(x)|^2\, dx$$

$$\le C \frac{d^{2(k+1)}}{|B|^2} \sum_{|\alpha|=k+1} \int_\Omega \Big(\int_B \int_0^1 |D^\alpha f(ty + (1-t)x)| t^k\, dt\, dy \Big)^2 dx$$

$$\le C \frac{d^{2(k+1)}}{|B|^2} \sum_{|\alpha|=k+1} \int_\Omega \Big(\int_B \int_0^1 |D^\alpha f(ty + (1-t)x)|^2\, dt\, dy \Big) \Big(\int_B \int_0^1 t^{2k}\, dt\, dy \Big)\, dx.$$

Therefore,

$$\int_\Omega |f(x) - Q_{k,B} f(x)|^2\, dx$$

$$\le C \frac{d^{2(k+1)}}{|B|} \sum_{|\alpha|=k+1} \int_\Omega \int_B \int_0^1 |D^\alpha f(ty + (1-t)x)|^2\, dt\, dy\, dx. \tag{5}$$

Now, for each α,

$$\int_\Omega \int_B \int_0^1 |D^\alpha f(ty+(1-t)x)|^2 dt\, dy\, dx$$
$$= \int_\Omega \int_B \int_0^{\frac12} |D^\alpha f(ty+(1-t)x)|^2 dt\, dy\, dx$$
$$+ \int_\Omega \int_B \int_{\frac12}^1 |D^\alpha f(ty+(1-t)x)|^2 dt\, dy\, dx =: I + II.$$

Let us call g_α the extension by zero of $D^\alpha f$ to $\mathbb{R}^n$. Then, by Fubini's theorem and two changes of variables we have

$$I \le \int_B \int_0^{\frac12} \int_{\mathbb{R}^n} |g_\alpha(ty+(1-t)x)|^2\, dx\, dt\, dy = \int_B \int_0^{\frac12} \int_{\mathbb{R}^n} |g_\alpha((1-t)x)|^2\, dx\, dt\, dy$$
$$= \int_B \int_0^{\frac12} \int_{\mathbb{R}^n} |g_\alpha(z)|^2 (1-t)^{-n}\, dz\, dt\, dy \le 2^{n-1}|B| \int_\Omega |D^\alpha f(z)|^2\, dz.$$

Analogously,

$$II \le \int_\Omega \int_{\frac12}^1 \int_{\mathbb{R}^n} |g_\alpha(ty+(1-t)x)|^2\, dy\, dt\, dx = \int_\Omega \int_{\frac12}^1 \int_{\mathbb{R}^n} |g_\alpha(ty)|^2\, dy\, dt\, dx$$
$$= \int_\Omega \int_{\frac12}^1 \int_{\mathbb{R}^n} |g_\alpha(z)|^2 t^{-n}\, dz\, dt\, dx \le 2^{n-1}|\Omega| \int_\Omega |D^\alpha f(z)|^2\, dz.$$

Therefore, replacing these bounds in (5) we obtain (3) for $\beta = 0$.

On the other hand, an elementary computation shows that

$$D^\beta Q_{k,B} f(x) = Q_{k-|\beta|,B}(D^\beta f)(x) \qquad \forall |\beta| \le k$$

and therefore, the estimate (3) for $|\beta| > 0$ follows from the case $\beta = 0$ applied to $D^\beta f$.

Important consequences of this result are the following error estimates for the L^2-projection onto $\mathcal{P}_m$.

Corollary 2.1. *Let $\Omega \subset \mathbb{R}^n$ be a domain with diameter d star-shaped with respect to a set of positive measure $B \subset \Omega$. Given an integer $m \ge 0$, let $P : L^2(\Omega) \to \mathcal{P}_m$ be the L^2-orthogonal projection. There exists a constant $C = C(j,n)$ such that, for $0 \le j \le m+1$, if $f \in H^j(\Omega)$, then*

$$\|f - Pf\|_{L^2(\Omega)} \le C \frac{|\Omega|^{1/2}}{|B|^{1/2}} d^j |\nabla^j f|_{L^2(\Omega)}.$$

Remark 2.1. Analogous results to Lemma 3 and its corollary hold for bounded Lipschitz domains because this kind of domains can be written as a finite union of star-shaped domains (see [25] for details).

The following result is fundamental in the analysis of mixed finite element approximations.

Lemma 2.2. *Let $\Omega \subset \mathbb{R}^n$ be a bounded domain. Given $f \in L^2(\Omega)$ there exists $\mathbf{v} \in H^1(\Omega)^n$ such that*

$$\operatorname{div} \mathbf{v} = f \quad \text{in } \Omega \tag{6}$$

and

$$\|\mathbf{v}\|_{H^1(\Omega)} \leq C \|f\|_{L^2(\Omega)} \tag{7}$$

with a constant C depending only on Ω.

Proof. Let $B \in \mathbb{R}^n$ be a ball containing Ω and ϕ be the solution of the boundary problem

$$\begin{cases} \Delta \phi = f & \text{in } B \\ \phi = 0 & \text{on } \partial B \end{cases} \tag{8}$$

It is known that ϕ satisfies the following a priori estimate (see for example [36])

$$\|\phi\|_{H^2(\Omega)} \leq C \|f\|_{L^2(\Omega)}$$

and therefore $\mathbf{v} = \nabla \phi$ satisfies (6) and (7).

Remark 2.2. To treat Neumann boundary conditions we would need the existence of a solution of $\operatorname{div}\mathbf{v} = f$ satisfying (7) and the boundary condition $\mathbf{v} \cdot \mathbf{n} = 0$ on $\partial\Omega$. Such a $\mathbf{v}$ can be obtained by solving a Neumann problem in Ω for smooth domains or convex polygonal or polyhedral domains. For more general domains, including arbitrary polygonal or polyhedral domains, the existence of $\mathbf{v}$ satisfying (6) and (7) can be proved in different ways. In fact $\mathbf{v}$ can be taken such that all its components vanish on $\partial\Omega$ (see for example [2, 7, 30]).

A usual technique to obtain error estimates for finite element approximations is to work in a reference element and then change variables to prove results for a general element. Let us introduce some notations and recall some basic estimates.

Fix a reference simplex $\widehat{T} \subset \mathbb{R}^n$. Given a simplex $T \subset \mathbb{R}^n$, there exists an invertible affine map $F : \widehat{T} \to T$, $F(\hat{x}) = A\hat{x} + b$, with $A \in \mathbb{R}^{n\times n}$ and $b \in \mathbb{R}^n$.

We call h_T the diameter of T and ρ_T the diameter of the largest ball inscribed in T (see Fig. 1). We will use the regularity assumption on the elements, namely, many of our estimates will depend on a constant σ such that

$$\frac{h_T}{\rho_T} \leq \sigma. \tag{9}$$

It is known that (see [19]), for the matrix norm associated with the euclidean vector norm, the following estimates hold:

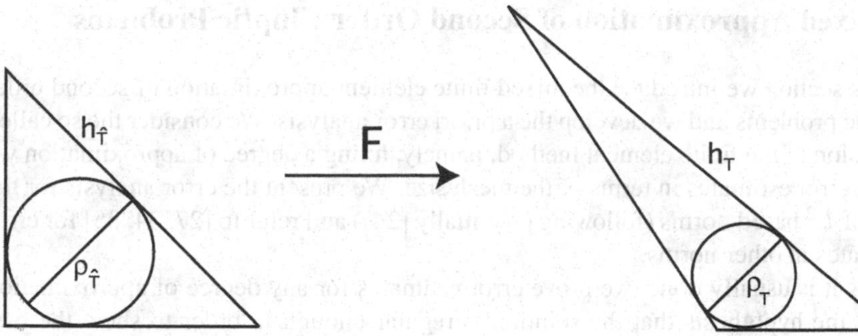

Fig. 1.

$$\|A\| \leq \frac{h_T}{\rho_{\widehat{T}}} \quad \text{and} \quad \|A^{-1}\| \leq \frac{h_{\widehat{T}}}{\rho_T}. \tag{10}$$

With any $\phi \in L^2(T)$ we associate $\hat{\phi} \in L^2(\widehat{T})$ in the usual way, namely,

$$\phi(x) = \hat{\phi}(\hat{x}) \tag{11}$$

where $x = F(\hat{x})$.

We end this section by recalling the so called inverse estimates which are a fundamental tool in finite element analysis. We give only a particular case which will be needed for our proofs (see for example [19] for more general inverse estimates).

Lemma 2.3. *Given a simplex T there exists a constant $C = C(\sigma, k, n, \widehat{T})$ such that, for any $p \in \mathcal{P}_k(T)$,*

$$\|\nabla p\|_{L^2(T)} \leq \frac{C}{h_T}\|p\|_{L^2(T)}.$$

Proof. Since $\mathcal{P}_k(\widehat{T})$ is a finite-dimensional space, all the norms defined on it are equivalent. In particular, there exists a constant $\widehat{C}$ depending on k and $\widehat{T}$ such that

$$\|\widehat{\nabla}\hat{p}\|_{L^2(\widehat{T})} \leq \widehat{C}\|\hat{p}\|_{L^2(\widehat{T})} \tag{12}$$

for any $\hat{p} \in \mathcal{P}_k(\widehat{T})$.

An easy computation shows that

$$\nabla p = A^{-T}\widehat{\nabla}\hat{p}$$

where A^{-T} is the transpose matrix of A^{-1}. Therefore, using the bound for $\|A^{-1}\|$ given in (10) together with (12) and (9) we have

$$\int_T |\nabla p|^2\,dx = \int_{\widehat{T}} |A^{-T}\widehat{\nabla}\hat{p}|^2 |\det A|\,d\hat{x} \leq \|A^{-1}\|^2 \int_{\widehat{T}} |\widehat{\nabla}\hat{p}|^2 |\det A|\,d\hat{x}$$

$$\leq \widehat{C}\frac{h_{\widehat{T}}^2}{\rho_T^2}\int_{\widehat{T}} |\hat{p}|^2 |\det A|\,d\hat{x} = \widehat{C}\frac{h_{\widehat{T}}^2}{\rho_T^2}\int_T |p|^2\,dx \leq \widehat{C}\sigma^2\frac{h_{\widehat{T}}^2}{h_T^2}\int_T |p|^2\,dx.$$

3 Mixed Approximation of Second Order Elliptic Problems

In this section we introduce the mixed finite element approximation of second order elliptic problems and we develop the a priori error analysis. We consider the so called h-version of the finite element method, namely, fixing a degree of approximation we prove error estimates in terms of the mesh size. We present the error analysis for the case of L^2 based norms (following essentially [24]) and refer to [27, 34, 35] for error estimates in other norms.

As it is usually done, we prove error estimates for any degree of approximation under the hypothesis that the solution is regular enough in order to show the best possible order of a method. However, the reader has to be aware that, in practice, for polygonal or polyhedral domains (which is the case considered here!) the solution is in general not smooth due to singularities at the angles and therefore the order of convergence is limited by the regularity of the solution of each particular problem considered. On the other hand, for domains with smooth boundary where the solutions might be very regular, a further error analysis considering the approximation of the boundary is needed.

Consider the elliptic problem

$$\begin{cases} -\operatorname{div}(a\nabla p) = f & \text{in } \Omega \\ p = 0 & \text{on } \partial\Omega \end{cases} \tag{13}$$

where $\Omega \subset \mathbb{R}^n$ is a polyhedral domain and $a = a(x)$ is a function bounded by above and below by positive constants.

In many applications the variable of interest is

$$\mathbf{u} = -a\nabla p$$

and then, it could be desirable to use a mixed finite element method which approximates $\mathbf{u}$ and p simultaneously. With this purpose, problem (13) is decomposed into a first order system as follows:

$$\begin{cases} \mathbf{u} + a\nabla p = 0 & \text{in } \Omega \\ \operatorname{div}\mathbf{u} = f & \text{in } \Omega \\ p = 0 & \text{on } \partial\Omega. \end{cases} \tag{14}$$

To write an appropriate weak formulation of this problem we introduce the space

$$H(\operatorname{div},\Omega) = \{\mathbf{v} \in L^2(\Omega)^n : \operatorname{div}\mathbf{v} \in L^2(\Omega)\}$$

which is a Hilbert space with norm given by

$$\|\mathbf{v}\|^2_{H(\operatorname{div},\Omega)} = \|\mathbf{v}\|^2_{L^2(\Omega)} + \|\operatorname{div}\mathbf{v}\|^2_{L^2(\Omega)}.$$

Defining $\mu(x) = 1/a(x)$, the first equation in (14) can be rewritten as

$$\mu\,\mathbf{u} + \nabla p = 0 \ \text{ in } \Omega.$$

Multiplying by test functions and integrating by parts we obtain the standard weak mixed formulation of problem (14), namely,

$$\begin{cases} \displaystyle\int_\Omega \mu \mathbf{u} \cdot \mathbf{v}\, dx - \int_\Omega p \operatorname{div} \mathbf{v}\, dx = 0 & \forall \mathbf{v} \in H(\operatorname{div}, \Omega) \\ \displaystyle\int_\Omega q \operatorname{div} \mathbf{u}\, dx = \int_\Omega f q\, dx & \forall q \in L^2(\Omega). \end{cases} \tag{15}$$

Observe that the Dirichlet boundary condition is implicit in the weak formulation (i.e., it is the type of condition usually called natural). Instead, Neumann boundary conditions would have to be imposed on the space (essential conditions). This is exactly opposite to what happens in the case of standard formulations.

The weak formulation (15) involves the divergence of the solution and of the test functions but not arbitrary first derivatives. This fact allows us to work on the space $H(\operatorname{div}, \Omega)$ instead of the smaller $H^1(\Omega)^n$ and this will be important for the finite element approximation because piecewise polynomials vector functions do not need to have both components continuous to be in $H(\operatorname{div}, \Omega)$, but only their normal component.

In order to define finite element approximations to the solution $(\mathbf{u}, p)$ of (15) we need to introduce finite-dimensional subspaces of $H(\operatorname{div}, \Omega)$ and $L^2(\Omega)$ made of piecewise polynomial functions.

For simplicity we will consider the case of triangular elements (or its generalizations to higher dimensions) and the associated Raviart–Thomas spaces which are the best-known spaces for this problem. This family of spaces was introduced in [44] in the 2-d case, while its extension to three dimensions was first considered in [41]. Since no essential technical difficulties arise in the general case, we prefer to present the spaces and the analysis of their properties in the general n-dimensional case (although, of course, we are mainly interested in the cases $n = 2$ and $n = 3$). Below we will comment and give references on different variants of spaces.

First we introduce the local spaces, analyze their properties and construct the Raviart–Thomas interpolation.

Given a simplex $T \in \mathbb{R}^n$, the local Raviart–Thomas space [44, 41] of order $k \geq 0$ is defined by

$$\mathcal{RT}_k(T) = \mathcal{P}_k(T)^n + x\,\mathcal{P}_k(T) \tag{16}$$

In the following lemma we give some basic properties of the spaces $\mathcal{RT}_k(T)$. We denote with F_i, $i = 1, \cdots, n+1$, the faces of a simplex T and with $\mathbf{n}_i$ their corresponding exterior normals.

Lemma 3.1. *(a)* $\dim \mathcal{RT}_k(T) = n\binom{k+n}{k} + \binom{k+n-1}{k}$.
(b) If $\mathbf{v} \in \mathcal{RT}_k(T)$ *then,* $\mathbf{v} \cdot \mathbf{n}_i \in \mathcal{P}_k(F_i)$ *for* $i = 1, \cdots, n+1$.
(c) If $\mathbf{v} \in \mathcal{RT}_k(T)$ *is such that* $\operatorname{div} \mathbf{v} = 0$ *then,* $\mathbf{v} \in \mathcal{P}_k^n$.

Proof. Any $\mathbf{v} \in \mathcal{RT}_k(T)$ can be written as

$$\mathbf{v} = \mathbf{w} + x \sum_{|\alpha|=k} a_\alpha x^\alpha \tag{17}$$

with $\mathbf{w} \in \mathcal{P}_k^n$.

Recall that $\dim \mathcal{P}_k = \binom{k+n}{k}$ and that the number of multi-indeces α such that $|\alpha| = k$ is $\binom{k+n-1}{k}$. Then, (a) follows from (17).

Now, the face F_i is on a hyperplane of equation $x \cdot \mathbf{n}_i = s$ with $s \in \mathrm{I\!R}$. Therefore, if $\mathbf{v} = \mathbf{w} + x\,p$ with $\mathbf{w} \in \mathcal{P}_k^n$ and $p \in \mathcal{P}_k$, we have

$$\mathbf{v} \cdot \mathbf{n}_i = \mathbf{w} \cdot \mathbf{n}_i + x \cdot \mathbf{n}_i\, p = \mathbf{w} \cdot \mathbf{n}_i + s\, p \in \mathcal{P}_k$$

which proves (b).

Finally, if $\operatorname{div} \mathbf{v} = 0$ we take the divergence in the expression (17) and conclude easily that $a_\alpha = 0$ for all α and therefore (c) holds.

Our next goal is to construct an interpolation operator

$$\Pi_T : H^1(T)^n \to \mathcal{RT}_k$$

which will be fundamental for the error analysis. We fix k and to simplify notation we omit the index k in the operator.

For simplicity we define the interpolation for functions in $H^1(T)^n$ although it is possible (and necessary in many cases!) to do the same construction for less regular functions. Indeed, the reader who is familiar with fractional order Sobolev spaces and trace theorems will realize that the degrees of freedom defining the interpolation are well defined for functions in $H^s(T)^n$, with $s > 1/2$.

The local interpolation operator is defined in the following lemma.

Lemma 3.2. *Given $\mathbf{v} \in H^1(T)^n$, where $T \in \mathrm{I\!R}^n$ is a simplex, there exists a unique $\Pi_T \mathbf{v} \in \mathcal{RT}_k(T)$ such that*

$$\int_{F_i} \Pi_T \mathbf{v} \cdot \mathbf{n}_i\, p_k\, ds = \int_{F_i} \mathbf{v} \cdot \mathbf{n}_i\, p_k\, ds \;\; \forall p_k \in \mathcal{P}_k(F_i), \;\; i = 1, \cdots, n+1 \qquad (18)$$

and, if $k \geq 1$,

$$\int_T \Pi_T \mathbf{v} \cdot \mathbf{p}_{k-1}\, dx = \int_T \mathbf{v} \cdot \mathbf{p}_{k-1}\, dx \;\; \forall \mathbf{p}_{k-1} \in \mathcal{P}_{k-1}^n(T). \qquad (19)$$

Proof. First, we want to see that the number of conditions defining $\Pi_T \mathbf{v}$ equals the dimension of $\mathcal{RT}_k(T)$. This is easily verified for the case $k = 0$, so let us consider the case $k \geq 1$.

Since $\dim \mathcal{P}_k(F_i) = \binom{k+n-1}{k}$, the number of conditions in (18) is

$$\#\,\text{of faces} \times \dim \mathcal{P}_k(F_i) = (n+1)\binom{k+n-1}{k}.$$

On the other hand, the number of conditions in (19) is

$$\dim \mathcal{P}_{k-1}^n(T) = n\binom{k+n-1}{k-1}.$$

Then, the total number of conditions defining $\Pi_T \mathbf{v}$ is

$$(n+1)\binom{k+n-1}{k} + n\binom{k+n-1}{k-1}.$$

Therefore, in view of (a) of lemma (3.1), we have to check that

$$n\binom{k+n}{k} + \binom{k+n-1}{k} = (n+1)\binom{k+n-1}{k} + n\binom{k+n-1}{k-1}$$

or equivalently,

$$\binom{k+n}{k} = \binom{k+n-1}{k} + \binom{k+n-1}{k-1}$$

which can be easily verified.

Therefore, in order to show the existence of $\Pi_T \mathbf{v}$, it is enough to prove uniqueness. So, take $\mathbf{v} \in \mathcal{RT}_k(T)$ such that

$$\int_{F_i} \mathbf{v} \cdot \mathbf{n}_i \, p_k \, ds = 0 \;\; \forall p_k \in \mathcal{P}_k(F_i)\,, \;\; i = 1, \cdots, n+1 \tag{20}$$

and

$$\int_T \mathbf{v} \cdot \mathbf{p}_{k-1} \, dx = 0 \;\; \forall \mathbf{p}_{k-1} \in \mathcal{P}^n_{k-1}(T). \tag{21}$$

From (b) of Lemma 3.1 and (20) it follows that $\mathbf{v} \cdot \mathbf{n}_i = 0$ on F_i. Then, using now (21) we have

$$\int_T (\operatorname{div} \mathbf{v})^2 \, dx = -\int_T \mathbf{v} \cdot \nabla(\operatorname{div} \mathbf{v}) \, dx = 0$$

because $\nabla(\operatorname{div} \mathbf{v}) \in \mathcal{P}^n_{k-1}(T)$. Consequently $\operatorname{div} \mathbf{v} = 0$ and so, from (c) of Lemma 3.1 we know that $\mathbf{v} \in \mathcal{P}^n_k(T)$.

Therefore, for each $i = 1, \cdots, n+1$, the component $\mathbf{v} \cdot \mathbf{n}_i$ is a polynomial of degree k on T which vanishes on F_i. Therefore, calling λ_i the barycentric coordinates associated with T (i.e., $\lambda_i(x) = 0$ on F_i), we have

$$\mathbf{v} \cdot \mathbf{n}_i = \lambda_i q_{k-1}$$

with $q_{k-1} \in \mathcal{P}_{k-1}(T)$. But, from (21) we know that

$$\int_T \mathbf{v} \cdot \mathbf{n}_i \, p_{k-1} \, dx = 0 \;\; \forall p_{k-1} \in \mathcal{P}_{k-1}(T)$$

and choosing $p_{k-1} = q_{k-1}$ we obtain

$$\int_T \lambda_i q^2_{k-1} \, dx = 0.$$

Therefore, since λ_i does not change sign on T, it follows that $q_{k-1} = 0$ and consequently $\mathbf{v} \cdot \mathbf{n}_i = 0$ in T for $i = 1, \cdots, n+1$. In particular, there are n linearly independent directions in which $\mathbf{v}$ has vanishing components and, therefore, $\mathbf{v} = 0$ as we wanted to see.

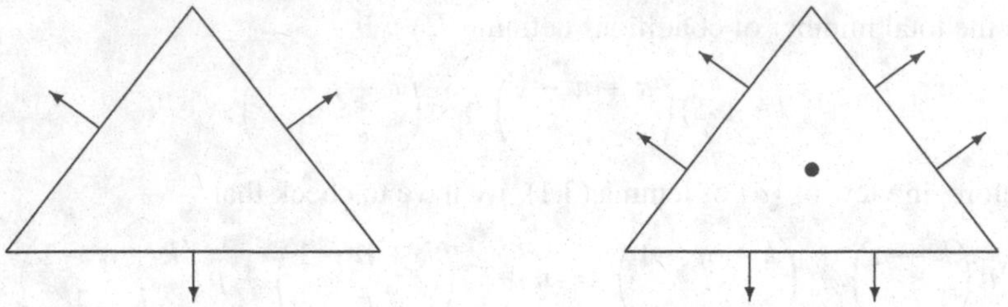

Fig. 2. Degrees of freedom for $\mathcal{RT}_0$ and $\mathcal{RT}_1$ in $\mathbb{R}^2$

Figure 2 shows the degrees of freedom defining Π_T for $k = 0$ and $k = 1$ in the 2-d case. The arrows indicate normal components values and the filled circle, moments of the components of $\mathbf{v}$ (and so it corresponds to two degrees of freedom).

To obtain error estimates for the mixed finite element approximations we need to know the approximation properties of the Raviart–Thomas interpolation Π_T. The analysis given in [44, 49] makes use of general standard arguments for polynomial-preserving operators (see [19]). The main difference with the error analysis for Lagrange interpolation is that here we have to use an appropriate transformation, known as the Piola transform, which preserves the degrees of freedom defining $\Pi_T \mathbf{v}$.

The Piola transform is defined as follows. Given two domains $\widehat{\Omega}, \Omega \subset \mathbb{R}^n$ and a smooth bijective map $F : \widehat{\Omega} \to \Omega$, let DF be the Jacobian matrix of F and $J := \det DF$. Assume that J does not vanish at any point, then, we define for $\hat{\mathbf{v}} \in L^2(\widehat{\Omega})^n$

$$\mathbf{v}(x) = \frac{1}{|J(\hat{x})|} \widehat{D}F(\hat{x}) \hat{\mathbf{v}}(\hat{x})$$

where $x = F(\hat{x})$. Here and in what follows, the hat over differential operators indicates that the derivatives are taken with respect to $\hat{x}$.

We recall that scalar functions are transformed as indicated in (11) (we are using the same notation for the transformation of vector and scalar functions since no confusion is possible).

In the particular case that F is an affine map given by $A\hat{x} + b$ we have $J = \det A$ and

$$\mathbf{v}(x) = \frac{1}{|J|} A \hat{\mathbf{v}}(\hat{x}). \tag{22}$$

In the next lemma we give some fundamental properties of the Piola transform. For simplicity, we prove the results only for affine transformations, which is the useful case for our purposes. However, it is important to remark that analogous results hold for general transformations and this is important, for example, to work with general quadrilateral elements.

Lemma 3.3. *If* $\mathbf{v} \in H(\text{div}, T)$ *and* $\phi \in H^1(T)$ *then*

$$\int_T \text{div}\, \mathbf{v}\, \phi \, dx = \int_{\widehat{T}} \widehat{\text{div}}\, \hat{\mathbf{v}}\, \hat{\phi}\, d\hat{x}\,, \tag{23}$$

$$\int_T \mathbf{v} \cdot \nabla\phi \, dx = \int_{\widehat{T}} \hat{\mathbf{v}} \cdot \hat{\nabla}\hat{\phi} \, d\hat{x} \tag{24}$$

and

$$\int_{\partial T} \mathbf{v} \cdot \mathbf{n} \, \phi \, ds = \int_{\partial\widehat{T}} \hat{\mathbf{v}} \cdot \hat{\mathbf{n}} \, \hat{\phi} \, d\hat{s}. \tag{25}$$

Proof. From the definition of the Piola transform (22) we have

$$D\mathbf{v}(x) = \frac{1}{|J|} A D(\hat{\mathbf{v}} \circ F^{-1})(x) = \frac{1}{|J|} A \widehat{D}\hat{\mathbf{v}}(\hat{x}) DF^{-1}(x) = \frac{1}{|J|} A \widehat{D}\hat{\mathbf{v}}(\hat{x}) A^{-1}.$$

Then,

$$\operatorname{div} \mathbf{v} = \operatorname{tr} D\mathbf{v} = \frac{1}{|J|} \operatorname{tr}(A \widehat{D}\hat{\mathbf{v}} A^{-1}) = \frac{1}{|J|} \operatorname{tr} \widehat{D}\hat{\mathbf{v}} = \frac{1}{|J|} \widehat{\operatorname{div}} \hat{\mathbf{v}}$$

and therefore (23) follows by a change of variable.

To prove (24) recall that

$$\nabla\phi = A^{-T} \hat{\nabla}\hat{\phi}.$$

Then,

$$\int_T \mathbf{v} \cdot \nabla\phi \, dx = \int_{\widehat{T}} A\hat{\mathbf{v}} \cdot A^{-T} \hat{\nabla}\hat{\phi} \, d\hat{x} = \int_{\widehat{T}} \hat{\mathbf{v}} \cdot \hat{\nabla}\hat{\phi} \, d\hat{x}.$$

Finally, (25) follows from (23) and (24) applying the divergence theorem.

Remark 3.1. The integral over ∂T in the previous lemma has to be understood as a duality product between $\mathbf{v} \cdot \mathbf{n} \in H^{-\frac{1}{2}}(\partial T)$ and $\phi \in H^{\frac{1}{2}}(\partial T)$.

We can now prove the invariance of the Raviart–Thomas interpolation under the Piola transform.

Lemma 3.4. *Given a simplex $T \in \mathbb{R}^n$ and $\mathbf{v} \in H^1(T)^n$ we have*

$$\Pi_{\widehat{T}} \hat{\mathbf{v}} = \widehat{\Pi_T \mathbf{v}}. \tag{26}$$

Proof. We have to check that $\widehat{\Pi_T \mathbf{v}}$ satisfies the conditions defining $\Pi_{\widehat{T}} \hat{\mathbf{v}}$, namely,

$$\int_{\widehat{F}_i} \widehat{\Pi_T \mathbf{v}} \cdot \hat{\mathbf{n}}_i \, \hat{p}_k \, d\hat{s} = \int_{\widehat{F}_i} \hat{\mathbf{v}} \cdot \hat{\mathbf{n}}_i \, \hat{p}_k \, d\hat{s} \;\; \forall \hat{p}_k \in \mathcal{P}_k(\widehat{F}_i), \;\; i = 1, \cdots, n+1, \tag{27}$$

where $\widehat{F}_i = F^{-1}(F_i)$, and

$$\int_{\widehat{T}} \widehat{\Pi_T \mathbf{v}} \cdot \hat{\mathbf{p}}_{k-1} \, d\hat{x} = \int_{\widehat{T}} \hat{\mathbf{v}} \cdot \hat{\mathbf{p}}_{k-1} \, d\hat{x} \;\; \forall \hat{\mathbf{p}}_{k-1} \in \mathcal{P}_{k-1}^n(\widehat{T}). \tag{28}$$

Given $\hat{p}_k \in \mathcal{P}_k(\widehat{F}_i)$ we have

$$\int_{\widehat{F}_i} \hat{\mathbf{v}} \cdot \hat{\mathbf{n}}_i \, \hat{p}_k \, d\hat{s} = \int_{F_i} \mathbf{v} \cdot \mathbf{n}_i \, p_k \, ds. \tag{29}$$

Indeed, this follows from (25) by a density argument. We can not apply (25) directly because the function obtained by extending p_k by zero to the other faces of T is not in $H^{\frac{1}{2}}(\partial T)$ and, therefore, it is not the restriction to the boundary of a function $\phi \in H^1(T)$. However, we can take a sequence of functions $q_j \in C_0^\infty(F_i)$ such that $q_j \to p_k$ in $L^2(F_i)$ and, since the extension by zero to ∂T of q_j is in $H^{\frac{1}{2}}(\partial T)$, there exists $\phi_j \in H^1(T)$ such that the restriction of ϕ_j to F_i is equal to q_j. Therefore, applying (25) we obtain,

$$\int_{\widehat{F}_i} \hat{\mathbf{v}} \cdot \hat{\mathbf{n}}_i \, \hat{q}_j \, d\hat{s} = \int_{F_i} \mathbf{v} \cdot \mathbf{n}_i \, q_j \, ds$$

and therefore, since $\mathbf{v} \cdot \mathbf{n}_i \in L^2(F_i)$, we can pass to the limit to obtain (29). Analogously we have

$$\int_{\widehat{F}_i} \widehat{\Pi_T \mathbf{v}} \cdot \hat{\mathbf{n}}_i \, \hat{p}_k \, d\hat{s} = \int_{F_i} \Pi_T \mathbf{v} \cdot \mathbf{n}_i \, p_k \, ds.$$

and therefore (27) follows from condition (18) in the definition of $\Pi_T \mathbf{v}$.

To check (28) observe that, for $\hat{\mathbf{p}}_{k-1} \in \mathcal{P}_{k-1}^n(\widehat{T})$, we have

$$\begin{aligned}
\int_{\widehat{T}} \widehat{\Pi_T \mathbf{v}} \cdot \hat{\mathbf{p}}_{k-1} \, d\hat{x} &= \int_T |J| A^{-1} \Pi_T \mathbf{v} \cdot |J| A^{-1} \mathbf{p}_{k-1} |J|^{-1} \, dx \\
&= \int_T \Pi_T \mathbf{v} \cdot |J| A^{-T} A^{-1} \mathbf{p}_{k-1} \, dx \\
&= \int_T \mathbf{v} \cdot |J| A^{-T} A^{-1} \mathbf{p}_{k-1} \, dx = \int_{\widehat{T}} \hat{\mathbf{v}} \cdot \hat{\mathbf{p}}_{k-1} \, d\hat{x}
\end{aligned}$$

where we have used condition (19) and that $|J| A^{-T} A^{-1} \mathbf{p}_{k-1} \in \mathcal{P}_{k-1}^n(T)$.

We can now prove the optimal order error estimates for the Raviart–Thomas interpolation.

Theorem 3.1. *There exists a constant C depending on k, n and the regularity constant σ such that, for any $\mathbf{v} \in H^m(T)^n$ and $1 \le m \le k+1$,*

$$\|\mathbf{v} - \Pi_T \mathbf{v}\|_{L^2(T)} \le C h_T^m \|\nabla^m \mathbf{v}\|_{L^2(T)}. \tag{30}$$

Proof. First we prove an estimate on the reference element $\widehat{T}$. We will denote with $\widehat{C}$ a generic constant which depends only on k, n and $\widehat{T}$. For each face $\widehat{F}_i$ of $\widehat{T}$ let $\{p_j^i\}_{1 \le j \le N}$ be a basis of $\mathcal{P}_k(\widehat{F}_i)$ and let $\{\mathbf{p}_m\}_{1 \le m \le M}$ be a basis of $\mathcal{P}_{k-1}^n(\widehat{T})$. Then, associated with this basis we can introduce the Lagrange-type basis of $\mathcal{RT}_k(\widehat{T})$, $\{\phi_j^i, \psi_m\}$ defined by

$$\int_{\widehat{F}_i} \phi_j^i \cdot \mathbf{n}_i \, p_s^r = \delta_{ir} \delta_{js}, \qquad \int_{\widehat{T}} \phi_j^i \cdot \mathbf{p}_m = 0\,,$$

$$\forall \quad i, r = 1, \cdots, n+1, \quad j, s = 1, \cdots, N, \quad m = 1, \cdots, M$$

and

$$\int_{\widehat{T}} \psi_m \cdot \mathbf{p}_\ell = \delta_{m\ell}, \quad \psi_m \cdot \mathbf{n}_i = 0$$

$$\forall \quad m, \ell = 1, \cdots, M, \quad i = 1, \cdots, n+1.$$

Then,

$$\Pi_{\widehat{T}} \hat{\mathbf{v}}(\hat{x}) = \sum_{i=1}^{n+1} \sum_{j=1}^{N} \Big(\int_{\widehat{F}_i} \hat{\mathbf{v}} \cdot \mathbf{n}_i p_j^i \Big) \phi_j^i(\hat{x}) + \sum_{m=1}^{M} \Big(\int_{\widehat{T}} \hat{\mathbf{v}} \cdot \mathbf{p}_m \Big) \psi_m(\hat{x}).$$

Now, from the trace theorem (1) on $\widehat{T}$ we have

$$\Big| \int_{\widehat{F}_i} \hat{\mathbf{v}} \cdot \mathbf{n}_i p_j^i \Big| \le \widehat{C} \|\hat{\mathbf{v}}\|_{H^1(\widehat{T})}.$$

Clearly, we also have

$$\Big| \int_{\widehat{T}} \hat{\mathbf{v}} \cdot \mathbf{p}_m \Big| \le \widehat{C} \|\hat{\mathbf{v}}\|_{L^2(\widehat{T})}.$$

In both estimates the constant $\widehat{C}$ depends on bounds for the polynomials p_j^i and $\mathbf{p}_m$ and then, it depends only on k, n and $\widehat{T}$.

Therefore, using now that $\|\phi_j^i\|_{L^2(\widehat{T})}$ and $\|\psi_m\|_{L^2(\widehat{T})}$ are also bounded by a constant $\widehat{C}$ we obtain

$$\|\Pi_{\widehat{T}} \hat{\mathbf{v}}\|_{L^2(\widehat{T})} \le \widehat{C} \|\hat{\mathbf{v}}\|_{H^1(\widehat{T})}. \tag{31}$$

Using now the relation (26) and making a change of variables we have

$$\int_T |\Pi_T \mathbf{v}|^2 \, dx = \int_{\widehat{T}} |J|^{-2} |A \Pi_{\widehat{T}} \hat{\mathbf{v}}|^2 |J| \, d\hat{x} \le |J|^{-1} \|A\|^2 \int_{\widehat{T}} |\Pi_{\widehat{T}} \hat{\mathbf{v}}|^2 \, d\hat{x}.$$

Then, using the bound for $\|A\|$ (10) and (31) we obtain

$$\int_T |\Pi_T \mathbf{v}|^2 \, dx \le |J|^{-1} \frac{h_T^2}{\rho_{\widehat{T}}^2} \Big\{ \int_{\widehat{T}} |\hat{\mathbf{v}}|^2 \, d\hat{x} + \int_{\widehat{T}} |\widehat{D} \hat{\mathbf{v}}|^2 \, d\hat{x} \Big\} \tag{32}$$

but, since $\hat{\mathbf{v}} = |J| A^{-1} \mathbf{v}$ and $\widehat{D} \hat{\mathbf{v}} = |J| A^{-1} D \mathbf{v} A$, using the bounds for $\|A\|$ and $\|A^{-1}\|$ (10),

$$|\hat{\mathbf{v}}| \le |J| \frac{h_{\widehat{T}}}{\rho_T} |\mathbf{v}| \qquad \text{and} \qquad |\widehat{D} \hat{\mathbf{v}}| \le |J| \frac{h_{\widehat{T}}}{\rho_T} \frac{h_T}{\rho_{\widehat{T}}} |D \mathbf{v}|$$

and so, it follows from (32), changing variables again, that

$$\|\Pi_T \mathbf{v}\|_{L^2(T)}^2 \le \widehat{C} \Big\{ \frac{h_T^2}{\rho_T^2} \|\mathbf{v}\|_{L^2(T)}^2 + \frac{h_T^4}{\rho_T^2} \|D \mathbf{v}\|_{L^2(T)}^2 \Big\}.$$

Therefore, from the regularity hypothesis (9) we obtain

$$\|\Pi_T \mathbf{v}\|_{L^2(T)} \le C\Big\{\|\mathbf{v}\|_{L^2(T)} + h_T\|D\mathbf{v}\|_{L^2(T)}\Big\} \tag{33}$$

where the constant depends only on $\widehat{T}$, k, n and the regularity constant σ.

Now we use a standard argument. Since $\mathcal{P}_k^n(T) \subset \mathcal{RT}_k(T)$ we know that $\Pi_T \mathbf{q} = \mathbf{q}$ for all $\mathbf{q} \in \mathcal{P}_k^n(T)$ and then

$$\begin{aligned}\|\mathbf{v} - \Pi_T \mathbf{v}\|_{L^2(T)} &= \|\mathbf{v} - \mathbf{q} - \Pi_T(\mathbf{v} - \mathbf{q})\|_{L^2(T)} \\ &\le C\{\|\mathbf{v} - \mathbf{q}\|_{L^2(T)} + h_T\|D(\mathbf{v} - \mathbf{q})\|_{L^2(T)}\}\end{aligned}$$

where the constant depends on that in (33). Therefore, we conclude the proof applying Lemma 2.1.

Let us now introduce the global Raviart–Thomas finite element spaces. Assume that we have a family of triangulations $\{\mathcal{T}_h\}$ of Ω, i.e., $\Omega = \cup_{T\in\mathcal{T}_h} T$, such that the intersection of two elements in $\mathcal{T}_h$ is either empty, or a vertex, or a common edge or face and h is a measure of the mesh-size, namely, $h = \max_{T\in\mathcal{T}_h} h_T$.

We assume that the family of triangulations is regular, i.e., for any $T \in \mathcal{T}_h$ and any h, the regularity condition (9) is satisfied with a uniform σ.

Associated with the triangulation $\mathcal{T}_h$ we introduce the global space

$$\mathcal{RT}_k(\mathcal{T}_h) = \{\mathbf{v} \in H(\mathrm{div}, \Omega) : \mathbf{v}|_T \in \mathcal{RT}_k(T)\ \forall T \in \mathcal{T}_h\} \tag{34}$$

When no confusion arises we will drop the $\mathcal{T}_h$ from the definition and call $\mathcal{RT}_k$ the global space. A fundamental tool in the error analysis is the operator

$$\Pi_h : H(\mathrm{div}, \Omega) \cap \prod_{T\in\mathcal{T}_h} H^1(T)^n \longrightarrow \mathcal{RT}_k$$

defined by

$$\Pi_h \mathbf{v}|_T = \Pi_T \mathbf{v} \qquad \forall T \in \mathcal{T}_h.$$

We have to check that $\Pi_h \mathbf{v} \in \mathcal{RT}_k$. Since by definition $\Pi_T \mathbf{v} \in \mathcal{RT}_k(T)$, it only remains to see that $\Pi_h \mathbf{v} \in H(\mathrm{div}, \Omega)$.

First we observe that a piecewise polynomial vector function is in $H(\mathrm{div}, \Omega)$ if and only if it has continuous normal component across the elements (this can be verified by applying the divergence theorem). But, since $\mathbf{v} \in H(\mathrm{div}, \Omega)$, the continuity of the normal component of $\Pi_h \mathbf{v}$ follows from (b) of Lemma 3.1 in view of the degrees of freedom (18) in the definition of Π_T.

The finite element space for the approximation of the scalar variable p is the standard space of, not necessarily continuous, piecewise polynomials of degree k, namely,

$$\mathcal{P}_k^d(\mathcal{T}_h) = \{q \in L^2(\Omega) : q|_T \in \mathcal{P}_k(T) :\ \forall T \in \mathcal{T}_h\} \tag{35}$$

where the d stands for "discontinuous". Also in this case we will write only $\mathcal{P}_k^d$ when no confusion arises. Observe that, since no derivative of the scalar variable appears in the weak form, we do not require any continuity in the approximation space for this variable.

In the following lemma we give two fundamental properties for the error analysis.

Lemma 3.5. *The operator Π_h satisfies*

$$\int_\Omega \operatorname{div}(\mathbf{v} - \Pi_h \mathbf{v})\, q\, dx = 0 \tag{36}$$

$\forall \mathbf{v} \in H(\operatorname{div}, \Omega) \cap \prod_{T \in \mathcal{T}_h} H^1(T)^n$ *and* $\forall q \in \mathcal{P}_k^d$. *Moreover,*

$$\operatorname{div} \mathcal{RT}_k = \mathcal{P}_k^d. \tag{37}$$

Proof. Using (18) and (19) it follows that, for any $\mathbf{v} \in H^1(T)^n$ and any $q \in \mathcal{P}_k(T)$,

$$\int_T \operatorname{div}(\mathbf{v} - \Pi_T \mathbf{v}) q\, dx = -\int_T (\mathbf{v} - \Pi_T \mathbf{v}) \cdot \nabla q\, dx + \int_{\partial T} (\mathbf{v} - \Pi_T \mathbf{v}) \cdot \mathbf{n}\, q = 0$$

thus, (36) holds.

It is easy to see that div $\mathcal{RT}_k \subset \mathcal{P}_k^d$. In order to see the other inclusion recall that from Lemma 2.2 we know that div : $H^1(\Omega)^n \to L^2(\Omega)$ is surjective. Therefore, given $q \in \mathcal{P}_k^d$ there exists $\mathbf{v} \in H^1(\Omega)^n$ such that div $\mathbf{v} = q$. Then, it follows from (36) that div $\Pi_h \mathbf{v} = q$ and so (37) is proved.

Introducing the orthogonal L^2-projection $P_h : L^2(\Omega) \to \mathcal{P}_k^d$, properties (36) and (37) can be summarized in the following commutative diagram

$$\begin{array}{ccccc} H^1(\Omega)^n & \xrightarrow{\operatorname{div}} & L^2(\Omega) & & \\ \Pi_h \big\downarrow & & \big\downarrow P_h & & \\ \mathcal{RT}_k & \xrightarrow{\operatorname{div}} & \mathcal{P}_k^d & \longrightarrow & 0 \end{array} \tag{38}$$

where, to simplify notation, we have replaced $H(\operatorname{div}, \Omega) \cap \prod_{T \in \mathcal{T}_h} H^1(T)^n$ by its subspace $H^1(\Omega)^n$.

Our next goal is to give error estimates for the mixed finite element approximation of problem (13), namely, $(\mathbf{u}_h, p_h) \in \mathcal{RT}_k \times \mathcal{P}_k^d$ defined by

$$\begin{cases} \displaystyle\int_\Omega \mu\, \mathbf{u}_h \cdot \mathbf{v}\, dx - \int_\Omega p_h \operatorname{div} \mathbf{v}\, dx = 0 & \forall \mathbf{v} \in \mathcal{RT}_k \\ \displaystyle\int_\Omega q \operatorname{div} \mathbf{u}_h\, dx = \int_\Omega f q\, dx & \forall q \in \mathcal{P}_k^d. \end{cases} \tag{39}$$

It is important to remark that, although we are considering the particular case of the Raviart–Thomas spaces on simplicial elements, the error analysis only makes use of the fundamental commutative diagram property (38) and of the approximation properties of the projections Π_h and P_h. Therefore, similar results can be obtained for other finite element spaces.

Lemma 3.6. *If* $\mathbf{u}$ *and* $\mathbf{u}_h$ *are the solutions of (15) and (39) then,*

$$\|\mathbf{u} - \mathbf{u}_h\|_{L^2(\Omega)} \le (1 + \|a\|_{L^\infty(\Omega)} \|\mu\|_{L^\infty(\Omega)}) \|\mathbf{u}_h - \Pi_h \mathbf{u}\|_{L^2(\Omega)}.$$

Proof. Subtracting (39) from (15) we obtain the error equations

$$\int_\Omega \mu\,(\mathbf{u}-\mathbf{u}_h)\cdot\mathbf{v}\,dx - \int_\Omega (p-p_h)\,\mathrm{div}\,\mathbf{v}\,dx = 0 \ \ \forall \mathbf{v}\in \mathcal{RT}_k \tag{40}$$

and,

$$\int_\Omega q\ \mathrm{div}(\mathbf{u}-\mathbf{u}_h)\,dx = 0 \ \ \forall q\in \mathcal{P}_k^d. \tag{41}$$

Using (36) and (41) we obtain

$$\int_\Omega q\ \mathrm{div}(\Pi_h\mathbf{u}-\mathbf{u}_h)\,dx = 0 \ \ \forall q\in \mathcal{P}_k^d$$

and, since (37) holds, we can take $q=\mathrm{div}(\Pi_h\mathbf{u}-\mathbf{u}_h)$ to conclude that

$$\mathrm{div}(\Pi_h\mathbf{u}-\mathbf{u}_h)=0.$$

Therefore, taking $\mathbf{v}=\Pi_h\mathbf{u}-\mathbf{u}_h$ in (40) we obtain

$$\int_\Omega \mu\,(\mathbf{u}-\mathbf{u}_h)\cdot(\Pi_h\mathbf{u}-\mathbf{u}_h)\,dx = 0$$

and so,

$$\|\Pi_h\mathbf{u}-\mathbf{u}_h\|_{L^2(\Omega)}^2 \le \|a\|_{L^\infty(\Omega)}\int_\Omega \mu\,(\Pi_h\mathbf{u}-\mathbf{u})(\Pi_h\mathbf{u}-\mathbf{u}_h)\,dx$$

$$\le \|a\|_{L^\infty(\Omega)}\|\mu\|_{L^\infty(\Omega)}\|\Pi_h\mathbf{u}-\mathbf{u}\|_{L^2(\Omega)}\|\Pi_h\mathbf{u}-\mathbf{u}_h\|_{L^2(\Omega)}$$

and we conclude the proof by using the triangle inequality.

As a consequence, we have the following optimal order error estimate for the approximation of the vector variable $\mathbf{u}$.

Theorem 3.2. *If the solution* $\mathbf{u}$ *of problem (14) belongs to* $H^m(\Omega)^n$, $1\le m\le k+1$, *there exists a constant* C *depending on* $\|a\|_{L^\infty(\Omega)}$, $\|\mu\|_{L^\infty(\Omega)}$, k, n *and the regularity constant* σ, *such that*

$$\|\mathbf{u}-\mathbf{u}_h\|_{L^2(\Omega)} \le Ch^m\|\nabla^m\mathbf{u}\|_{L^2(\Omega)}.$$

Proof. The result is an immediate consequence of Lemma 3.6 and Theorem 3.1.

In the next theorem we obtain error estimates for the scalar variable p. We will use that

$$\|\mathbf{v}-\Pi_h\mathbf{v}\|_{L^2(\Omega)} \le Ch\|\mathbf{v}\|_{H^1(\Omega)} \ \ \forall\mathbf{v}\in H^1(\Omega) \tag{42}$$

which follows from a particular case of Theorem 3.1. In particular,

$$\|\Pi_h\mathbf{v}\|_{L^2(\Omega)} \le C\|\mathbf{v}\|_{H^1(\Omega)}. \tag{43}$$

Lemma 3.7. *If* $(\mathbf{u}, p)$ *and* $(\mathbf{u}_h, p_h)$ *are the solutions of (15) and (39), there exists a constant* C *depending on* $\|a\|_{L^\infty(\Omega)}$, $\|\mu\|_{L^\infty(\Omega)}$, k, n *and the regularity constant* σ, *such that*

$$\|p - p_h\|_{L^2(\Omega)} \leq C\{\|p - P_h p\|_{L^2(\Omega)} + \|\mathbf{u} - \Pi_h \mathbf{u}\|_{L^2(\Omega)}\}. \tag{44}$$

Proof. From (37) we know that for any $q \in \mathcal{P}_k^d$ there exists $\mathbf{w}_h \in \mathcal{RT}_k$ such that $\operatorname{div} \mathbf{w}_h = q$. Moreover, it is easy to see that $\mathbf{w}_h$ can be taken such that

$$\|\mathbf{w}_h\|_{L^2(\Omega)} \leq C\|q\|_{L^2(\Omega)}. \tag{45}$$

Indeed, recall that $\mathbf{w}_h = \Pi_h \mathbf{w}$ where $\mathbf{w} \in H^1(\Omega)$ satisfies $\operatorname{div} \mathbf{w} = q$ and $\|\mathbf{w}\|_{H^1(\Omega)} \leq C\|q\|_{L^2(\Omega)}$ (from Lemma 2.2 we know that such a $\mathbf{w}$ exists). Then, (45) follows from (43).

Now, from the error equation (40) we have

$$\int_\Omega (P_h p - p_h) \operatorname{div} \mathbf{v}\, dx = \int_\Omega (\mathbf{u} - \mathbf{u}_h)\mathbf{v}\, dx \qquad \forall \mathbf{v} \in \mathcal{RT}_k$$

and so, taking $\mathbf{v} \in V_h$ such that $\operatorname{div} \mathbf{v} = P_h p - p_h$ and

$$\|\mathbf{v}\|_{L^2(\Omega)} \leq C\|P_h p - p_h\|_{L^2(\Omega)},$$

we obtain

$$\|P_h p - p_h\|_{L^2(\Omega)}^2 \leq C\|\mathbf{u} - \mathbf{u}_h\|_{L^2(\Omega)}\|P_h p - p_h\|_{L^2(\Omega)}$$

which combined with Lemma 3.6 and the triangular inequality yields (44).

As a consequence, we obtain an error estimate for the approximation of the scalar variable p.

Theorem 3.3. *If the solution* $(\mathbf{u}, p)$ *of problem 14 belongs to* $H^m(\Omega)^n \times H^m(\Omega)$, $1 \leq m \leq k+1$, *there exists a constant* C *depending on* $\|a\|_{L^\infty(\Omega)}$, $\|\mu\|_{L^\infty(\Omega)}$, k, n *and the regularity constant* σ, *such that*

$$\|p - p_h\|_{L^2(\Omega)} \leq Ch^m\{\|\nabla^m \mathbf{u}\|_{L^2(\Omega)} + \|\nabla^m p\|_{L^2(\Omega)}\}. \tag{46}$$

Proof. The result follows immediately from Theorem 3.2, Lemma 3.7 and the error estimates for the L^2-projection given in (2.1).

For the case in which Ω is a convex polygon or a smooth domain and the coefficient a is smooth enough to have the a priori estimate

$$\|p\|_{H^2(\Omega)} \leq C_0\|f\|_{L^2(\Omega)} \tag{47}$$

we also obtain a higher order error estimate for $\|P_h p - p_h\|_{L^2(\Omega)}$ using a duality argument.

Lemma 3.8. *If $a \in W^{1,\infty}(\Omega)$ and (47) holds, there exists a constant C depending on $\|a\|_{W^{1,\infty}(\Omega)}$, $\|\mu\|_{L^\infty(\Omega)}$, k, n, C_0 and the regularity constant σ such that*

$$\|P_h p - p_h\|_{L^2(\Omega)} \le Ch\{\|\mathbf{u} - \mathbf{u}_h\|_{L^2(\Omega)} + \|\operatorname{div}(\mathbf{u} - \mathbf{u}_h)\|_{L^2(\Omega)}\}. \tag{48}$$

Proof. We use a duality argument. Let ϕ be the solution of

$$\begin{cases} \operatorname{div}(a\nabla\phi) = P_h p - p_h & \text{in } \Omega \\ \phi = 0 & \text{on } \partial\Omega. \end{cases}$$

Using (36), (37), (40), (41), and (42) we have,

$$\begin{aligned}
\|P_h p - p_h\|^2_{L^2(\Omega)} &= \int_\Omega (P_h p - p_h)\operatorname{div}(a\nabla\phi)\,dx \\
&= \int_\Omega (P_h p - p_h)\operatorname{div}\Pi_h(a\nabla\phi)\,dx = \int_\Omega (p - p_h)\operatorname{div}\Pi_h(a\nabla\phi)\,dx \\
&= \int_\Omega \mu(\mathbf{u} - \mathbf{u}_h)\cdot(\Pi_h(a\nabla\phi) - a\nabla\phi)\,dx + \int_\Omega (\mathbf{u} - \mathbf{u}_h)\cdot\nabla\phi\,dx \\
&= \int_\Omega \mu(\mathbf{u} - \mathbf{u}_h)\cdot(\Pi_h(a\nabla\phi) - a\nabla\phi)\,dx - \int_\Omega \operatorname{div}(\mathbf{u} - \mathbf{u}_h)(\phi - P_h\phi)\,dx \\
&\le C\|\mathbf{u} - \mathbf{u}_h\|_{L^2(\Omega)} h\|\phi\|_{H^2(\Omega)} + C\|\operatorname{div}(\mathbf{u} - \mathbf{u}_h)\|_{L^2(\Omega)} h\|\phi\|_{H^1(\Omega)}
\end{aligned}$$

where for the last inequality we have used that $a \in W^{1,\infty}(\Omega)$. The proof concludes by using the a priori estimate (47) for ϕ.

Theorem 3.4. *If $a \in W^{1,\infty}(\Omega)$, (47) holds, $\mathbf{u} \in H^{k+1}(\Omega)^n$ and $f \in H^{k+1}(\Omega)$, there exists a constant C depending on $\|a\|_{W^{1,\infty}(\Omega)}$, $\|\mu\|_{L^\infty(\Omega)}$, k, n, C_0 and the regularity constant σ such that*

$$\|P_h p - p_h\|_{L^2(\Omega)} \le Ch^{k+2}\{\|\nabla^{k+1}\mathbf{u}\|_{L^2(\Omega)} + \|\nabla^{k+1} f\|_{L^2(\Omega)}\}. \tag{49}$$

Proof. The second equation in (39) can be written as $\operatorname{div}\mathbf{u}_h = P_h f$. Then we have

$$\operatorname{div}(\mathbf{u} - \mathbf{u}_h) = f - P_h f$$

and, therefore, the theorem follows from Theorem 3.2 and Lemma 3.8 and the error estimates for the L^2-projection given in (2.1).

The estimate for $\|P_h p - p_h\|_{L^2(\Omega)}$ given by this theorem is important because it can be used to construct superconvergent approximations of p, i.e., approximations which converge at a higher order than p_h (see for example [11, 48])

For the sake of clarity we have presented the error analysis for the Raviart–Thomas spaces which were the first ones introduced for the mixed approximation of second order elliptic problems. However, as we mentioned above, the analysis makes use only of the existence of a projection Π_h satisfying the commutative diagram property and on approximation properties of Π_h and of the L^2-projection on the finite element space used to approximate the scalar variable p.

For the particular case of the Raviart–Thomas spaces the regularity assumption (9) can be replaced by the weaker "maximum angle condition" (see [1] for $k = 0$ and $n = 2, 3$, [28] for $k = 1$ and $n = 2$ and [29] for general $k \geq 0$ and $n = 2, 3$).

The Raviart–Thomas spaces were constructed in order to approximate both vector and scalar variables with the same order. However, if one is more interested in the approximation of the vector variable $\mathbf{u}$, one can try to use different order approximations for each variable in order to reduce the degrees of freedom (thus reducing the computational cost) while preserving the same order of convergence for $\mathbf{u}$ provided by the $\mathcal{RT}_k$ spaces. This is the main idea to define the following spaces which were introduced by Brezzi, Douglas and Marini [16]. Although with this choice the order of convergence for p is reduced, estimate (49) allows to improve it by a post-processing of the computed solution [16].

In the examples below, we will define the local spaces for each variable. It is not difficult to check that the degrees of freedom defining the spaces approximating the vector variable guarantee the continuity of the normal component and therefore the global spaces are subspaces of $H(\text{div}, \Omega)$.

For $n = 2$, $k \geq 1$ and T a triangle, the space $\mathcal{BDM}_k(T)$ is defined in the following way:

$$\mathcal{BDM}_k(T) = \mathcal{P}_k^2(T) \tag{50}$$

and the corresponding space for the scalar variable is $\mathcal{P}_{k-1}(T)$.

Observe that

$$dim\, \mathcal{BDM}_k(T) = (k+1)(k+2).$$

For example, $dim\, \mathcal{BDM}_1(T) = 6$ and $dim\, \mathcal{BDM}_2(T) = 12$. Figure 3 shows the degrees of freedom for these two spaces. The arrows correspond to degrees of freedom of normal components while the circles indicate the internal degrees of freedom corresponding to the second and third conditions in the definition of Π_T below.

In what follows, ℓ_i, $i = 1, 2, 3$ are the sides of T, $b_T = \lambda_1 \lambda_2 \lambda_3$ is a "bubble" function and, for $\phi \in H^1(\Omega)$,

$$\mathbf{curl}\, \phi = \left(\frac{\partial \phi}{\partial y}, -\frac{\partial \phi}{\partial x} \right).$$

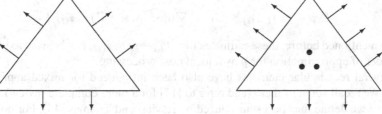

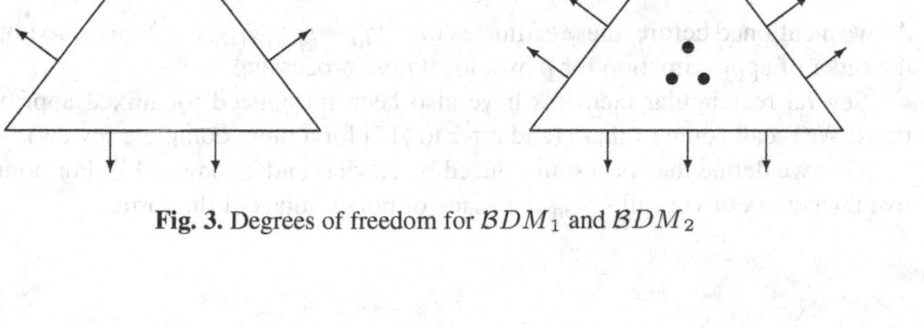

Fig. 3. Degrees of freedom for $\mathcal{BDM}_1$ and $\mathcal{BDM}_2$

The operator Π_T for this case is defined as follows:

$$\int_{\ell_i} \Pi_T \mathbf{v} \cdot \mathbf{n}_i p_k \, ds = \int_{\ell_i} \mathbf{v} \cdot \mathbf{n}_i p_k \, ds \ \forall p_k \in \mathcal{P}_k(\ell_i)\,, \ \ i = 1, 2, 3$$

$$\int_T \Pi_T \mathbf{v} \cdot \nabla p_{k-1} \, dx = \int_T \mathbf{v} \cdot \nabla p_{k-1} \, dx \ \forall p_{k-1} \in \mathcal{P}_{k-1}(T)$$

and, when $k \geq 2$

$$\int_T \Pi_T \mathbf{v} \cdot \mathbf{curl}\,(b_T p_{k-2}) \, dx = \int_T \mathbf{v} \cdot \mathbf{curl}\,(b_T p_{k-2}) \, dx \ \forall p_{k-2} \in \mathcal{P}_{k-2}(T).$$

The reader can check that all the conditions for convergence are satisfied in this case. Property (36) follows from the definition of Π_T and the proof of its existence is similar to that of Lemma 3.2. Consequently, the same arguments used for the Raviart–Thomas approximation provide the same error estimate for the approximation of $\mathbf{u}$ that we had in Theorem 3.2 while for p we have

$$\|p - p_h\|_{L^2(\Omega)} \leq C h^m \{\|\nabla^m \mathbf{u}\|_{L^2(\Omega)} + \|\nabla^m p\|_{L^2(\Omega)}\},$$

$1 \leq m \leq k$ and the estimate does not hold for $m = k + 1$ i.e., the best order of convergence is reduced in one with respect to the estimate obtained for the Raviart–Thomas approximation.

However, with the same argument used in Lemma 3.8 it can be proved that, for $k \geq 2$,

$$\|P_h p - p_h\|_{L^2(\Omega)} \leq C\{h\|\mathbf{u} - \mathbf{u}_h\|_{L^2(\Omega)} + h^2 \|\mathrm{div}(\mathbf{u} - \mathbf{u}_h)\|_{L^2(\Omega)}\},$$

indeed, since P_h is the orthogonal projection on $\mathcal{P}_{k-1}^d$ and $k - 1 \geq 1$, this follows by using that

$$\|\phi - P_h \phi\|_{L^2(\Omega)} \leq C h^2 \|\phi\|_{H^2(\Omega)} \tag{51}$$

in the last step of the proof of that lemma.

Therefore, for $k \geq 2$, we obtain the following result analogous to that in Theorem 3.4

$$\|P_h p - p_h\|_{L^2(\Omega)} \leq C h^{k+2} \{\|\nabla^{k+1} \mathbf{u}\|_{L^2(\Omega)} + \|\nabla^k f\|_{L^2(\Omega)}\}.$$

On the other hand, if $k = 1$, (51) does not hold (because in this case P_h is the projection over piecewise constant functions). Then, in this case we can prove only

$$\|P_h p - p_h\|_{L^2(\Omega)} \leq C h^2 \{\|\nabla \mathbf{u}\|_{L^2(\Omega)} + \|\nabla f\|_{L^2(\Omega)}\}.$$

As we mentioned before, these estimates for $\|P_h p - p_h\|_{L^2(\Omega)}$ can be used to improve the order of approximation for p by a local post-processing.

Several rectangular elements have also been introduced for mixed approximations. We recall some of them (and refer to [17] for a more complete review).

First we define the spaces introduced by Raviart and Thomas [44]. For nonnegative integers k, m we call $\mathcal{Q}_{k,m}$ the space of polynomials of the form

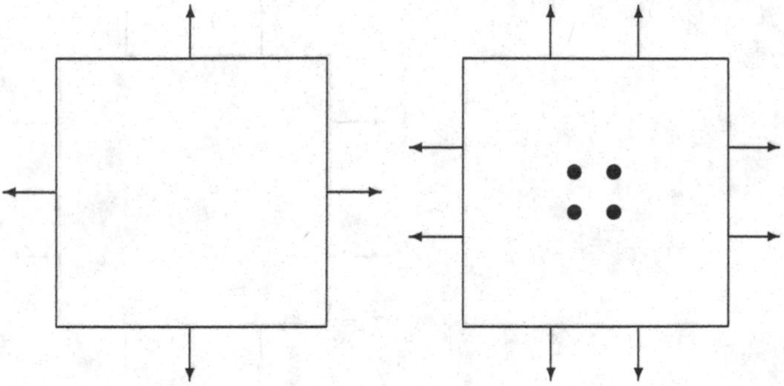

Fig. 4. Degrees of freedom for $\mathcal{RT}_0$ and $\mathcal{RT}_1$

$$q(x,y) = \sum_{i=0}^{k}\sum_{j=0}^{m} a_{ij}x^i y^j$$

then, the $\mathcal{RT}_k(R)$ space on a rectangle R is given by

$$\mathcal{RT}_k(R) = \mathcal{Q}_{k+1,k}(R) \times \mathcal{Q}_{k,k+1}(R)$$

and the space for the scalar variable is $\mathcal{Q}_k(R)$. It can be easily checked that

$$dim\,\mathcal{RT}_k(R) = 2(k+1)(k+2).$$

Figure 4 shows the degrees of freedom for $k = 0$ and $k = 1$.

Denoting with ℓ_i, $i = 1,2,3,4$ the four sides of R, the degrees of freedom defining the operator Π_T for this case are

$$\int_{\ell_i} \Pi_T \mathbf{v}\cdot\mathbf{n}_i p_k\,d\ell = \int_{\ell_i} \mathbf{v}\cdot\mathbf{n}_i p_k\,d\ell \;\forall p_k \in \mathcal{P}_k(\ell_i),\; i = 1,2,3,4$$

and (for $k \geq 1$)

$$\int_R \Pi_T\mathbf{v}\cdot\phi_k\,dx = \int_R \mathbf{v}\cdot\phi_k\,dx \;\;\forall\phi_k \in \mathcal{Q}_{k-1,k}(R)\times\mathcal{Q}_{k,k-1}(R).$$

Our last example in the 2D case are the spaces introduced by Brezzi, Douglas and Marini on rectangular elements. They are defined for $k \geq 1$ as

$$\mathcal{BDM}_k(R) = \mathcal{P}_k^2(R) + \langle\mathbf{curl}\,(x^{k+1}y)\rangle + \langle\mathbf{curl}\,(xy^{k+1})\rangle$$

and the associated scalar space is $\mathcal{P}_{k-1}(R)$. It is easy to see that

$$dim\,\mathcal{BDM}_k(R) = (k+1)(k+2)+2.$$

The degrees of freedom for $k = 1$ and $k = 2$ are shown in Fig. 5.

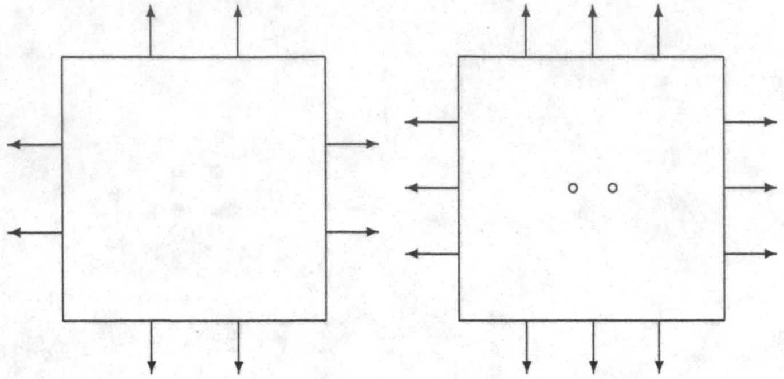

Fig. 5. Degrees of freedom for $\mathcal{BDM}_1$ and $\mathcal{BDM}_2$

The operator Π_T is defined by

$$\int_{\ell_i} \Pi_T \mathbf{v} \cdot \mathbf{n}_i p_k \, d\ell = \int_{\ell_i} \mathbf{v} \cdot \mathbf{n}_i p_k \, d\ell \;\; \forall p_k \in \mathcal{P}_k(\ell_i)\,, \;\; i = 1, 2, 3, 4$$

and (for $k \geq 2$)

$$\int_R \Pi_T \mathbf{v} \cdot \mathbf{p}_{k-2} \, dx = \int_R \mathbf{v} \cdot \mathbf{p}_{k-2} \, dx \;\; \forall \mathbf{p}_{k-2} \in \mathcal{P}^2_{k-2}(R).$$

The $\mathcal{RT}_k$ as well as the $\mathcal{BDM}_k$ spaces on rectangles have analogous properties to those on triangles. Therefore, the same error estimates obtained for triangular elements are valid in both cases.

More generally, one can consider general quadrilateral elements. Given a convex quadrilateral Q, the spaces are defined using the Piola transform from a reference rectangle R to Q. Let us define for example the Raviart–Thomas spaces $\mathcal{RT}_k(Q)$.

Let $R = [0,1] \times [0,1]$ be the reference rectangle and $F : R \to Q$ a bilinear transformation taking the vertices of R into the vertices of Q. Then, we define the local space $\mathcal{RT}_k(R)$ by using the Piola transform, i.e., if $x = F(\hat{x})$, DF is the Jacobian matrix of F and $J = |\det DF|$,

$$\mathcal{RT}_k(Q) = \{\mathbf{v} : Q \to \mathbb{R}^2 \, : \, \mathbf{v}(x) = \frac{1}{J(\hat{x})} DF(\hat{x}) \hat{\mathbf{v}}(\hat{x}) \;\text{ with }\; \hat{\mathbf{v}} \in \mathcal{RT}_k(R)\}.$$

Also in this case similar error estimates to those obtained for triangular elements can be proved under appropriate regularity assumptions on the quadrilaterals. The analysis of this case is more technical and so we omit details and refer to [5, 37, 49].

3-d extensions of the spaces defined above have been introduced by Nédélec [41, 42] and by Brezzi, Douglas, Durán and Fortin [14]. For tetrahedral elements the spaces are defined in an analogous way, although the construction of the operator Π_T requires a different analysis (we refer to [41] for the extension of the $\mathcal{RT}_k$ spaces and to [42, 14] for the extension of the $\mathcal{BDM}_k$ spaces). In the case of 3-d rectangular elements, the extensions of $\mathcal{RT}_k$ are again defined in an analogous way [41] and the extensions of $\mathcal{BDM}_k$ [14] can be defined for a 3-d rectangle R by

$$\begin{aligned}\mathcal{BDDF}_k(R) = \mathcal{P}_k^3 &+ \langle\{\mathbf{curl}\,(0,0,xy^{i+1}z^{k-i}),\ i=0,\dots,k\}\rangle \\ &+\langle\{\mathbf{curl}\,(0,x^{k-i}yz^{i+1},0),\ i=0,\dots,k\}\rangle \\ &+\langle\{\mathbf{curl}\,(x^{i+1}y^{k-i}z,0,0),\ i=0,\dots,k\}\rangle\end{aligned}$$

where now we are using the usual notation $\mathbf{curl}\,\mathbf{v}$ for the rotational of a 3-d vector field $\mathbf{v}$.

All the convergence results obtained in 2-d can be extended for the 3-d spaces mentioned here. Other families of spaces, in both 2 and 3 dimensions which are intermediate between the $\mathcal{RT}$ and the $\mathcal{BDM}$ spaces were introduced and analyzed by Brezzi, Douglas, Fortin and Marini [15].

Finally, we refer to [10] for the case of general isoparametric hexahedral elements.

4 A Posteriori Error Estimates

In this section we present an a posteriori error analysis for the mixed finite element approximation of second order elliptic problems. For simplicity, we will assume that the restriction of the coefficient a in (13) to any element of the triangulation is constant. If not, higher order terms corresponding to the approximation of a arise in the estimates.

For simplicity, we prove the results for the approximations obtained by the Raviart–Thomas spaces and in the 2-d case. However, simple variants of the method can be applied for mixed approximations in other spaces, in particular, for all the spaces described in the previous section.

We introduce error estimators of the residual type for both scalar and vector variables and prove that the error is bounded by a constant times the estimator plus a term which is of higher order (i.e., what is usually called "reliability" of the estimator). We also prove that the estimator is less than or equal a constant times the error. This last estimate (usually called "efficiency" of the estimator) is local, more precisely, the error in one element T can be bounded below by the estimators in the same triangle plus the estimators in the elements sharing a side with T.

It is well known that several mixed methods are related to nonconforming finite element approximations (see [6]). In particular the lowest order Raviart–Thomas method corresponds to the nonconforming linear elements of Crouzeix–Raviart (see also [40]).

A posteriori error estimates were obtained first for the Crouzeix–Raviart method by using a Helmoltz type decomposition of the error (see [23]). The same technique has been applied for mixed finite element approximations in [4, 18]. In [4] only the vector variable is estimated while in [18] both variables are estimated, but to estimate the scalar variable the a priori estimate (47) was assumed to hold. In particular, this hypothesis excludes nonconvex polygonal domains. We refer also to [3, 39] for related results.

Our analysis for the vector variable follows the approach of [4, 18], while for the scalar variable we present a new argument which does not require the a priori estimate (47).

We will use the following well-known approximation result. We denote with $\mathcal{P}_{k+1}^c$ the standard continuous piecewise polynomials of degree $k+1$. For any $\phi \in H^1(\Omega)$ there exists $\phi_h \in \mathcal{P}_{k+1}^c$ such that

$$\|\phi - \phi_h\|_{L^2(\ell)} \leq C|\ell|^{1/2}\|\nabla\phi\|_{L^2(\widetilde{T})} \tag{52}$$

and,

$$\|\phi - \phi_h\|_{L^2(T)} \leq C|T|^{1/2}\|\nabla\phi\|_{L^2(\widetilde{T})} \tag{53}$$

where $\widetilde{T}$ is the union of all the elements sharing a vertex with T (we can take for example the Clément approximation [21] or any variant of it (see for example [37, 47]).

We will use the notation $\mathbf{curl}\,\phi$ introduced in the previous section for $\phi \in H^1(\Omega)$ and for $\mathbf{v} \in H^1(\Omega)^2$ we define

$$\operatorname{rot}\mathbf{v} = \frac{\partial v_2}{\partial x} - \frac{\partial v_1}{\partial y}.$$

Also, for a field $\mathbf{v}$ such that its restriction $\mathbf{v}|_T$ to each $T \in \mathcal{T}_h$ belongs to $H^1(T)^2$ we will denote with $\operatorname{rot}_h\mathbf{v}$ the function such that its restriction to T is given by $\operatorname{rot}(\mathbf{v}|_T)$.

For an element T, let E_T be the set of edges of T and t be the unit tangent on ℓ oriented clockwise. For an interior side ℓ, $[\![\mathbf{u}_h\cdot t]\!]_\ell$ denotes the jump of the tangential component of $\mathbf{u}_h$, namely, if T_1 and T_2 are the triangles sharing ℓ, and t_1 and t_2 the corresponding unit tangent vectors on ℓ then

$$[\![\mathbf{u}_h \cdot t]\!]_\ell = \mathbf{u}_h|_{T_1} \cdot t_1 - \mathbf{u}_h|_{T_2} \cdot t_1 = \mathbf{u}_h|_{T_1} \cdot t_1 + \mathbf{u}_h|_{T_2} \cdot t_2.$$

We define

$$J_\ell = \begin{cases} [\![\mathbf{u}_h \cdot t]\!]_\ell & \text{if} \quad \ell \not\subset \partial\Omega \\ 2\mathbf{u}_h \cdot t & \text{if} \quad \ell \subset \partial\Omega. \end{cases}$$

We now introduce the estimator for the vector variable and prove the efficiency and reliability of this estimator.

The local error estimator is defined by

$$\eta_{vect,T}^2 = |T|\|\operatorname{rot}_h\mathbf{u}_h\|_{L^2(T)}^2 + \sum_{\ell\in E_T} |\ell|\|J_\ell\|_{L^2(\ell)}^2$$

and the global one by,

$$\eta_{vect}^2 = \sum_{T\in\mathcal{T}_h} \eta_{vect,T}^2.$$

The key point to prove the reliability of the estimator is to decompose the error by using a generalized Helmholtz decomposition given in the next lemma.

Lemma 4.1. *If the domain Ω is simply connected and $\mathbf{v} \in L^2(\Omega)^2$, then there exist $\psi \in H_0^1(\Omega)$ and $\phi \in H^1(\Omega)$ such that*

$$\mathbf{v} = a\nabla\psi + \mathbf{curl}\,\phi \tag{54}$$

and

$$\|\nabla\phi\|_{L^2(\Omega)} + \|\nabla\psi\|_{L^2(\Omega)} \leq C\|\mathbf{v}\|_{L^2(\Omega)} \tag{55}$$

with a constant C depending only on a.

Proof. To obtain this decomposition we solve the problem

$$\operatorname{div}(a\nabla\psi) = \operatorname{div}\mathbf{v}$$

with $\psi \in H_0^1(\Omega)$, namely, ψ satisfies

$$\int_\Omega a\nabla\psi \cdot \nabla\xi = \int_\Omega \mathbf{v} \cdot \nabla\xi \qquad \forall\xi \in H_0^1(\Omega).$$

In particular, choosing $\xi = \psi$ we obtain

$$\|\nabla\psi\|_{L^2(\Omega)} \leq C\|\mathbf{v}\|_{L^2(\Omega)}. \tag{56}$$

Now, since

$$\operatorname{div}(\mathbf{v} - a\nabla\psi) = 0,$$

and the domain is simply connected, there exists $\phi \in H^1(\Omega)$ such that (54) holds.

Moreover, observe that (55) follows easily from (56) and (54).

Theorem 4.1. *If Ω is simply connected and the restriction of a to any $T \in \mathcal{T}_h$ is constant, there exists a constant C_1 such that*

$$\|\mathbf{u} - \mathbf{u}_h\|_{L^2(\Omega)} \leq C_1\{\eta_{vect} + h\|f - P_h f\|_{L^2(\Omega)}\}. \tag{57}$$

Proof. For $\phi \in H^1(\Omega)$ we have

$$\int_\Omega \mu\mathbf{u} \cdot \mathbf{curl}\,\phi\,dx = -\int_\Omega \nabla p \cdot \mathbf{curl}\,\phi\,dx = 0..$$

Analogously, for $\phi_h \in \mathcal{P}_{k+1}^c$, $\mathbf{curl}\,\phi_h \in \mathcal{RT}_k$ and therefore, using the first equation in (39),

$$\int_\Omega \mu\mathbf{u}_h \cdot \mathbf{curl}\,\phi_h\,dx = 0.$$

Then,

$$\int_\Omega \mu\,(\mathbf{u}-\mathbf{u}_h)\cdot\mathbf{curl}\,\phi\,dx = -\int_\Omega \mu\,\mathbf{u}_h\cdot\mathbf{curl}\,(\phi-\phi_h)\,dx$$

$$= -\sum_T \Big\{ \int_T \mathrm{rot}_h(\mu\mathbf{u}_h)\,(\phi-\phi_h)\,dx + \int_{\partial T} \mu\mathbf{u}_h\cdot t\,(\phi-\phi_h)\,ds \Big\}$$

$$= -\sum_T \Big\{ \int_T \mathrm{rot}_h(\mu\mathbf{u}_h)\,(\phi-\phi_h)\,dx + \frac{1}{2}\sum_{\ell\in E_T}\int_\ell J_\ell\,(\phi-\phi_h)\,ds \Big\}.$$

Then, if $\phi_h \in \mathcal{P}^c_{k+1}$ is an approximation of ϕ satisfying (52) and (53), applying the Schwarz inequality we obtain

$$\int_\Omega \mu\,(\mathbf{u}-\mathbf{u}_h)\cdot\mathbf{curl}\,\phi\,dx \le C\eta_{vect}|\phi|_{1,\Omega}. \tag{58}$$

On the other hand, if $\psi\in H^1_0(\Omega)$ we have

$$\int_\Omega \mu\,(\mathbf{u}-\mathbf{u}_h)\cdot a\,\nabla\psi\,dx = \int_\Omega (\mathbf{u}-\mathbf{u}_h)\cdot\nabla\psi\,dx$$

$$= -\int_\Omega \mathrm{div}(\mathbf{u}-\mathbf{u}_h)\,\psi\,dx = -\int_\Omega (f-P_hf)\,\psi\,dx = -\int_\Omega (f-P_hf)\,(\psi-P_h\psi)\,dx$$

and, therefore, using that

$$\|\psi-P_h\psi\|_{L^2(\Omega)} \le Ch\|\nabla\psi\|_{L^2(\Omega)},$$

which follows immediately from Corollary 2.1, we obtain

$$\int_\Omega (\mathbf{u}-\mathbf{u}_h)\cdot\nabla\psi\,dx \le Ch\|f-P_hf\|_{L^2(\Omega)}\|\nabla\psi\|_{L^2(\Omega)}. \tag{59}$$

Using now Lemma 4.1 for $\mathbf{v}=\mathbf{u}-\mathbf{u}_h$ we have

$$\mathbf{u}-\mathbf{u}_h = a\nabla\psi + \mathbf{curl}\,\phi$$

with $\psi\in H^1_0(\Omega)$ and $\phi\in H^1(\Omega)$ such that

$$\|\nabla\phi\|_{L^2(\Omega)} + \|\nabla\psi\|_{L^2(\Omega)} \le C\|\mathbf{u}-\mathbf{u}_h\|_{L^2(\Omega)}. \tag{60}$$

Then,

$$\|\mathbf{u}-\mathbf{u}_h\|^2_{L^2(\Omega)} \le C\Big\{\int_\Omega \mu(\mathbf{u}-\mathbf{u}_h)\cdot\mathbf{curl}\,\phi\,dx + \int_\Omega (\mathbf{u}-\mathbf{u}_h)\cdot\nabla\psi\,dx\Big\}$$

and therefore (57) follows immediately from (58), (59) and (60).

To prove the efficiency we will use a well-known argument of Verfürth [50, 52]. In our case this argument will make use of the following lemma.

Lemma 4.2. *Given a triangle T and functions $q_T\in L^2(T)$ and, for each side ℓ of T, $p_\ell\in L^2(\ell)$, there exists $\phi\in\mathcal{P}_{k+3}(T)$ such that*

$$\begin{cases} \int_T \phi\, r\, dx = \int_T q_T\, r\, dx & \forall r \in \mathcal{P}_k(T), \\ \int_\ell \phi\, s\, dx = \int_\ell p_\ell\, s\, dx & \forall s \in \mathcal{P}_{k+1}(\ell)\ \forall \ell \in E_T, \\ \phi = 0 & \textit{at the vertices of } T. \end{cases} \tag{61}$$

Moreover,

$$\|\nabla\phi\|_{L^2(T)} \le C\{|T|^{-\frac{1}{2}}\|q_T\|_{L^2(T)} + \sum_{\ell\in E_T} |\ell|^{-\frac{1}{2}}\|p_\ell\|_{L^2(\ell)}\}. \tag{62}$$

Proof. The number of conditions is

$$\dim \mathcal{P}_k(T) + 3\dim \mathcal{P}_{k+1}(\ell) = \frac{(k+2)(k+1)}{2} + 3(k+2) = \frac{(k+2)(k+7)}{2}$$

while the dimension of the subspace of $\mathcal{P}_{k+3}$ of polynomials vanishing at the vertices of T is

$$\dim \mathcal{P}_{k+3}(T) - 3 = \frac{(k+4)(k+5)}{2} - 3 = \frac{(k+2)(k+7)}{2}.$$

Therefore, (61) is a square system and so it is enough to show the uniqueness. So, assume that

$$\begin{cases} \int_T \phi\, r\, dx = 0 & \forall r \in \mathcal{P}_k(T) \\ \int_\ell \phi\, s\, dx = 0 & \forall s \in \mathcal{P}_{k+1}(\ell)\ \forall \ell \in E_T \\ \phi = 0 & \text{at the vertices of } T. \end{cases} \tag{63}$$

Since ϕ vanishes at the vertices of ℓ, it follows from the second condition in (63) that $\phi = 0$ on the sides of T. Then,

$$\phi = \lambda_1\lambda_2\lambda_3\, r \qquad \text{with } r \in \mathcal{P}_k$$

and, therefore, it follows from the first condition in (63) that $\phi = 0$.

We will call D_T the union of T with the triangles sharing a side with it.

Theorem 4.2. *If the restriction of a to any $T \in \mathcal{T}_h$ is constant, there exists a constant C_2 such that, for any $T \in \mathcal{T}_h$,*

$$\eta_{vect,T} \le C_2\|\mathbf{u} - \mathbf{u}_h\|_{L^2(D_T)}. \tag{64}$$

Proof. We apply Lemma 4.2 on T and its neighbors T_i, $i = 1, 2, 3$ (we assume that T does not have a side on $\partial\Omega$, trivial modifications are needed if this is not the case). In this way we can construct $\phi \in H_0^1(D_T)$ vanishing at the vertices of T and T_i, $i = 1, 2, 3$ and such that

$$\int_T \phi\, r\, dx = -\int_T |T|\, \mathrm{rot}_h(\mu\mathbf{u}_h)\, r\, dx \qquad \forall r \in \mathcal{P}_k(T) \tag{65}$$

$$\int_\ell \phi\, s\, dx = -\int_\ell |\ell| J_\ell\, s\, dx \qquad \forall s \in \mathcal{P}_{k+1}(\ell), \forall \ell \in E_T, \tag{66}$$

$$\int_{T_i} \phi\, r\, dx = 0 \qquad \forall r \in \mathcal{P}_k(T_i) \tag{67}$$

and

$$\int_\ell \phi\, s\, dx = 0 \qquad \forall s \in \mathcal{P}_{k+1}(\ell) \quad \text{on the other two sides of} \quad T_i. \tag{68}$$

Since ϕ vanishes at the boundary of D_T we can extend it by zero to obtain a function $\phi \in H_0^1(\Omega)$. Then,

$$\int_\Omega \mu\, (\mathbf{u} - \mathbf{u}_h) \cdot \mathbf{curl}\, \phi\, dx = -\sum_T \Big\{ \int_T \mathrm{rot}_h(\mu \mathbf{u}_h)\, \phi\, dx + \frac{1}{2} \sum_{\ell \in E_T} \int_\ell J_\ell\, \phi\, ds \Big\}. \tag{69}$$

But,

$$\mathrm{rot}_h(\mu \mathbf{u}_h)|_T \in \mathcal{P}_k(T) \quad \text{and} \quad J_\ell \in \mathcal{P}_{k+1}(\ell),$$

therefore, we can take $r = \mathrm{rot}\,(\mu \mathbf{u}_h)$ and, for each $\ell \in E_T$, $s = J_\ell$ in (65) and (66) respectively to obtain

$$\int_T \mathrm{rot}_h(\mu \mathbf{u}_h)\, \phi\, dx = -|T| \|\mathrm{rot}_h \mathbf{u}_h\|^2_{L^2(T)}$$

and

$$\sum_{\ell \in E_T} \int_\ell J_\ell\, \phi\, ds = -\sum_{\ell \in E_T} |\ell| \|J_\ell\|^2_{L^2(\ell)}.$$

Analogously, using now (66), (67), (68), we obtain

$$\int_{T_i} \mathrm{rot}_h(\mu \mathbf{u}_h)\, \phi\, dx = 0, \quad i = 1, 2, 3$$

and

$$\sum_{\tilde\ell \in E_{T_i}} \int_{\tilde\ell} J_{\tilde\ell}\, \phi\, ds = -|\ell| \|J_\ell\|^2_{L^2(\ell)}, \quad i = 1, 2, 3$$

where $\ell = T \cap T_i$.

Therefore, recalling that ϕ vanishes outside D_T, it follows from (69) that

$$\eta^2_{vect,T} = \int_{D_T} \mu\, (\mathbf{u} - \mathbf{u}_h) \cdot \mathbf{curl}\, \phi\, dx$$

and so,

$$\eta^2_{vect,T} \le C \|\mathbf{u} - \mathbf{u}_h\|_{L^2(\Omega)} \|\nabla \phi\|_{L^2(D_T)}.$$

But, using (62) we have

$$\|\nabla \phi\|_{L^2(D_T)} \le C\{|T|^{\frac{1}{2}} \|\mathrm{rot}_h(\mu \mathbf{u}_h)\|_{L^2(T)} + \sum_{\ell \in E_T} |\ell|^{\frac{1}{2}} \|J_\ell\|_{L^2(\ell)}\} \le C \eta_{vect,T}$$

and therefore (64) holds.

To estimate the error in the scalar variable p we introduce the local estimator

$$\eta_{sc,T}^2 = |T| \|\nabla_h p_h + \mu \mathbf{u}_h\|_{L^2(T)}^2 + \sum_{\ell \in E_T} |\ell| \| [\![p_h]\!]_\ell \|_{L^2(\ell)}^2$$

where $[\![p_h]\!]_\ell$ denotes the jump of p_h across the side ℓ if ℓ is an interior side or $[\![p_h]\!]_\ell = 2p_h$ if $\ell \subset \partial\Omega$ and, for a function q such that its restriction to each $T \in \mathcal{T}_h$ belongs to $H^1(T)$ we denote with $\nabla_h q$ the function such that its restriction to T is given by $\nabla(q|_T)$.

Then, the global estimator is defined as usual by

$$\eta_{sc}^2 = \sum_{T \in \mathcal{T}_h} \eta_{sc,T}^2 \, .$$

The next lemma shows that the error in the scalar variable is bounded by η_{sc} plus the error in the vector variable.

Apart from (30) we will use the following error estimate which can be obtained in a similar way.

If ℓ is a side of an element T we have

$$\|(\mathbf{v} - \Pi_T \mathbf{v}) \cdot \mathbf{n}\|_{L^2(\ell)} \leq C |\ell|^{\frac{1}{2}} \|\nabla \mathbf{v}\|_{L^2(T)} . \tag{70}$$

Lemma 4.3. *There exists a constant C such that*

$$\|p - p_h\|_{L^2(\Omega)} \leq C\{\eta_{sc} + \|\mathbf{u} - \mathbf{u}_h\|_{L^2(\Omega)}\} .$$

Proof. By Lemma 2.2 we know that there exists $\mathbf{v} \in H^1(\Omega)^2$ such that

$$\operatorname{div} \mathbf{v} = p - p_h \tag{71}$$

and

$$\|\mathbf{v}\|_{H^1(\Omega)} \leq C \|p - p_h\|_{L^2(\Omega)} \tag{72}$$

with a constant C depending only on the domain.

Then,

$$\begin{aligned}
\|p - p_h\|_{L^2(\Omega)}^2 &= \int_\Omega (p - p_h) \operatorname{div} \mathbf{v} \, dx \\
&= \int_\Omega (p - p_h) \operatorname{div}(\mathbf{v} - \Pi_h \mathbf{v}) \, dx + \int_\Omega (p - p_h) \operatorname{div} \Pi_h \mathbf{v} \, dx \\
&= \int_\Omega (p - p_h) \operatorname{div}(\mathbf{v} - \Pi_h \mathbf{v}) \, dx \\
&\quad - \int_\Omega \mu (\mathbf{u} - \mathbf{u}_h) \cdot (\mathbf{v} - \Pi_h \mathbf{v}) \, dx + \int_\Omega \mu (\mathbf{u} - \mathbf{u}_h) \cdot \mathbf{v} \, dx .
\end{aligned} \tag{73}$$

But using that

$$\int_\Omega p \operatorname{div}(\mathbf{v} - \Pi_h \mathbf{v})\, dx - \int_\Omega \mu \mathbf{u} \cdot (\mathbf{v} - \Pi_h \mathbf{v})\, dx = 0$$

and integrating by parts on each element we have

$$\begin{aligned}
\int_\Omega (p - p_h) \operatorname{div}(\mathbf{v} - \Pi_h \mathbf{v})\, dx &- \int_\Omega \mu(\mathbf{u} - \mathbf{u}_h) \cdot (\mathbf{v} - \Pi_h \mathbf{v})\, dx \\
&= \sum_{T \in \mathcal{T}_h} \Big\{ \int_T \nabla_h p_h \cdot (\mathbf{v} - \Pi_h \mathbf{v})\, dx - \int_{\partial T} p_h (\mathbf{v} - \Pi_h \mathbf{v}) \cdot \mathbf{n}\, ds \\
&\quad + \int_T \mu \mathbf{u}_h \cdot (\mathbf{v} - \Pi_h \mathbf{v})\, dx \Big\} \\
&= \sum_{T \in \mathcal{T}_h} \Big\{ \int_T (\nabla_h p_h + \mu \mathbf{u}_h) \cdot (\mathbf{v} - \Pi_h \mathbf{v})\, dx \\
&\quad - \frac{1}{2} \sum_{\ell \in E_T} \int_\ell [[p_h]]_\ell (\mathbf{v} - \Pi_h \mathbf{v}) \cdot \mathbf{n}\, ds \Big\}.
\end{aligned}$$

Therefore, the Lemma follows from this equality and (73) using the Schwarz inequality and the error estimates (30) and (70).

Using now the results for the vector variable we obtain the following a posteriori error estimate for the scalar variable.

Theorem 4.3. *If Ω is simply connected and the restriction of a to any $T \in \mathcal{T}_h$ is constant, there exists a constant C_3 such that*

$$\|p - p_h\|_{L^2(\Omega)} \le C_3 \{\eta_{sc} + \eta_{vect} + h \|f - P_h f\|_{L^2(\Omega)}\}. \tag{74}$$

Proof. This result follows immediately from Theorem 4.1 and Lemma 4.3.

To prove the efficiency of η_{sc} we first prove that the jumps involved in the definition of the estimator can be bounded by the error plus the other part of the estimator.

Lemma 4.4. *There exists a constant C such that*

$$|\ell|^{\frac{1}{2}} \| [[p_h]]_\ell \|_{L^2(\ell)} \le C \{ \|p - p_h\|_{L^2(D_\ell)} + |\ell| \|\mathbf{u} - \mathbf{u}_h\|_{L^2(D_\ell)} + |\ell| \|\nabla_h p_h + \mu \mathbf{u}_h\|_{L^2(D_\ell)} \}$$

where D_ℓ is the union of the triangles sharing ℓ.

Proof. If $\ell \in E_T$, it follows from the regularity assumption on the meshes that $|\ell| \sim h_T \sim |T|^{\frac{1}{2}}$. Now, since p is continuous we have

$$\| [[p_h]]_\ell \|_{L^2(\ell)} = \| [[p_h - p]]_\ell \|_{L^2(\ell)}$$

and so, applying the trace inequality (1) and the standard scaling argument, we obtain

$$\begin{aligned}|\ell|^{\frac{1}{2}} \| [[p_h]]_\ell \|_{L^2(\ell)} &\leq C\{\|p_h - p\|_{L^2(D_\ell)} + |\ell| \|\nabla_h(p_h - p)\|_{L^2(D_\ell)}\} \\ &\leq C\{\|p_h - p\|_{L^2(D_\ell)} + |\ell| \|\nabla_h p_h + \mu \mathbf{u}\|_{L^2(D_\ell)}\} \\ &\leq C\{\|p_h - p\|_{L^2(D_\ell)} + |\ell| \|\nabla_h p_h + \mu \mathbf{u}_h\|_{L^2(D_\ell)} \\ &\quad + |\ell| \|\mu(\mathbf{u} - \mathbf{u}_h)\|_{L^2(D_\ell)}\}\end{aligned}$$

concluding the proof because μ is bounded.

Now, in order to bound $\|\nabla_h p_h + \mu \mathbf{u}\|_{L^2(T)}$ by the error we will use again the argument of Verfürth.

Lemma 4.5. *There exists a constant C such that*

$$|T|^{\frac{1}{2}} \|\nabla_h p_h + \mu \mathbf{u}_h\|_{L^2(T)} \leq C\{|T|^{\frac{1}{2}} \|\mathbf{u} - \mathbf{u}_h\|_{L^2(T)} + \|p - p_h\|_{L^2(T)}\}. \quad (75)$$

Proof. Using again that

$$\int_\Omega \mu \mathbf{u} \cdot \mathbf{v} \, dx - \int_\Omega p \operatorname{div} \mathbf{v} \, dx = 0 \qquad \forall \mathbf{v} \in H_0^1(\Omega)^2$$

we have, for any $\mathbf{v} \in H_0^1(T)^2$,

$$\begin{aligned}\int_T \mu (\mathbf{u} - \mathbf{u}_h) \cdot \mathbf{v} \, dx - \int_T (p - p_h) \operatorname{div} \mathbf{v} \, dx &= - \int_T \mu \, \mathbf{u}_h \cdot \mathbf{v} \, dx + \int_T p_h \operatorname{div} \mathbf{v} \, dx \\ = - \int_T \mu \, \mathbf{u}_h \cdot \mathbf{v} \, dx - \int_T \nabla_h p_h \cdot \mathbf{v} \, dx &= - \int_T (\nabla_h p_h + \mu \mathbf{u}_h) \cdot \mathbf{v} \, dx.\end{aligned}$$

Choosing now $\mathbf{v} = -b_T(\nabla_h p_h + \mu \mathbf{u}_h)$, with $b_T \in \mathcal{P}_3(T)$ vanishing at the boundary and equal to one at the barycenter of T, we obtain

$$\int_T \mu (\mathbf{u} - \mathbf{u}_h) \cdot \mathbf{v} \, dx - \int_T (p - p_h) \operatorname{div} \mathbf{v} \, dx = \int_T |\nabla_h p_h + \mu \mathbf{u}_h|^2 b_T \, dx. \quad (76)$$

But, since $\nabla_h p_h + \mu \mathbf{u}_h \in \mathcal{P}_{k+1}(T)$, a standard argument (equivalence of norms in a reference element and an affine change of variables) gives

$$\int_T |\nabla_h p_h + \mu \mathbf{u}_h|^2 \, dx \leq C \int_T |\nabla_h p_h + \mu \mathbf{u}_h|^2 b_T \, dx,$$

which together with (76) and the Schwarz inequality yields

$$\|\nabla_h p_h + \mu \mathbf{u}_h\|_{L^2(T)}^2 \leq C\{\|\mathbf{u} - \mathbf{u}_h\|_{0,T} \|\mathbf{v}\|_{L^2(T)} + \|p - p_h\|_{L^2(T)} \|\nabla \mathbf{v}\|_{L^2(T)}\} \quad (77)$$

but, since b_T is bounded by a constant independent of T we know that

$$\|\mathbf{v}\|_{L^2(T)} \leq C \|\nabla_h p_h + \mu \mathbf{u}_h\|_{L^2(T)}$$

and, using the inverse inequality given in Lemma 2.3,

$$\|\nabla \mathbf{v}\|_{L^2(T)} \leq C |T|^{-\frac{1}{2}} \|\nabla_h p_h + \mu \mathbf{u}_h\|_{L^2(T)}$$

and, therefore, (75) follows from (77).

Collecting the lemmas we can prove the efficiency of the estimator η_{sc}.

Theorem 4.4. *If the restriction of a to any $T \in \mathcal{T}_h$ is constant, there exists a constant C_4 such that, for any $T \in \mathcal{T}_h$,*

$$\eta_{sc,T} \le C_4\{|T|^{\frac{1}{2}}\|\mathbf{u} - \mathbf{u}_h\|_{L^2(D_T)} + \|p - p_h\|_{L^2(D_T)}\}. \tag{78}$$

Proof. This result is an immediate consequence of Lemmas 4.4 and 4.5.

Putting together the results for both estimators we have the following a posteriori error estimate for the mixed finite element approximation.

We define

$$\eta_T^2 = \eta_{vect,T}^2 + \eta_{sc,T}^2 \quad \text{and} \quad \eta^2 = \sum_{T \in \mathcal{T}_h} \eta_T^2.$$

Theorem 4.5. *If Ω is simply connected and the restriction of a to any $T \in \mathcal{T}_h$ is constant, there exist constants C_5 and C_6 such that*

$$\eta_T \le C_5\{\|\mathbf{u} - \mathbf{u}_h\|_{L^2(D_T)} + \|p - p_h\|_{L^2(D_T)}\}$$

and

$$\|\mathbf{u} - \mathbf{u}_h\|_{L^2(\Omega)} + \|p - p_h\|_{L^2(\Omega)} \le C_6\{\eta + h\|f - P_h f\|_{L^2(\Omega)}\}.$$

Proof. This result is an immediate consequence of Theorems 4.1, 4.2, 4.3 and 4.4.

5 The General Abstract Setting

The problem considered in the previous sections is a particular case of a general class of problems that we are going to analyze in this section. The theory presented here was first developed by Brezzi [13]. Some of the ideas were also introduced for particular problems by Babuška [9] and by Crouzeix and Raviart [22]. We also refer the reader to [32, 31] and to the books [17, 45, 37].

Let V and Q be two Hilbert spaces and suppose that $a(\,,\,)$ and $b(\,,\,)$ are continuous bilinear forms on $V \times V$ and $V \times Q$ respectively, i.e.,

$$|a(u, v)| \le \|a\|\|u\|_V\|v\|_V \quad \forall u \in V, \forall v \in V$$

and

$$|b(v, q)| \le \|b\|\|v\|_V\|q\|_Q \quad \forall v \in V, \ \forall q \in Q.$$

Consider the problem: given $f \in V'$ and $g \in Q'$ find $(u, p) \in V \times Q$ solution of

$$\begin{cases} a(u, v) + b(v, p) = \langle f, v \rangle & \forall v \in V \\ b(u, q) = \langle g, q \rangle & \forall q \in Q \end{cases} \tag{79}$$

where $\langle\, .\, ,\, .\, \rangle$ denotes the duality product between a space and its dual one.

For example, the mixed formulation of second order elliptic problems considered in the previous sections can be written in this way with

$$V = H(\text{div}, \Omega), \quad Q = L^2(\Omega)$$

and

$$a(\mathbf{u}, \mathbf{v}) = \int_\Omega \mu \mathbf{u} \cdot \mathbf{v} \, dx, \quad b(\mathbf{v}, p) = \int_\Omega p \, \text{div} \, \mathbf{v} \, dx \, .$$

The general problem (79) can be written in the standard way

$$c((u,p),(v,q)) = \langle f, v \rangle + \langle g, q \rangle \;\; \forall (v,q) \in V \times Q \tag{80}$$

where c is the continuous bilinear form on $V \times Q$ defined by

$$c((u,p),(v,q)) = a(u,v) + b(v,p) + b(u,q).$$

However, the bilinear form is not coercive and therefore the usual finite element error analysis can not be applied.

We will give sufficient conditions (indeed, they are also necessary although we are not going to prove it here, we refer to [17, 37]) on the forms a and b for the existence and uniqueness of a solution of problem (79). Below, we will also show that their discrete version ensures the stability and optimal order error estimates for the Galerkin approximations. These results were obtained by Brezzi [13] (see also [17] were more general results are proved).

Introducing the continuous operators $A : V \to V'$, $B : V \to Q'$ and its adjoint $B^* : Q \to V'$ defined by,

$$\langle Au, v \rangle_{V' \times V} = a(u,v)$$

and

$$\langle Bv, q \rangle_{Q' \times Q} = b(v,q) = \langle v, B^* q \rangle_{V \times V'}$$

problem (79) can also be written as

$$\begin{cases} Au + B^* p = f & \text{in } V' \\ \qquad\quad Bu = g & \text{in } Q'. \end{cases} \tag{81}$$

Let us introduce $W = KerB \subset V$ and, for $g \in Q'$,

$$W(g) = \{ v \in V : Bv = g \}.$$

Now, if $(u,p) \in V \times Q$ is a solution of (79) then, it is easy to see that u is a solution of the problem

$$u \in W(g), \quad a(u,v) = \langle f, v \rangle \quad \forall v \in W. \tag{82}$$

We will find conditions under which both problems (79) and (82) are equivalent, in the sense that for a solution $u \in W(g)$ of (82) there exists a unique $p \in Q$ such that (u,p) is a solution of (79).

In what follows we will use the following well-known result of functional analysis. Given a Hilbert space V and $S \subset V$ we define $S^0 \subset V'$ by

$$S^0 = \{ L \in V' : \langle L, v \rangle = 0, \;\; \forall v \in S \}.$$

Theorem 5.1. *Let V_1 and V_2 be Hilbert spaces and $A : V_1 \longrightarrow V_2'$ be a continuous linear operator. Then,*

$$(Ker\,A)^0 = \overline{Im\,A^*} \tag{83}$$

and

$$(Ker\,A^*)^0 = \overline{Im\,A}. \tag{84}$$

Proof. It is easy to see that $Im\,A^* \subset (Ker\,A)^0$ and that $(Ker\,A)^0$ is a closed subspace of V_1'. Therefore

$$\overline{Im\,A^*} \subset (Ker\,A)^0 .$$

Suppose now that there exists $L_0 \in V_1'$ such that $L_0 \in (Ker\,A)^0 \setminus \overline{Im\,A^*}$. Then, by the Hahn-Banach theorem there exists a linear continuous functional defined on V_1' which vanishes on $\overline{Im\,A^*}$ and is different from zero on L_0. In other words, using the standard identification between V_1'' and V_1, there exists $v_0 \in V_1$ such that

$$\langle L_0, v_0\rangle \neq 0 \quad \text{and} \quad \langle L, v_0\rangle = 0 \quad \forall L \in \overline{Im\,A^*}.$$

In particular, for all $v \in V_2$

$$\langle A v_0, v\rangle = \langle v_0, A^* v\rangle = 0$$

and so $v_0 \in Ker\,A$ which, since $L_0 \in (Ker\,A)^0$, contradicts $\langle L_0, v_0\rangle \neq 0$. Therefore, $(Ker\,A)^0 \subset \overline{Im\,A^*}$ and so (83) holds. Finally, (84) is an immediate consequence of (83) because $(A^*)^* = A$.

Lemma 5.1. *The following properties are equivalent:*

(a) There exists $\beta > 0$ such that

$$\sup_{v \in V} \frac{b(v,q)}{\|v\|_V} \geq \beta \|q\|_Q \quad \forall q \in Q. \tag{85}$$

(b) B^* is an isomorphism from Q onto W^0 and,

$$\|B^* q\|_{V'} \geq \beta \|q\|_Q \quad \forall q \in Q. \tag{86}$$

(c) B is an isomorphism from $W^\perp$ onto Q' and,

$$\|Bv\|_{Q'} \geq \beta \|v\|_V \quad \forall v \in W^\perp. \tag{87}$$

Proof. Assume that (a) holds then, (86) is satisfied and so B^* is injective. Moreover $Im\,B^*$ is a closed subspace of V', indeed, suppose that $B^* q_n \rightarrow w$ then, it follows from (86) that

$$\|B^*(q_n - q_m)\|_{V'} \geq \beta \|q_n - q_m\|_Q$$

and, therefore, $\{q_n\}$ is a Cauchy sequence and so it converges to some $q \in Q$ and, by continuity of B^*, $w = B^* q \in Im\,B^*$. Consequently, using (83) we obtain that $Im\,B^* = W^0$ and therefore (b) holds.

Now, we observe that W^0 can be isometrically identified with $(W^\perp)'$. Indeed, denoting with $P^\perp : V \to W^\perp$ the orthogonal projection, for any $g \in (W^\perp)'$ we define $\tilde{g} \in W^0$ by $\tilde{g} = g \circ P^\perp$ and it is easy to check that $g \to \tilde{g}$ is an isometric bijection from $(W^\perp)'$ onto W^0 and then, we can identify these two spaces. Therefore (b) and (c) are equivalent.

Corollary 5.1. *If the form b satisfies (85) then, problems (79) and (82) are equivalent, that is, there exists a unique solution of (79) if and only if there exists a unique solution of (82).*

Proof. If (u, p) is a solution of (79) we know that u is a solution of (82). It rests only to check that for a solution u of (82) there exists a unique $p \in Q$ such that $B^*p = f - Au$ but, this follows from (b) of the previous lemma since, as it is easy to check, $f - Au \in W^0$.

Now we can prove the fundamental existence and uniqueness theorem for problem (79).

Lemma 5.2. *If there exists $\alpha > 0$ such that a satisfies*

$$\sup_{v \in W} \frac{a(u,v)}{\|v\|_V} \geq \alpha \|u\|_V \quad \forall u \in W \tag{88}$$

$$\sup_{u \in W} \frac{a(u,v)}{\|u\|_V} \geq \alpha \|v\|_V \quad \forall v \in W \tag{89}$$

then, for any $g \in W'$ there exists $w \in W$ such that

$$a(w,v) = \langle g, v \rangle \qquad \forall v \in W$$

and moreover

$$\|w\|_W \leq \frac{1}{\alpha} \|g\|_{W'}. \tag{90}$$

Proof. Considering the operators

$$A : W \to W' \text{ and } A^* : W \to W'$$

defined by

$$\langle Au, v \rangle_{W' \times W} = a(u,v) \text{ and } \langle u, A^*v \rangle_{W \times W'} = a(u,v),$$

conditions (88) and (89) can be written as

$$\|Au\|_{W'} \geq \alpha \|u\|_W \quad \forall u \in W \tag{91}$$

and

$$\|A^*v\|_{W'} \geq \alpha \|v\|_W \quad \forall v \in W \tag{92}$$

respectively. Therefore, it follows from (89) that

$$Ker\, A^* = \{0\}.$$

Then, from (84), we have

$$(Ker\, A^*)^0 = \overline{Im\, A}$$

and so

$$\overline{Im\, A} = W'.$$

Using now (91) and the same argument used in (85) to prove that $Im\, B^*$ is closed, we can show that $Im\, A$ is a closed subspace of W' and consequently $Im\, A = W'$ as we wanted to show. Finally (90) follows immediately from (91).

Theorem 5.2. *If a satisfies (88) and (89), and b satisfies (85) then, there exists a unique solution $(u,p) \in V \times Q$ of problem (79) and moreover,*

$$\|u\|_V \leq \frac{1}{\alpha}\|f\|_{V'} + \frac{1}{\beta}\left(1 + \frac{\|a\|}{\alpha}\right)\|g\|_{Q'} \tag{93}$$

and,

$$\|p\|_Q \leq \frac{1}{\beta}\left(1 + \frac{\|a\|}{\alpha}\right)\|f\|_{V'} + \frac{\|a\|}{\beta^2}\left(1 + \frac{\|a\|}{\alpha}\right)\|g\|_{Q'}. \tag{94}$$

Proof. First we show that there exists a solution u of problem (82). Since (85) holds we know from Lemma 5.1 that there exists a unique $u_0 \in W^\perp$ such that $Bu_0 = g$ and

$$\|u_0\|_V \leq \frac{1}{\beta}\|g\|_{Q'} \tag{95}$$

then, the existence of u solution of (82) is equivalent to the existence of $w = u - u_0 \in W$ such that

$$a(w,v) = \langle f, v\rangle - a(u_0, v) \;\; \forall v \in W$$

but, from Lemma 5.2, it follows that such a w exists and moreover,

$$\|w\|_V \leq \frac{1}{\alpha}\{\|f\|_{V'} + \|a\|\|u_0\|_V\} \leq \frac{1}{\alpha}\{\|f\|_{V'} + \frac{\|a\|}{\beta}\|g\|_{Q'}\}$$

where we have used (95).

Therefore, $u = w + u_0$ is a solution of (82) and satisfies (93).

Now, from Corollary 5.1 it follows that there exists a unique $p \in Q$ such that (u,p) is a solution of (79). On the other hand, from Lemma 5.1 it follows that (86) holds and using it, it is easy to check that

$$\|p\|_Q \leq \frac{1}{\beta}\{\|f\|_{V'} + \|a\|\|u\|_V\}$$

which combined with (93) yields (94). Finally, the uniqueness of solution follows from (93) and (94).

Assume now that we have two families of subspaces $V_h \subset V$ and $Q_h \subset Q$. The Galerkin approximation $(u_h, p_h) \in V_h \times Q_h$ to the solution $(u,p) \in V \times Q$ of problem (79), is defined by

$$\begin{cases} a(u_h, v) + b(v, p_h) = \langle f, v \rangle & \forall v \in V_h \\ b(u_h, q) = \langle g, q \rangle & \forall q \in Q_h. \end{cases} \tag{96}$$

For the error analysis it is convenient to introduce the associated operator $B_h : V_h \to Q_h'$ defined by

$$\langle B_h v, q \rangle_{Q_h' \times Q_h} = b(v, q)$$

and the subsets of V_h, $W_h = Ker\, B_h$ and

$$W_h(g) = \{v \in V_h : B_h v = g \text{ in } Q_h'\}$$

where g is restricted to Q_h.

In order to have the Galerkin approximation well defined we need to know that there exists a unique solution $(u_h, p_h) \in V_h \times Q_h$ of problem (96). In view of Theorem 5.2, this will be true if there exist $\alpha^* > 0$ and $\beta^* > 0$ such that

$$\sup_{v \in W_h} \frac{a(u,v)}{\|v\|_V} \geq \alpha^* \|u\|_V \quad \forall u \in W_h \tag{97}$$

$$\sup_{u \in W_h} \frac{a(u,v)}{\|u\|_V} \geq \alpha^* \|v\|_V \quad \forall v \in W_h \tag{98}$$

and,

$$\sup_{v \in V_h} \frac{b(v,q)}{\|v\|_V} \geq \beta^* \|q\|_Q \quad \forall q \in Q_h. \tag{99}$$

In fact, (98) follows from (97) since W_h is finite dimensional.

Now, we can prove the fundamental general error estimates due to Brezzi [13].

Theorem 5.3. *If the forms a and b satisfy (97), (98) and (99), problem (96) has a unique solution and there exists a constant C, depending only on α^*, β^*, $\|a\|$ and $\|b\|$ such that the following estimates hold. In particular, if the constants α^* and β^* are independent of h then, C is independent of h.*

$$\|u - u_h\|_V + \|p - p_h\|_Q \leq C\{\inf_{v \in V_h} \|u - v\|_V + \inf_{q \in Q_h} \|p - q\|_Q\} \tag{100}$$

and, when $Ker\, B_h \subset Ker\, B$,

$$\|u - u_h\|_V \leq C \inf_{v \in V_h} \|u - v\|_V. \tag{101}$$

Proof. From Theorem 5.2, there exists a unique solution $(u_h, p_h) \in V_h \times Q_h$ of (96).

On the other hand, given $(v, q) \in V_h \times Q_h$, we have

$$a(u_h - v, w) + b(w, p_h - q) = a(u - v, w) + b(w, p - q) \quad \forall w \in V_h \tag{102}$$

and

$$b(u_h - v, r) = b(u - v, r) \quad \forall r \in Q_h. \tag{103}$$

Now, for fixed (v, q) , the right hand sides of (102) and (103) define linear functionals on V_h and Q_h which are continuous with norms bounded by

$$\|a\|\|u - v\|_V + \|b\|\|p - q\|_Q \quad \text{and} \quad \|b\|\|u - v\|_V$$

respectively. Then, it follows from Theorem 5.2 that, for any $(v, q) \in V_h \times Q_h$,

$$\|u_h - v\|_V + \|p_h - q\|_Q \leq C\{\|u - v\|_V + \|p - q\|_Q\}$$

and therefore (100) follows by the triangular inequality.

On the other hand, we know that $u_h \in W_h(g)$ is a solution of

$$a(u_h, v) = \langle f, v\rangle \quad \forall v \in W_h \tag{104}$$

and, since $W_h \subset W$, subtracting (104) from (82) we have,

$$a(u - u_h, v) = 0 \quad \forall v \in W_h. \tag{105}$$

Now, for $w \in W_h(g)$, $u_h - w \in W_h$ and so from (97) and (105) we have

$$\alpha^* \|u_h - w\|_V \leq \sup_{v \in W_h} \frac{a(u_h - w, v)}{\|v\|_V} = \sup_{v \in W_h} \frac{a(u - w, v)}{\|v\|_V} \leq \|a\|\|u - w\|_V$$

and therefore,

$$\|u - u_h\|_V \leq \left(1 + \frac{\|a\|}{\alpha^*}\right) \inf_{w \in W_h(g)} \|u - w\|_V.$$

To conclude the proof we will see that, if (99) holds then,

$$\inf_{w \in W_h(g)} \|u - w\|_V \leq \left(1 + \frac{\|b\|}{\beta^*}\right) \inf_{v \in V_h} \|u - v\|_V. \tag{106}$$

Given $v \in V_h$, from Lemma 5.1 we know that there exists a unique $z \in W_h^{\perp}$ such that

$$b(z, q) = b(u - v, q) \quad \forall q \in Q_h$$

and

$$\|z\|_V \leq \frac{\|b\|}{\beta^*} \|u - v\|_V$$

thus, $w = z + v \in V_h$ satisfies $B_h w = g$, that is, $w \in W_h(g)$. But

$$\|u - w\|_V \leq \|u - v\|_V + \|z\|_V \leq (1 + \frac{\|b\|}{\beta^*})\|u - v\|_V$$

and so (106) holds.

In the applications, a very useful criterion to check the inf–sup condition (99) is the following result due to Fortin [32].

Theorem 5.4. *Assume that (85) holds. Then, the discrete inf–sup condition (99) holds with a constant* $\beta^* > 0$ *independent of* h*, if and only if, there exists an operator*

$$\Pi_h : V \longrightarrow V_h$$

such that

$$b(v - \Pi_h v, q) = 0 \quad \forall v \in V\,,\ \forall q \in Q_h \tag{107}$$

and,

$$\|\Pi_h v\|_V \le C\|v\|_V \quad \forall v \in V \tag{108}$$

with a constant $C > 0$ *independent of* h.

Proof. Assume that such an operator Π_h exists. Then, from (107), (108) and (85) we have, for $q \in Q_h$,

$$\beta\|q\|_Q \le \sup_{v\in V} \frac{b(v,q)}{\|v\|_V} = \sup_{v\in V} \frac{b(\Pi_h v,q)}{\|v\|_V} \le C \sup_{v\in V} \frac{b(\Pi_h v,q)}{\|\Pi_h v\|_V}$$

and therefore, (99) holds with $\beta^* = \beta/C$.

Conversely, suppose that (99) holds with β^* independent of h. Then, from (87) we know that for any $v \in V$ there exists a unique $v_h \in W_h^\perp$ such that

$$b(v_h, q) = b(v,q) \quad \forall q \in Q_h$$

and,

$$\|v_h\|_V \le \frac{\|b\|}{\beta^*}\|v\|_V.$$

Therefore, $\Pi_h v = v_h$ defines the required operator.

Remark 5.1. In practice, it is sometimes enough to show the existence of the operator Π_h on a subspace $S \subset V$, where the exact solution belongs, verifying (107) and (108) for $v \in S$ and the norm on the right hand side of (108) replaced by a strongest norm (that of the space S). This is in some cases easier because the explicit construction of the operator Π_h requires regularity assumptions which do not hold for a general function in V. For example, in the problem analyzed in the previous sections we have constructed this operator on a subspace of $V = H(\text{div}, \Omega)$ because the degrees of freedom defining the operator do not make sense in $H(\text{div}, T)$, indeed, we need more regularity for $\mathbf{v}$ (for example $\mathbf{v} \in H^1(T)^n$) in order to have the integral of the normal component of $\mathbf{v}$ against a polynomial on a face F of T well defined. It is possible to show the existence of Π_h defined on $H(\text{div}, \Omega)$ satisfying (107) and (108) (see [32, 46]). However, as we have seen, this is not really necessary to obtain optimal error estimates.

References

1. G. Acosta and R. G. Durán, The maximum angle condition for mixed and non conforming elements: Application to the Stokes equations, *SIAM J. Numer. Anal.* 37, 18–36, 2000.
2. G. Acosta, R. G. Durán and M. A. Muschietti, Solutions of the divergence operator on John domains, *Adv. Math.* 206, 373–401, 2006.
3. M. Ainsworth, Robust a posteriori error estimation for nonconforming finite element approximations, *SIAM J. Numer. Anal.* 42, 2320–2341, 2005.
4. A. Alonso, Error estimator for a mixed method, *Numer. Math.* 74, 385–395, 1996.
5. D. N. Arnold, D. Boffi and R. S. Falk, Quadrilateral $H(\mathrm{div})$ finite elements, *SIAM J. Numer. Anal.* 42, 2429–2451, 2005.
6. D. N. Arnold and F. Brezzi, Mixed and nonconforming finite element methods implementation, postprocessing and error estimates, *R.A.I.R.O., Modél. Math. Anal. Numer.* 19, 7–32, 1985.
7. D. N. Arnold, L. R. Scott and M. Vogelius, Regular inversion of the divergence operator with Dirichlet boundary conditions on a polygon, *Ann. Scuola Norm. Sup. Pisa* Cl. Sci-Serie IV, XV, 169–192, 1988.
8. I. Babuška, Error bounds for finite element method, *Numer. Math. 16*, 322–333, 1971.
9. I. Babuška, The finite element method with lagrangian multipliers, *Numer. Math.*, 20, 179–192, 1973.
10. A. Bermúdez, P. Gamallo, M. R. Nogueiras and R. Rodríguez, Approximation of a structural acoustic vibration problem by hexhaedral finite elements, *IMA J. Numer. Anal. 26*, 391–421, 2006.
11. J. H. Bramble and J. M. Xu, Local post-processing technique for improving the accuracy in mixed finite element approximations, *SIAM J. Numer. Anal.* 26, 1267–1275, 1989.
12. S. Brenner and L. R. Scott, *The Mathematical Analysis of Finite Element Methods*, Springer, Berlin Heidelberg New York, 1994.
13. F. Brezzi, On the existence, uniqueness and approximation of saddle point problems arising from lagrangian multipliers, *R.A.I.R.O. Anal. Numer.* 8, 129–151, 1974.
14. F. Brezzi, J. Douglas, R. Durán and M. Fortin, Mixed finite elements for second order elliptic problems in three variables, *Numer. Math.* 51, 237–250, 1987.
15. F. Brezzi, J. Douglas, M. Fortin and L. D. Marini, Efficient rectangular mixed finite elements in two and three space variables, *Math. Model. Numer. Anal.* 21, 581–604, 1987.
16. F. Brezzi, J. Douglas and L. D. Marini, Two families of mixed finite elements for second order elliptic problems, *Numer. Math. 47*, 217–235, 1985.
17. F. Brezzi, and M. Fortin, *Mixed and Hybrid Finite Element Methods*, Springer, Berlin Heidelberg New York, 1991.
18. C. Carstensen, A posteriori error estimate for the mixed finite element method, *Math. Comp.* 66, 465–476, 1997.
19. P. G. Ciarlet, *The Finite Element Method for Elliptic Problems*, North Holland, 1978.
20. P. G. Ciarlet, *Mathematical Elasticity, Volume 1. Three-Dimensional Elasticity*, North Holland, 1988.
21. P. Clément, Approximation by finite element function using local regularization, *RAIRO* R-2 77–84, 1975.
22. M. Crouzeix and P. A. Raviart, Conforming and non-conforming finite element methods for solving the stationary Stokes equations, *R.A.I.R.O. Anal. Numer.* 7, 33–76, 1973.
23. E. Dari, R. G. Durán, C. Padra and V. Vampa, A posteriori error estimators for nonconforming finite element methods, *Math. Model. Numer. Anal.* 30, 385–400, 1996.

24. J. Douglas and J. E. Roberts, Global estimates for mixed methods for second order elliptic equations, *Math. Comp.* 44, 39–52, 1985.
25. T. Dupont, and L. R. Scott, Polynomial approximation of functions in Sobolev spaces, *Math. Comp.* 34, 441–463, 1980.
26. R. G. Durán, On polynomial Approximation in Sobolev Spaces, *SIAM J. Numer. Anal.* 20, 985–988, 1983.
27. R. G. Durán, Error Analysis in L^p for Mixed Finite Element Methods for Linear and quasilinear elliptic problems, *R.A.I.R.O. Anal. Numér* 22, 371–387, 1988.
28. R. G. Durán, Error estimates for anisotropic finite elements and applications *Proceedings of the International Congress of Mathematicians*, 1181–1200, 2006.
29. R. G. Durán and A. L. Lombardi, Error estimates for the Raviart-Thomas interpolation under the maximum angle condition, preprint, http://mate.dm.uba.ar/ rduran/papers/dl3.pdf
30. R. G. Durán and M. A. Muschietti, An explicit right inverse of the divergence operator which is continuous in weighted norms, *Studia Math.* 148, 207–219, 2001.
31. R. S. Falk and J. Osborn, Error estimates for mixed methods, *R.A.I.R.O. Anal. Numer.* 4, 249–277, 1980.
32. M. Fortin, An analysis of the convergence of mixed finite element methods, *R.A.I.R.O. Anal. Numer.* 11, 341–354, 1977.
33. E. Gagliardo, Caratterizzazioni delle tracce sulla frontiera relative ad alcune classi di funzioni in n variabili, *Rend. Sem. Mat. Univ. Padova* 27, 284–305, 1957.
34. L. Gastaldi and R. H. Nochetto, Optimal L^∞-error estimates for nonconforming and mixed finite element methods of lowest order, *Numer. Math. 50*, 587–611, 1987.
35. L. Gastaldi and R. H. Nochetto, On L^∞- accuracy of mixed finite element methods for second order elliptic problems, *Mat. Aplic. Comp.* 7, 13–39, 1988.
36. D. Gilbarg and N. S. Trudinger, *Elliptic Partial Differential Equations of Second Order*, Springer, Berlin Heidelberg New York, 1983.
37. V. Girault and P. A. Raviart, *Element Methods for Navier–Stokes Equations*, Springer, Berlin Heidelberg New York, 1986.
38. P. Grisvard, *Elliptic Problems in Nonsmooth Domain*, Pitman, Boston, 1985.
39. C. Lovadina and R. Stenberg, Energy norm a posteriori error estimates for mixed finite element methods, *Math. Comp.* 75, 1659–1674, 2006.
40. L. D. Marini, An inexpensive method for the evaluation of the solution of the lowest order Raviart–Thomas mixed method, *SIAM J. Numer. Anal.* 22, 493–496, 1985.
41. J. C. Nédélec, Mixed finite elements in $\mathbb{R}^3$, *Numer. Math.* 35, 315–341, 1980.
42. J. C. Nédélec, A new family of mixed finite elements in $\mathbb{R}^3$, *Numer. Math.* 50, 57–81, 1986.
43. L. E. Payne and H. F. Weinberger, An optimal Poincaré inequality for convex domains, *Arch. Rat. Mech. Anal.* 5, 286–292, 1960.
44. P. A. Raviart and J. M. Thomas, A mixed finite element method for second order elliptic problems, *Mathematical Aspects of the Finite Element Method*, (I. Galligani, E. Magenes, eds.), *Lectures Notes in Mathematics*, vol. 606, Springer, Berlin Heidelberg New York, 1977.
45. J. E. Roberts and J. M. Thomas, *Mixed and Hybrid Methods in Handbook of Numerical Analysis*, Vol. II (P. G. Ciarlet and J. L. Lions, eds.), Finite Element Methods (Part 1), North Holland, 1989.
46. J. Schöberl, Commuting quasi-interpolation operators for mixed finite elements, Preprint ISC-01-10-MATH, Texas A&M University, 2001.
47. L. R. Scott and S. Zhang, Finite element interpolation of non-smooth functions satisfying boundary conditions, *Math. Comp.* 54, 483–493, 1990.

48. R. Stenberg, Postprocessing schemes for some mixed finite elements, *RAIRO, Model. Math. Anal. Numer.* 25, 151–167, 1991.
49. J. M. Thomas, Sur l'Analyse Numérique des Méthodes d'Éléments Finis Hybrides et Mixtes, Thèse de Doctorat d'Etat, Université Pierre et Marie Curie, Paris, 1977.
50. R. Verfürth, A posteriori error estimators for the Stokes equations, *Numer. Math.* 55, 309–325, 1989.
51. R. Verfürth, A note on polynomial approximation in Sobolev spaces, *RAIRO M2AN* 33, 715–719, 1999.
52. R. Verfürth, *A Review of A Posteriori Error Estimation and Adaptive Mesh-Refinement Techniques*, Wiley, New York, 1996.

Finite Elements for the Stokes Problem

Daniele Boffi[1], Franco Brezzi[2], and Michel Fortin[3]

[1] Dipartimento di Matematica "F. Casorati", Università degli studi di Pavia, Via Ferrata 1, 27100 Pavia, Italy
daniele.boffi@unipv.it

[2] Istituto Universitario di Studi Superiori (IUSS) and I.M.A.T.I.–C.N.R., Via Ferrata 3, 27100 Pavia Pavia, Italy
brezzi@imati.cnr.it

[3] Département de Mathématiques et de Statistique, Pavillon Alexandre-Vachon, Université Laval, 1045, Avenue de la Mé decine, Quebec G1V 0A6, Canada
mfortin@mat.ulaval.ca

1 Introduction

Given a domain $\Omega \subset \mathbb{R}^n$, the Stokes problem models the motion of an incompressible fluid occupying Ω and can be written as the following system of variational equations,

$$\begin{cases} 2\mu \displaystyle\int_\Omega \underline{\underline{\varepsilon}}(\underline{u}) : \underline{\underline{\varepsilon}}(\underline{v})\, dx - \int_\Omega p \operatorname{div} \underline{v}\, dx = \int_\Omega \underline{f} \cdot \underline{v}\, dx, \quad \forall \underline{v} \in V, \\ \displaystyle\int_\Omega q \operatorname{div} \underline{u}\, dx = 0, \quad \forall q \in Q, \end{cases} \tag{1}$$

where $V = (H_0^1(\Omega))^n$ and Q is the subspace of $L^2(\Omega)$ consisting of functions with zero mean value on Ω. In this formulation $\underline{u}$ is the velocity of the fluid and p its pressure. A similar problem arises for the displacement of an incompressible elastic material.

An elastic material, indeed, can be modeled by the following variational equation, λ and μ being the Lamé coefficients

$$2\mu \int_\Omega \underline{\underline{\varepsilon}}(\underline{u}) : \underline{\underline{\varepsilon}}(\underline{v})\, dx + \lambda \int_\Omega \operatorname{div} \underline{u} \operatorname{div} \underline{v}\, dx = \int_\Omega \underline{f} \cdot \underline{v}\, dx, \quad \forall v \in V. \tag{2}$$

The case where λ is large (or equivalently when $\nu = \lambda/2(\lambda+\mu)$ approaches $1/2$) can be considered as an approximation of (1) by a penalty method. The limiting case is exactly (1) up to the fact that $\underline{u}$ is a displacement instead of a velocity. Problems where λ is large are quite common and correspond to almost incompressible materials.

It is also worth recalling that, defining $A\underline{u} = \underline{\mathrm{div}}\,\underline{\underline{\varepsilon}}(\underline{u})$, that in 2D reads

$$A\underline{u} = \begin{cases} \dfrac{\partial^2 u_1}{\partial x_1^2} + \dfrac{1}{2}\dfrac{\partial}{\partial x_2}\left(\dfrac{\partial u_1}{\partial x_2} + \dfrac{\partial u_2}{\partial x_1}\right), \\ \dfrac{\partial^2 u_2}{\partial x_2^2} + \dfrac{1}{2}\dfrac{\partial}{\partial x_1}\left(\dfrac{\partial u_1}{\partial x_2} + \dfrac{\partial u_2}{\partial x_1}\right), \end{cases} \tag{3}$$

we have $2\mu A\underline{u} = \mu\Delta\underline{u} + \mu\,\underline{\mathrm{grad}}\,\mathrm{div}\,\underline{u}$. Problems (1) and (2) are then respectively equivalent to

$$\begin{cases} -2\mu A\underline{u} + \underline{\mathrm{grad}}\,p = \mu\Delta\underline{u} + \underline{\mathrm{grad}}\,p = \underline{f}, \\ \mathrm{div}\,\underline{u} = 0, \\ \underline{u}|_\Gamma = 0, \end{cases} \tag{4}$$

and

$$-2\mu A\underline{u} - \lambda\,\underline{\mathrm{grad}}\,\mathrm{div}\,\underline{u} = -\mu\Delta\underline{u} - (\lambda + \mu)\,\underline{\mathrm{grad}}\,\mathrm{div}\,\underline{u} = \underline{f}. \tag{5}$$

Remark 1.1. The problems described above are, of course, physically unrealistic, as they involve body forces and homogeneous Dirichlet boundary conditions. The aim of doing so is to avoid purely technical difficulties and implies no loss of generality. The results obtained will be valid, unless otherwise stated, for all acceptable boundary conditions.

To approximate the Stokes problem, two approaches follow quite naturally from the preceding considerations. *The first* one is to use system (1) and to discretize $\underline{u}$ and p by standard (or less standard) finite element spaces. *The second* one is to use formulation (2) with λ large as a penalty approximation to system (1).

It rapidly became clear that both these approaches could yield strange results. In particular, the first one often led to nonconvergence of the pressure and the second one to a *locking mechanism*, the numerical solution being uniformly zero, or unnaturally small for big λ.

For velocity–pressure approximations, empirical cures were found by [46], [45] and others. At about the same time some elements using discontinuous pressure fields were shown to work properly [31], [35] from the mathematical point of view.

For the penalty method, the cure was found in selective or reduced integration procedures. This consisted in evaluating terms like $\int_\Omega \mathrm{div}\,\underline{u}\,\mathrm{div}\,\underline{v}\,dx$ by quadrature formulas of low order. This *sometimes* led to good results.

It was finally stated [50], even if the result was implicit in earlier works [8], that the analysis underlying the two approaches must be the same. Penalty methods are often equivalent to some mixed methods. In such cases, the penalty method works if and only if the associated mixed method works [9].

2 The Stokes Problem as a Mixed Problem

2.1 Mixed Formulation

We shall describe in this section how the Stokes problem (1) can be analyzed in the general framework of mixed methods. Defining $V = (H_0^1(\Omega))^n$, $\tilde{Q} = L^2(\Omega)$, and

$$a(\underline{u}, \underline{v}) = 2\mu \int_\Omega \underline{\underline{\varepsilon}}(\underline{u}) : \underline{\underline{\varepsilon}}(\underline{v})\, dx, \tag{6}$$

$$b(\underline{v}, q) = -\int_\Omega q \operatorname{div} \underline{v}\, dx, \tag{7}$$

problem (1) can clearly be written in the form: find $\underline{u} \in V$ and $p \in \tilde{Q}$ such that

$$\begin{cases} a(\underline{u}, \underline{v}) + b(\underline{v}, p) = (\underline{f}, \underline{v}), & \forall \underline{v} \in V, \\ b(\underline{u}, q) = 0, & \forall q \in \tilde{Q}, \end{cases} \tag{8}$$

which is a mixed problem. Indeed, it can be observed that p is the Lagrange multiplier associated with the incompressibility constraint.

Remark 2.1. It is apparent, from the definition (7) of $b(\cdot,\cdot)$ and from the boundary conditions of the functions in V, that p, if exists, is defined up to a constant. Therefore, we introduce the space

$$Q = L^2(\Omega)/\mathbb{R}, \tag{9}$$

where two elements $q_1, q_2 \in L^2(\Omega)$ are identified if their difference is constant. It is not difficult to show that Q is isomorphic to the subspace of $L^2(\Omega)$ consisting of functions with zero mean value on Ω.

With this choice, our problem reads: find $\underline{u} \in V$ and $p \in Q$ such that

$$\begin{cases} a(\underline{u}, \underline{v}) + b(\underline{v}, p) = (\underline{f}, \underline{v}), & \forall \underline{v} \in V, \\ b(\underline{u}, q) = 0, & \forall q \in Q. \end{cases} \tag{10}$$

Let us check that our problem is well-posed. With standard procedure, we can introduce the following operators

$$B = -\operatorname{div} : (H_0^1(\Omega))^n \to L^2(\Omega)/\mathbb{R} \tag{11}$$

and

$$B^t = \underline{\operatorname{grad}} : L^2(\Omega)/\mathbb{R} \to (H^{-1}(\Omega))^n. \tag{12}$$

It can be shown (see, e.g., [64]) that

$$\operatorname{Im} B = Q \cong \left\{ q \in L^2(\Omega) : \int_\Omega q\, dx = 0 \right\}, \tag{13}$$

hence the operator B has a continuous lifting, and the continuous inf–sup condition is fulfilled. We also notice that, with our definition of the space Q, the kernel $\ker B^t$ reduces to zero.

The bilinear form $a(\cdot,\cdot)$ is coercive on V (see [32, 64]), whence the ellipticity in the kernel also will follow (i.e., A is invertible on $\ker B$).

We state the well-posedness of problem (10) in the following theorem.

Theorem 2.1. *Let* $\underline{f}$ *be given in* $(H^{-1}(\Omega))^n$*. Then there exists a unique* $(\underline{u}, p) \in V \times Q$ *solution to problem* (10) *which satisfies*

$$||\underline{u}||_V + ||p||_Q \le C||\underline{f}||_{H^{-1}}. \tag{14}$$

Now choosing an approximation $V_h \subset V$ and $Q_h \subset Q$ yields the discrete problem

$$\begin{cases} 2\mu \displaystyle\int_\Omega \underline{\underline{\varepsilon}}(\underline{u}_h) : \underline{\underline{\varepsilon}}(\underline{v}_h)\, dx - \int_\Omega p_h \operatorname{div} \underline{v}_h\, dx = \int_\Omega \underline{f} \cdot \underline{v}_h\, dx, & \forall \underline{v}_h \in V_h, \\ \displaystyle\int_\Omega q_h \operatorname{div} \underline{u}_h\, dx = 0, \quad \forall q_h \in Q_h. \end{cases} \tag{15}$$

The bilinear form $a(\cdot,\cdot)$ is coercive on V; hence, according to the general theory of mixed approximations, there is no problem for the *existence* of a solution $\{\underline{u}_h, p_h\}$ to problem (15), while we might have troubles with the *uniqueness* of p_h. We thus try to obtain estimates of the errors $||\underline{u} - \underline{u}_h||_V$ and $||p - p_h||_Q$.

First we observe that, in general, the discrete solution $\underline{u}_h$ *needs not be divergence-free*. Indeed, the bilinear form $b(\cdot,\cdot)$ defines a discrete divergence operator

$$B_h = -\operatorname{div}_h : V_h \to Q_h. \tag{16}$$

(It is convenient here to identify $Q = L^2(\Omega)/\mathbb{R}$ and $Q_h \subset Q$ with their dual spaces). In fact, we have

$$(\operatorname{div}_h \underline{u}_h, q_h)_Q = \int_\Omega q_h \operatorname{div} \underline{u}_h\, dx, \tag{17}$$

and, thus, $\operatorname{div}_h \underline{u}_h$ turns out to be the L^2-projection of $\operatorname{div} \underline{u}_h$ onto Q_h.

The discrete divergence operator coincides with the standard divergence operator if $\operatorname{div} V_h \subset Q_h$. Referring to the abstract setting, we see that obtaining error estimates requires a careful study of the properties of the operator $B_h = -\operatorname{div}_h$ and of its transpose that we denote by $\underline{\operatorname{grad}}_h$.

The first question is to characterize the kernel $\ker B_h^t = \ker(\underline{\operatorname{grad}}_h)$. It might happen that $\ker B_h^t$ contains nontrivial functions In these cases $\operatorname{Im} B_h = \operatorname{Im}(\operatorname{div}_h)$ will be *strictly smaller* than $Q_h = P_{Q_h}(\operatorname{Im} B)$; this may lead to pathologies. In particular, if we consider a modified problem, like the one that usually originates when dealing with nonhomogeneous boundary conditions, the strict inclusion $\operatorname{Im} B_h \subset Q_h$ may even imply troubles with the existence of the solution. This situations is made clearer with the following example.

Example 2.1. Let us consider problem (4) with *nonhomogeneous boundary conditions*, that is let $\underline{r}$ be such that

$$\underline{u}|_\Gamma = \underline{r}, \qquad \int_\Gamma \underline{r} \cdot \underline{n}\, ds = 0, \tag{18}$$

It is classical to reduce this case to a problem with homogeneous boundary conditions by first introducing a function $\underline{\tilde{u}} \in (H^1(\Omega))^n$ such that $\underline{\tilde{u}}|_\Gamma = \underline{r}$. Setting $\underline{u} = \underline{u}_0 + \underline{\tilde{u}}$ with $\underline{u}_0 \in (H_0^1(\Omega))^n$ we have to solve

$$\begin{cases} -2\mu A\underline{u}_0 + \underline{\text{grad}}\, p = \underline{f} + 2\mu A\underline{\tilde{u}} = \underline{\tilde{f}}, \\ \operatorname{div} \underline{u}_0 = -\operatorname{div} \underline{\tilde{u}} = g, \quad \underline{u}_0|_\Gamma = 0 \end{cases} \tag{19}$$

with A defined in (3). We thus find a problem with a constraint $B\underline{u}_0 = g$ where $g \neq 0$. It may happen that the associated discrete problem fails to have a solution, because $g_h = P_{Q_h} g$ does not necessarily belongs to $\operatorname{Im} B_h$, whenever $\ker B_h^t \not\subset \ker B^t$. Discretization where $\ker(\underline{\text{grad}}_h)$ is nontrivial *can therefore lead to ill-posed problems* in particular for some nonhomogeneous boundary conditions. Examples of such conditions can be found in [56, 57]. In general, any method that relies on extra compatibility conditions is a source of trouble when applied to more complicated (nonlinear, time-dependent, etc.) problems.

Let us now turn our attention to the study of the error estimates. Since the bilinear form $a(\cdot,\cdot)$ is coercive on V, we only have to deal with the inf–sup condition. The following proposition will be the starting point for the analysis of any finite element approximation of (10).

Proposition 2.1. *Let $(\underline{u}, p) \in V \times Q$ be the solution of* (10) *and suppose the following inf–sup condition holds true (with k_0 independent of h)*

$$\inf_{q_h \in Q_h} \sup_{\underline{v}_h \in V_h} \frac{\int_\Omega q_h \operatorname{div} \underline{v}_h \, dx}{||q_h||_Q ||\underline{v}_h||_V} \geq k_h \geq k_0 > 0. \tag{20}$$

Then there exists a unique $(\underline{u}_h, p_h) \in V_h \times Q_h$ solution to (15) *and the following estimate holds*

$$||\underline{u} - \underline{u}_h||_V + ||p - p_h||_Q \leq C \inf_{\underline{v}_h \in V_h,\, q_h \in Q_h} \{||\underline{u} - \underline{v}_h||_V + ||p - q_h||_Q\}. \tag{21}$$

Remark 2.2. Actually, as it has been already observed, the existence of the discrete solution $(\underline{u}_h, p_h)$ (when the right-hand side in the second equation of (10) is zero) is not a consequence of the inf–sup condition (20). However, we should not forget about the possible situation presented in Example 2.1.

Remark 2.3. We shall also meet cases in which the constant k_h is not bounded below by k_0. We shall then try to know precisely how it depends on h and to see whether a lower-order convergence can be achieved. When $\ker(\underline{\text{grad}}_h)$ is nontrivial, we are interested in a weaker form of (20)

$$\sup_{\underline{v}_h \in V_h} \frac{\int_\Omega q_h \operatorname{div} \underline{v}_h \, dx}{||\underline{v}_h||_V} \geq k_h \inf_{q \in \ker(\underline{\text{grad}}_h)} ||q_h - q||_{L^2(\Omega)}, \tag{22}$$

and in the dependence of k_h in terms of h.

Several ways have also been proposed to get a more direct and intuitive evaluation of how a finite element scheme can approximate divergence-free functions. One of them is the *constraint ratio*, that we denote by C_r, and which is defined as

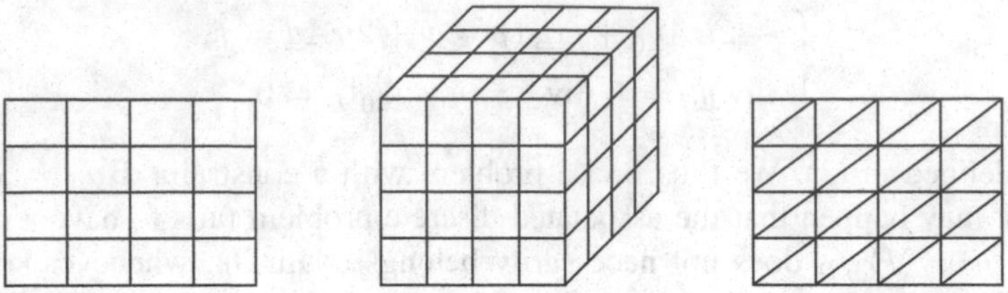

Fig. 1. Uniform meshes

$$C_r = \dim Q_h / \dim V_h. \tag{23}$$

It is, therefore, the ratio between the number of linearly independent constraints arising from the discrete divergence-free condition and the total number of degrees of freedom of the discrete velocity.

The value of C_r has no direct interpretation, *unless it is larger than* 1, which means that the number of constraints exceeds that of the variables. We then have a *locking phenomenon*.

Conversely, a small value of C_r implies a poor approximation of the divergence-free condition. It must however be emphasized that such a use of the constraint ratio has only a limited empirical value.

Another heuristic evaluation can be found by looking at the smallest representable vortex for a given mesh. This will be closely related to building a divergence-free basis (cf. Sect. 10). The idea behind this procedure [37] is that a discrete divergence-free function can be expressed as a sum of small vortices, that are, indeed, basis functions for $\ker B_h$. The size of the smallest vertices can be thought of as the equivalent of the smallest representable wavelenght in spectral methods.

In this context, we shall refer to a uniform mesh of n^2 rectangles, n^3 cubes or $2n^2$ triangles (Fig. 1). We must also quote the results of [67] who introduced a "patch test" to analyze similar problems. This patch test is only heuristic and does not yield a proof of stability. Moreover, such a test may be misleading in several cases.

3 Some Basic Examples

We start this section with some two-dimensional examples of possible choices for the spaces V_h and Q_h, namely the $P_1 - P_1$, $P_1 - P_0$ elements. These elements in general do not satisfy the inf–sup condition (20) and are not applicable in practice.

Then we present a complete analysis of the $P_2 - P_0$ element. Even though it might not be recommended to use this element because of its "unbalanced" approximation properties ($O(h^2)$ for V_h in the V-norm and only $O(h)$ for Q_h in the norm of Q), so that estimate (21) turns out to be suboptimal, the analysis of this element contains basic issues for getting familiar with the approximation of the Stokes problem. Moreover, the stability properties of this element will often be used as an intermediate step for the analysis of other, more efficient, elements.

Example 3.1. The $P_1 - P_1$ element

Let us consider a very simple case, that is, a P_1 continuous interpolation for both velocity and pressure, namely, using the notation of [24],

$$V_h = (\mathfrak{L}_1^1)^2 \cap V, \quad Q_h = \mathfrak{L}_1^1 \cap Q. \tag{24}$$

It is easy to check that if the number of triangles is large enough, then there exist nontrivial functions satisfying the *discrete* divergence-free condition. Thus no locking will occur and a solution can be computed. Indeed, this method would not provide an optimal approximation of the pressures by virtue of the unbalanced approximation properties of the discrete spaces (while Q_h achieves second order in L^2, V_h gives only first order in H^1). On the other hand, users of such methods (you can think of using also, for instance, $(P_2 - P_2)$, $(Q_1 - Q_1)$, etc.), soon became aware that their results were strongly mesh dependent. In particular, the computed pressures exhibited a very strange instability. This comes from the fact that for some meshes the kernel of the discrete gradient operator is nontrivial. This means that the solution obtained is determined only up to a given number of *spurious pressure modes*, [56, 57] and that, at best, some filtering will have to be done before accurate results are available. We shall come back later on to this phenomenon also named checkerboarding in Sect. 5. To better understand the nature of spurious pressure modes, the reader may check the results of Fig. 2 in which different symbols denote points where functions in $\ker(\underline{\mathrm{grad}}_h)$ must have equal values for a $(P_1 - P_1)$ approximation. In this case we have *three* spurious pressure modes. This also shows that there exists on this mesh one nontrivial discrete divergence-free function whereas a direct count would predict locking.

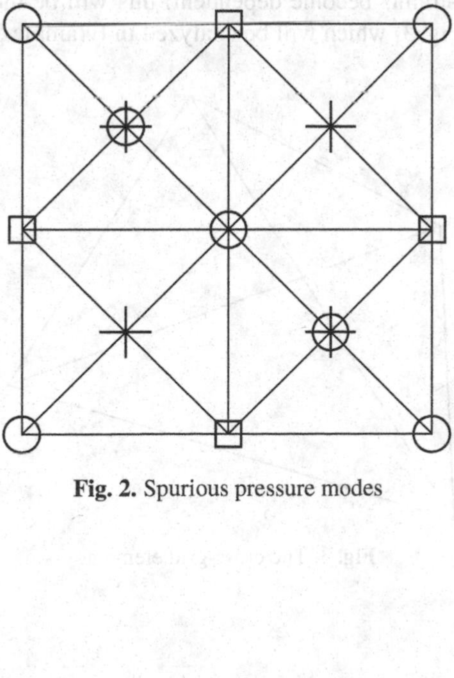

Fig. 2. Spurious pressure modes

Example 3.2. $P_1 - P_0$ approximation

This is probably the simplest element one can imagine for the approximation of an incompressible flow: one uses a standard P_1 approximation for the velocities and a piecewise constant approximation for the pressures. With the notation of [24] this would read

$$V_h = (\mathfrak{L}_1^1)^2 \cap V, \quad Q_h = \mathfrak{L}_0^0 \cap Q. \tag{25}$$

As the divergence of a P_1 velocity field is piecewise constant, this would lead to a divergence-free approximation. Moreover, this would give a well-balanced $O(h)$ approximation in estimate (21).

However, it is easy to see that such an element will not work for a general mesh. Indeed, consider a triangulation of a (simply connected) domain Ω and let us denote by

— t the number of triangles,
— v_I the number of internal vertices,
— v_B the number of boundary vertices.

We shall thus have $2v_I$ degrees of freedom (d.o.f.) for the space V_h (since the velocities vanish on the boundary) and $(t-1)$ d.o.f. for Q_h (because of the zero mean value of the pressures) leading to $(t-1)$ independent divergence-free constraints. By Euler's relations, we have

$$t = 2v_I + v_B - 2 \tag{26}$$

and thus

$$t - 1 > 2v_I \tag{27}$$

whenever $v_B > 3$. A function $\underline{u}_h \in V_h$ is thus overconstrained and a *locking phenomenon* is likely to occur: in general the only divergence-free discrete function is $\underline{u}_h \equiv 0$. When the mesh is built under certain restrictions, it is, however, possible that some linear constraints become dependent: this will be the case for the cross-grid macroelement (Fig. 3) which will be analyzed in Example 5.3.

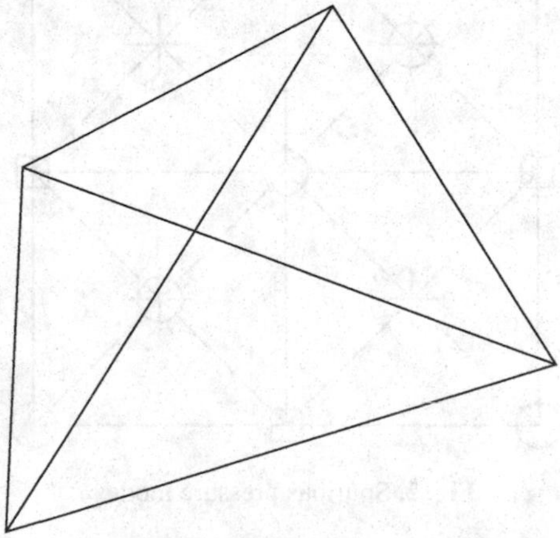

Fig. 3. The cross-grid element

Example 3.3. A stable approximation: the $P_2 - P_0$ element

Let us now move to the *stable* $P_2 - P_0$ element; namely, we use continuous piecewise quadratic vectors for the approximation of the velocities and piecewise constants for the pressures.

The discrete divergence-free condition can then be written as

$$\int_K \operatorname{div} \underline{u}_h \, dx = \int_{\partial K} \underline{u}_h \cdot \underline{n} \, ds = 0, \quad \forall K \in \mathcal{T}_h, \tag{28}$$

that is as a conservation of mass on every element. This is intuitively an approximation of $\operatorname{div} \underline{u}_h = 0$, directly related to the physical meaning of this condition. It is clear from error estimate (21) and standard approximation results that such an approximation will lead to the loss of one order of accuracy due to the poor approximation of the pressures. However, an augmented Lagrangian technique can be used, in order to recover a part of the accuracy loss (see Remark 3.2).

We are going to prove the following proposition.

Proposition 3.1. *The choice*

$$V_h = (\mathfrak{L}_2^1)^2 \cap V, \quad Q_h = \mathfrak{L}_0^0 \cap Q \tag{29}$$

fulfills the inf–sup condition (20).

Proof. Before giving the rigorous proof of Proposition 3.1 we are going to sketch the main argument.

If we try to check the inf–sup condition by building a Fortin operator Π_h, then, given $\underline{u}$, we have to build $\underline{u}_h = \Pi_h \underline{u}$ such that

$$\int_\Omega \operatorname{div}(\underline{u} - \underline{u}_h) q_h \, dx = 0, \quad \forall q_h \in Q_h. \tag{30}$$

Since q_h is constant on every element $K \in \mathcal{T}_h$, this is equivalent to

$$\int_K \operatorname{div}(\underline{u} - \underline{u}_h) \, dx = \int_{\partial K} (\underline{u} - \underline{u}_h) \cdot \underline{n} \, ds = 0. \tag{31}$$

This last condition would be satisfied if $\underline{u}_h$ could be built in the following way. Let us denote by M_i and e_i, $i = 1, 2, 3$, the vertices and the sides of the triangular element K (Fig. 4); the midside nodes are denoted by M_{ij}. We then define

$$\underline{u}_h(M_i) = \underline{u}(M_i), \quad i = 1, 2, 3 \tag{32}$$

$$\int_{e_i} \underline{u}_h \, ds = \int_{e_i} \underline{u} \, ds. \tag{33}$$

Condition (33) can be fulfilled by a correct choice of $\underline{u}_h(M_{ij})$. Moreover this construction can be done at element level as the choice of $\underline{u}_h(M_{ij})$ is compatible on adjacent elements (that is, with this definition, $\underline{u}_h$ turns out to be continuous).

Although this is the basic idea, some technicalities must be introduced before a real construction is obtained. Indeed, for $\underline{u} \in (H_0^1(\Omega))^2$, condition (32) has no sense.

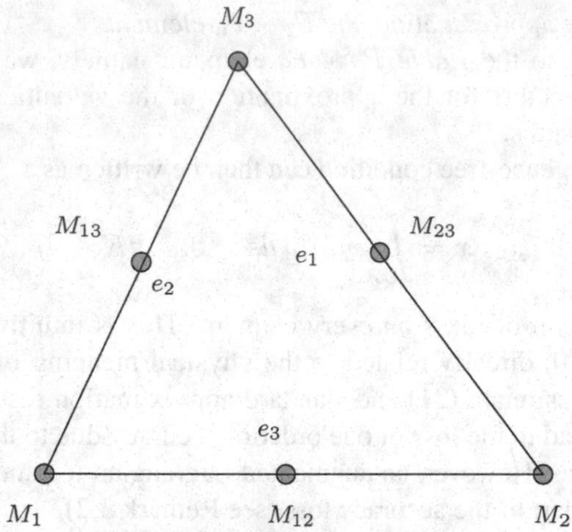

Fig. 4. Vertices, edges and midnodes

Let us then give a rigorous proof of Proposition 3.1. We denote by $\Pi_1 : V \to V_h$ the Clément interpolant [30]. We then have

$$\sum_K h_K^{2r-2} |\underline{v} - \Pi_1 \underline{v}|^2_{r,K} \leq c ||\underline{v}||^2_{1,\Omega}, \quad r = 0, 1. \tag{34}$$

Setting $r = 1$ and using the triangular inequality $||\Pi_1 \underline{v}|| \leq ||\underline{v} - \Pi_1 \underline{v}|| + ||\underline{v}||$ gives

$$||\Pi_1 \underline{v}||_V \leq c_1 ||\underline{v}||_V, \quad \forall \underline{v} \in V. \tag{35}$$

We now modify Π_1 in a suitable way. Let us define $\Pi_2 : V \to V_h$ in the following way:

$$\Pi_2 \underline{v}|_K(M) = 0, \quad \forall M \text{ vertex of } K, \tag{36}$$

$$\int_e \Pi_2 \underline{u}\, ds = \int_e \underline{u}\, ds, \quad \forall e \text{ edge of } K. \tag{37}$$

By construction Π_2 satisfies

$$\int_\Omega \operatorname{div}(\underline{v} - \Pi_2 \underline{v}) q_h\, dx = 0, \quad \forall \underline{v}_h \in V_h,\ q_h \in Q_h \tag{38}$$

and a scaling argument gives

$$|\Pi_2 \underline{v}|_{1,K} = |\widehat{\Pi_2 \underline{v}}|_{1,\hat{K}} < c(K, \theta_0) ||\hat{\underline{v}}||_{1,\hat{K}} \leq c(K, \theta_0)(h_K^{-1} |\underline{v}|_{0,K} + |\underline{v}|_{1,K}). \tag{39}$$

We can now define

$$\Pi_h \underline{u} = \Pi_1 \underline{u} + \Pi_2(\underline{u} - \Pi_1 \underline{u}) \tag{40}$$

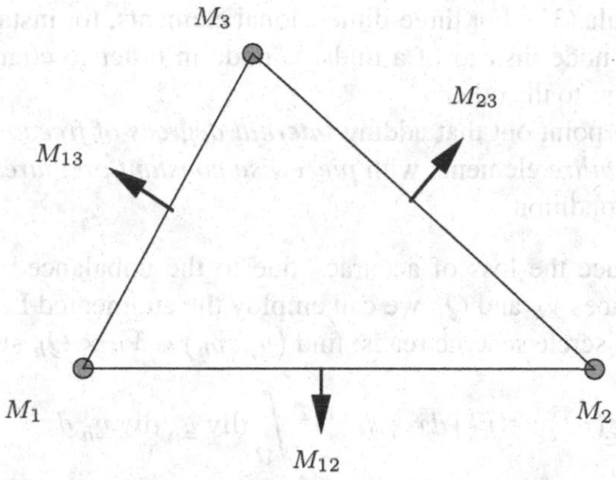

Fig. 5. Reduced $P_2 - P_0$ element

and observe that (39) and (34) imply

$$||\Pi_2(I - \Pi_1)\underline{u}||_V \leq c_2||\underline{u}||_V, \quad \forall \underline{v} \in V, \tag{41}$$

since

$$||\Pi_2(I - \Pi_1)\underline{v}||^2_{1,\Omega} = \sum_K ||\Pi_2(I - \Pi_1)\underline{v}||^2_{1,K} \tag{42}$$

$$\leq c \sum_K \{h_K^{-2}||(I - \Pi_1)\underline{v}||^2_{0,K} + |(I - \Pi_1)\underline{v}|^2_{1,K}\} \leq c||\underline{v}||^2_{1,\Omega}. \tag{43}$$

Hence Π_h is a Fortin operator and the proof is concluded.

The above proof can easily be extended to more general cases. It applies to the $Q_2 - P_0$ quadrilateral element provided the usual regularity assumptions on quadrilateral meshes are made. A simple modification will hold for elements in which only the normal component of velocity is used as a d.o.f. at the midside nodes [37], [33], [11]. Indeed, if only the normal component of $\underline{u}_h$ is used as a degree of freedom, the $P_2 - P_0$ element becomes the element of Fig. 5 in which, on each side, the normal component of $\underline{u}_h$ is quadratic, whereas the tangential component is only linear. In this case we can define $\Pi_2\underline{v}$ by setting

$$\int_e (\Pi_2\underline{v} \cdot \underline{n})\, ds = \int_e \underline{v} \cdot \underline{n}\, ds \tag{44}$$

The above proof applies directly. The same remark is valid for the $Q_2 - P_0$ quadrilateral element.

Remark 3.1. The philosophical idea behind the $P_2 - P_0$ element is that we need one degree of freedom per each interface (actually, the normal component of the velocity) in order to control the jump of the pressures. This is basically the meaning of the Green's formula (31). For three-dimensional elements, for instance, we would need some midface node instead of a midside node in order to control the normal flux from one element to the other.

In particular, we point out that adding *internal degrees of freedom* to the velocity space *cannot stabilize* elements with *piecewise constant pressures* which do not satisfy the inf–sup condition.

Remark 3.2. To reduce the loss of accuracy due to the unbalanced approximation properties of the spaces V_h and Q_h we can employ the augmented Lagrangian technique of [16]. The discrete scheme reads: find $(\underline{u}_h, p_h) \in V_h \times Q_h$ such that

$$\begin{aligned} \int_\Omega \underline{\underline{\varepsilon}}(\underline{u}_h) : \underline{\underline{\varepsilon}}(\underline{v}_h)\, dx + h^{-1/2} \int_\Omega \operatorname{div} \underline{u}_h \operatorname{div} \underline{v}_h\, dx \\ - \int_\Omega p_h \operatorname{div} \underline{v}_h\, dx = \int_\Omega \underline{f} \cdot \underline{v}_h\, dx \quad \forall \underline{v}_h \in V_h, \\ \int_\Omega q_h \operatorname{div} \underline{u}_h\, dx = 0 \quad \forall q_h \in Q_h. \end{aligned} \tag{45}$$

Following [16] we have the following error estimate

$$||\underline{u} - \underline{u}_h||_V + ||p - p_h||_Q \leq c h^{3/2} \inf_{\underline{v} \in V_h,\, q \in Q_h} (||\underline{u} - \underline{v}||_V + ||p - q||_Q) . \tag{46}$$

4 Standard Techniques for Checking the Inf–Sup Condition

We consider in this section standard techniques for the proof of the inf–sup stability condition (20) that can be applied to a large class of elements. For ease of presentation, in this section we develop the theory only and postpone the examples to Sects. 6 and 7, for two- and three-dimensional schemes, respectively. However, after the description of each technique, we list some schemes for which that technique applies too.

Of course, the first method consists in the direct estimate of the inf–sup constant. In order to do that, we need to construct explicitly the operator $\underline{\operatorname{grad}}_h : Q_h \to V_h$ satisfying

$$\int_\Omega q_h \operatorname{div} \underline{\operatorname{grad}}_h q_h\, dx = ||q_h||^2_Q, \tag{47}$$

$$||\underline{\operatorname{grad}}_h q_h||_V \leq c_h ||q_h||_Q, \tag{48}$$

for any $q_h \in Q_h$. If the constant c_h in (48) is bounded above, then the inf–sup condition (20) will hold true.

4.1 Fortin's Trick

An efficient way of proving the inf–sup condition (20) consists in using the technique presented in [36] which consists of building an interpolation operator Π_h as follows.

Proposition 4.1. *If there exists a linear operator $\Pi_h : V \longrightarrow V_h$ such that*

$$\int_\Omega \operatorname{div}(\underline{u} - \Pi_h \underline{u}) q_h \, dx = 0, \quad \forall \underline{v} \in V, \; q_h \in Q_h, \tag{49}$$

$$||\Pi_h \underline{u}||_V \leq c||\underline{u}||_V. \tag{50}$$

then the inf–sup condition (20) *holds true.*

Remark 4.1. Condition (49) is equivalent to $\ker(\underline{\mathrm{grad}}_h) \subset \ker(\underline{\mathrm{grad}})$. An element with this property will present no spurious pressure modes.

In several cases the operator Π_h can be constructed in two steps as it has been done for the P_2–P_0 element in Proposition 3.1. In general it will be enough to build two operators $\Pi_1, \Pi_2 \in \mathcal{L}(V, V_h)$ such that

$$||\Pi_1 \underline{v}||_V \leq c_1 ||\underline{v}||_V, \qquad \forall \underline{v} \in V, \tag{51}$$

$$||\Pi_2 (I - \Pi_1) \underline{v}||_V \leq c_2 ||\underline{v}||_V, \quad \forall \underline{v} \in V, \tag{52}$$

$$\int_\Omega \operatorname{div}(\underline{v} - \Pi_2 \underline{v}) q_h = 0, \qquad \forall \underline{v} \in V, \forall q_h \in Q_h, \tag{53}$$

where the constants c_1 and c_2 are independent of h. Then the operator Π_h satisfying (49) and (50) will be found as

$$\Pi_h \underline{u} = \Pi_1 \underline{u} + \Pi_2 (\underline{u} - \Pi_1 \underline{u}). \tag{54}$$

In many cases, Π_1 will be the Clément operator of [30] defined in $H^1(\Omega)$.

On the contrary, the choice of Π_2 will vary from one case to the other, according to the choice of V_h and Q_h. However, the common feature of the various choices for Π_2 will be the following one: the operator Π_2 is constructed on each element K in order to satisfy (53). In many cases it will be such that

$$||\Pi_2 \underline{v}||_{1,K} \leq c(h_K^{-1} ||\underline{v}||_{0,K} + |\underline{v}|_{1,K}). \tag{55}$$

We can summarize this results in the following proposition.

Proposition 4.2. *Let V_h be such that a "Clément's operator": $\Pi_1 : V \rightarrow V_h$ exists and satisfies* (34). *If there exists an operator $\Pi_2 : V \longrightarrow V_h$ such that* (53) *and* (55) *hold, then the operator Π_h defined by* (54) *satisfies* (49) *and* (50) *and therefore the discrete inf–sup condition* (20) *holds.*

Example 4.1. The construction of Fortin's operator has been used, for instance, for the sability proof of the $P_2 - P_0$ element (see Example 3.3).

4.2 Projection onto Constants

Following [19] we now consider a modified inf–sup condition.

$$\inf_{q_h \in Q_h} \sup_{\underline{v}_h \in V_h} \frac{\int_\Omega q_h \operatorname{div} \underline{v}_h \, dx}{||\underline{v}_h||_V ||q_h - \bar{q}_h||_Q} \geq k_0 > 0, \tag{56}$$

where $\bar{q}_h$ is the L^2-projection of q_h onto $\mathfrak{L}_0^0$ (that is, piecewise constant functions).

Proposition 4.3. *Let us suppose that the modified inf–sup condition* (56) *holds with* k_0 *independent of* h*. Assume moreover that* V_h *is such that, for any* $q_h \in \mathfrak{L}_0^0 \cap Q$,

$$\sup_{\underline{v}_h \in V_h} \frac{\int_\Omega q_h \operatorname{div} \underline{v}_h \, dx}{||\underline{v}_h||_V} \geq \gamma_0 ||q_h||_Q, \tag{57}$$

with γ_0 *independent of* h*. Then the inf–sup condition* (20) *holds true.*

Proof. For any $q_h \in Q_h$ one has

$$\begin{aligned} \sup_{\underline{v}_h \in V_h} \frac{b(\underline{v}_h, q_h)}{||\underline{v}_h||_V} &= \sup_{\underline{v}_h \in V_h} \left\{ \frac{b(\underline{v}_h, q_h - \bar{q}_h)}{||\underline{v}_h||_V} + \frac{b(\underline{v}_h, \bar{q}_h)}{||\underline{v}_h||_V} \right\} \\ &\geq \sup_{\underline{v}_h \in V_h} \frac{b(\underline{v}_h, \bar{q}_h)}{||\underline{v}_h||_V} - \sup_{\underline{v}_h \in V_h} \frac{b(\underline{v}_h, q_h - \bar{q}_h)}{||\underline{v}_h||_V} \\ &\geq \gamma_0 ||\bar{q}_h||_Q - ||q_h - \bar{q}_h||_0, \end{aligned} \tag{58}$$

which implies

$$\sup_{\underline{v}_h \in V_h} \frac{b(\underline{v}_h, q_h)}{||\underline{v}_h||_V} \geq \frac{k_0 \gamma_0}{1 + k_0} ||\bar{q}_h||_Q. \tag{59}$$

Putting together (56) and (59) proves the proposition.

Remark 4.2. In the case of *continous pressures* schemes, hypothesis (57) can be replaced with the following approximation assumption: for any $\underline{v} \in V$ there exists $\underline{v}^I \in V_h$ such that

$$||\underline{v} - \underline{v}^I||_{L^2(\Omega)} \leq c_1 h ||\underline{v}||_V, \quad ||\underline{v}^I||_V \leq c_2 ||\underline{v}||_V. \tag{60}$$

The details of the proof can be found in [19] when the mesh is quasiuniform. The quasiuniformity assumption is actually not needed, as it can be shown with an argument similar to the one which will be presented in the next subsection (see, in particular, Remark 4.3).

Example 4.2. The technique presented in this section will be used, for instance, for the stability proof of the generalized two-dimensional Hood–Taylor element (see Sect. 8.2 and Theorem 8.1).

4.3 Verfürth's Trick

Verfürth's trick [66] applies to *continuous pressures* approximations and is essentially based on two steps. The first step is quite general and can be summarized in the following lemma.

Lemma 4.1. *Let* Ω *be a bounded domain in* $\mathbb{R}^N$ *with Lipschitz continuous boundary. Let* $V_h \subset (H_0^1(\Omega))^2 = V$ *and* $Q_h \subset H^1(\Omega)$ *be closed subspaces. Assume that there exists a linear operator* Π_h^0 *from* V *into* V_h *and a constant* c *(independent of* h*) such that*

$$||\underline{v}_h - \Pi_h^0 \underline{v}||_r \leq c \sum_{K \in \mathcal{T}_h} \left(h_K^{2-2r} ||\underline{v}||_{1,K}^2 \right)^{1/2}, \quad \forall \underline{v} \in V, \; r = 0, 1. \tag{61}$$

Then there exist two positive constants c_1 and c_2 such that, for every $q_h \in Q_h$,

$$\sup_{\underline{v} \in V_h} \frac{\int_\Omega q_h \operatorname{div} \underline{v}_h \, dx}{||\underline{v}_h||_V} \geq c_1 ||q_h||_Q - c_2 \left(\sum_{K \in \mathcal{T}_h} h_K^2 ||\underline{\operatorname{grad}}\, q_h||_{0,K}^2 \right)^{1/2}. \tag{62}$$

Proof. Given $q_h \in Q_h$, let $\bar{\underline{v}} \in V$ be such that

$$\frac{\int_\Omega q_h \operatorname{div} \bar{\underline{v}} \, dx}{||\bar{\underline{v}}||_V ||q_h||_Q} \geq \beta > 0, \tag{63}$$

where β is the continuous inf–sup constant. Then,

$$\begin{aligned}
\sup_{\underline{v}_h \in V_h} \frac{\int_\Omega q_h \operatorname{div} \underline{v}_h \, dx}{||\underline{v}_h||_V} &\geq \frac{\int_\Omega q_h \operatorname{div} \Pi_h^0 \bar{\underline{v}} \, dx}{||\Pi_h^0 \bar{\underline{v}}||_V} \geq \frac{1}{2c} \frac{\int_\Omega q_h \operatorname{div} \Pi_h^0 \bar{\underline{v}} \, dx}{||\bar{\underline{v}}||_V} \\
&= \frac{1}{2c} \frac{\int_\Omega q_h \operatorname{div} \bar{\underline{v}} \, dx}{||\bar{\underline{v}}||_V} + \frac{1}{2c} \frac{\int_\Omega q_h \operatorname{div}(\Pi_h^0 \bar{\underline{v}} - \bar{\underline{v}}) \, dx}{||\bar{\underline{v}}||_V} \\
&\geq \frac{\beta}{4c} ||q_h||_Q - \frac{1}{2c} \frac{\int_\Omega \underline{\operatorname{grad}}\, q_h \cdot (\Pi_h^0 \bar{\underline{v}} - \bar{\underline{v}}) \, dx}{||\bar{\underline{v}}||_V} \\
&\geq \frac{\beta}{4c} ||q_h||_Q - \left(\frac{1}{2} \sum_{K \in \mathcal{T}_h} h_K^2 ||\underline{\operatorname{grad}}\, q_h||_{0,K}^2 \right)^{1/2}.
\end{aligned} \tag{64}$$

Remark 4.3. Indeed, via a scaling argument, it can be shown that the last term in the right-hand side of equation (62) is equivalent to $||q_h - \bar{q}_h||_0$, where $\bar{q}_h$ denotes, as in the previous subsection, the L^2-projection onto the piecewise constants.

We are now in the position of stating the main result of this subsection. Note that Verfürth's trick consists in proving a kind of inf–sup condition where the zero norm of q_h is substituted by $h|q_h|_1$.

Proposition 4.4. *Suppose the hypotheses of Lemma 4.1 hold true. Assume, moreover, that there exists a constant c_3 such that, for every $q_h \in Q_h$,*

$$\sup_{\underline{v}_h \in V_h} \frac{\int_\Omega q_h \operatorname{div} \underline{v}_h}{||\underline{v}_h||_V} \geq c_3 \left(\sum_{K \in \mathcal{T}_h} h_K^2 |q_h|_{1,K}^2 \right)^{1/2}. \tag{65}$$

Then the standard inf–sup condition (20) *holds true.*

Proof. Let us multiply (62) by c_3 and (65) by c_2 and sum up the two equations. We have

$$(c_3 + c_2) \sup_{\underline{v}_h \in V_h} \frac{\int_\Omega q_h \operatorname{div} \underline{v}_h \, dx}{||\underline{v}_h||_V} \geq c_1 c_3 ||q_h||_Q, \tag{66}$$

that is, the inf–sup condition (20).

Example 4.3. The Verfürth's trick has been designed for the stability analysis of the Hood–Taylor method. It will be used for this purpose in Sect. 8.2 (see Theorem 8.1).

4.4 Space and Domain Decomposition Techniques

Sometimes the spaces V_h and Q_h decomposes into the sum (direct or not) of subspaces for which it might be easier to prove an inf–sup condition. This is the case, for instance, when a *domain decomposition* technique is employed. Some of the results we are going to present, can be viewed as a particular case of the macroelement technique which will be introduced in Sect. 4.5.

The next result has been presented and proved in [41].

Proposition 4.5. *Suppose that Ω can be decomposed as the union of disjoint subdomains with Lipschitz continuous boundaries*

$$\Omega = \bigcup_{r=1}^{R} \Omega_r \tag{67}$$

and that such decomposition is compatible with the mesh in the sense that each element of the mesh is contained in one subdomain. We make use of the following notation:

$$\begin{aligned} V_{0,r} &= \{\underline{v} \in V_h : \underline{v} = 0 \text{ in } \Omega \setminus \Omega_r\}, \\ Q_{0,r} &= \{q \in Q_h : \int_{\Omega_r} q\, dx = 0\}, \\ K &= \{q \in Q : q|_{\Omega_r} \text{ is constant}, \ r = 1, \dots, R\}. \end{aligned} \tag{68}$$

Suppose, moreover, that the spaces $V_{0,r}$ and $Q_{0,r}$ satisfy the following inf–sup condition

$$\inf_{q_h \in Q_{0,r}} \sup_{\underline{v}_h \in V_{0,r}} \frac{\int_{\Omega_r} q_h \operatorname{div} \underline{v}_h \, dx}{\|q_h\|_Q \|\underline{v}_h\|_V} \geq k_r > 0, \tag{69}$$

with k_r independent of h $(r = 1, \dots, R)$ and that the following inf–sup condition between V_h and K holds true

$$\inf_{q_h \in K} \sup_{\underline{v}_h \in V_h} \frac{\int_{\Omega} q_h \operatorname{div} \underline{v}_h \, dx}{\|q_h\|_Q \|\underline{v}_h\|_V} \geq k_K > 0, \tag{70}$$

with k_K independent of h. Then the spaces V_h and Q_h satisfy the inf–sup condition (20).

Sometimes it is not possible (or it is not the best choice) to partition Ω into *disjoint* subdomains. Let us describe the case of two overlapping subdomains. The following proposition can be checked by a direct computation.

Proposition 4.6. *Let Ω be the union of two subdomains Ω_1 and Ω_2 with Lipschitz continuous boundaries. With the notation of the previous proposition, suppose that the spaces $V_{0,r}$ and $Q_{0,r}$ satisfy the inf–sup conditions*

$$\inf_{q_h \in Q_{0,r}} \sup_{\underline{v}_h \in V_{0,r}} \frac{\int_{\Omega_r} q_h \operatorname{div} \underline{v}_h \, dx}{\|q_h\|_Q \|\underline{v}_h\|_V} \geq k_r > 0, \tag{71}$$

for $r = 1, 2$. Then the spaces V_h and Q_h satisfy the condition

$$\inf_{q_h \in Q_h} \sup_{\underline{v}_h \in V_h} \frac{\int_\Omega q_h \operatorname{div} \underline{v}_h \, dx}{\|q_h - \overline{q}_h\|_Q \|\underline{v}_h\|_V} \geq \frac{1}{\sqrt{2}} \min(k_1, k_2), \tag{72}$$

where, as in Sect. 4.2, we have denoted by $\overline{q}_h$ the L^2 projection of q_h onto the space $\mathfrak{L}_0^0$.

Another useful technique for proving the inf–sup condition can be found in [54]. This result is quite general; in particular, the decomposition of the spaces V_h and Q_h does not rely on a decomposition of the domain Ω. In [54] the following proposition is stated for a two-subspaces decomposition, but it obviously extends to more general situations.

Proposition 4.7. *Let Q_1 and Q_2 be subspaces of Q_h such that*

$$Q_h = Q_1 + Q_2. \tag{73}$$

If V_1, V_2 are subspaces of V_h and α_1, α_2 positive constants such that

$$\inf_{q_h \in Q_i} \sup_{\underline{v}_h \in V_i} \frac{\int_\Omega q_h \operatorname{div} \underline{v}_h \, dx}{\|q_h\|_Q \|\underline{v}_h\|_V} \geq \alpha_i, \quad i = 1, 2 \tag{74}$$

and β_1, β_2 are nonnegative constants such that

$$\begin{aligned} \left| \int_\Omega q_1 \operatorname{div} \underline{v}_2 \, dx \right| &\leq \beta_1 \|q_1\|_Q \|\underline{v}_2\|_V, \quad q_1 \in Q_1, \ \forall \underline{v}_2 \in V_2, \\ \left| \int_\Omega q_2 \operatorname{div} \underline{v}_1 \, dx \right| &\leq \beta_2 \|q_2\|_Q \|\underline{v}_1\|_V, \quad q_2 \in Q_2, \ \forall \underline{v}_1 \in V_1, \end{aligned} \tag{75}$$

with

$$\beta_1 \beta_2 < \alpha_1 \alpha_2, \tag{76}$$

then the inf–sup condition (20) *holds true with k_0 depending only on α_i, β_i, $i = 1, 2$.*

Remark 4.4. Condition (76) is trivially true, for instance, when $\beta_1 \beta_2 = 0$ and $\alpha_1 \alpha_2 > 0$.

Example 4.4. Most of the technique presented in this section can be seen as a particular case of the macroelement technique (see Sect. 4.5) Proposition 4.6 will be used in Theorem 8.1 for the stability proof of the Hood–Taylor scheme.

4.5 Macroelement Technique

In this section we present a technique introduced by Stenberg (see [59, 60, 62, 61, 63]) which, under suitable hypotheses, reduces the matter of checking the inf–sup condition (20) to an algebraic problem. We refer also to [18] for related results in a somewhat different setting.

The present technique is based on a decomposition of the triangulation $\mathcal{T}_h$ into disjoint macroelements, where we refer to a *macroelement* as an open polygon (resp., polyhedron in $\mathbb{R}^3$) which is the union of adjacent elements.

Let us introduce some notation.

A macroelement M is said to be *equivalent* to a reference macroelement $\hat{M}$ if there exists a mapping $F_M : \hat{M} \to M$ such that

1. F_M is continuous and invertible;
2. $F_M(\hat{M}) = M$;
3. If $\hat{M} = \cup \hat{K}_j$, where K_j, $j = 1,\dots,m$ are the elements defining $\hat{M}$, then $K_j = F_M(\hat{K}_j)$, $j = 1,\dots,m$, are the elements of M;
4. $F_M|_{\hat{K}_j} = F_{K_j} \circ F^{-1}_{\hat{K}_j}$, $j = 1,\dots,m$, where F_K denotes the affine mapping from the reference element to a generic element K.

We denote by $\mathcal{E}_{\hat{M}}$ the equivalence class of $\hat{M}$. We now introduce the discrete spaces associated with V_h and Q_h on the generic macroelement M (n is the dimension of Ω).

$$\begin{aligned} V_{0,M} &= \{\underline{v} \in (H^1_0(M))^n : \underline{v} = \underline{w}|_M \text{ with } \underline{w} \in V_h\}, \\ Q_{0,M} &= \left\{p \in L^2(\Omega) : \int_M p\,dx = 0,\ p = q|_M \text{ with } q \in Q_h\right\}. \end{aligned} \tag{77}$$

We finally introduce a space which corresponds to the kernel of B_h^t on the macroelement M.

$$K_M = \left\{p \in Q_{0,M} : \int_M p \operatorname{div} \underline{v}\, dx = 0,\ \forall \underline{v} \in V_{0,m}\right\}. \tag{78}$$

The *macroelements condition* reads

$$K_M = \{0\}, \tag{79}$$

that is, the analogous (at a macroelement level) of the necessary condition for the discrete Stokes problem to be well-posed that the kernel of B_h^t reduces to the zero function.

Proposition 4.8. *Suppose that each triangulation $\mathcal{T}_h$ can be decomposed into disjoint macroelements belonging to a* fixed *number (independent of h) of equivalence classes $\mathcal{E}_{\hat{M}_i}$, $i = 1,\dots,n$. Suppose, moreover, that the pair $V_h - \mathfrak{L}^0_0/\mathbb{R}$ is a stable Stokes element, that is,*

$$\inf_{q_h \in \mathfrak{L}^0_0/\mathbb{R}} \sup_{\underline{v}_h \in V_h} \frac{\int_\Omega q_h \operatorname{div} \underline{v}_h\, dx}{||q_h||_Q ||\underline{v}_h||_V} \geq \beta > 0, \tag{80}$$

with β independent of h. Then the macroelement condition (79) *(for every $M \in \mathcal{E}_{\hat{M}_i}$, $i = 1,\dots n$) implies the inf–sup condition* (20).

Proof. We do not enter the technical details of the proof, for which we refer to [59]. The basic arguments of the proof are sketched in Remark 4.5.

Remark 4.5. The macroelement condition (79) is strictly related to the *patch test* commonly used in the engineering practice (cf., e.g., [67]). However, the count of the degrees of freedom is clearly insufficient by itself. Hence, let us point out how the hypotheses of Proposition 4.8 are important.

Hypothesis (79) (the macroelement condition) implies, via a compactness argument, that a discrete inf–sup condition holds true between the spaces $V_{0,M}$ and $Q_{0,M}$. The *finite* number of equivalent macroelements classes is sufficient to conclude that the corresponding inf–sup constants are uniformly bounded below by a positive number.

Then, we are basically in the situation of the domain decomposition technique of Sect. 4.4. We now use hypothesis (80) to control the constant functions on each macroelement and to conclude the proof.

Remark 4.6. Hypothesis (80) is satisfied in the two-dimensional case whenever V_h contains piecewise quadratic functions (see Sect. 3). In the three-dimensional case things are not so easy (to control the constants we need extra degrees of freedom on the faces, as observed in Remark 3.1. For this reason, let us state the following proposition which can be proved with the technique of Sect. 4.2 (see Remark 4.2) and which applies to the case of *continuous pressures* approximations.

Proposition 4.9. *Let us make the same assumptions as in Proposition 4.8 with* (80) *replaced by the condition of Remark 4.2 (see* (60)*). Then, provided* $Q_h \subset C^0(\Omega)$, *the inf–sup condition* (70) *holds true.*

Remark 4.7. The hypothesis that the macroelement partition of $\mathcal{T}_h$ is *disjoint* can be weakened, in the spirit of Proposition 4.6, by requiring that each element K of $\mathcal{T}_h$ belongs at most to a finite number N of macroelements with N independent of h.

Example 4.5. The macroelement technique can be used in order to prove the stability of several schemes. Among those, we recall the $Q_2 - P_1$ element (see Sect. 6.4) and the three-dimensional generalized Hood–Taylor scheme (see Theorem 8.2).

4.6 Making Use of the Internal Degrees of Freedom

This subsection presents a general framework providing a general tool for the analysis of finite element approximations to incompressible materials problems.

The basic idea has been used several times on particular cases, starting from [31] for discontinuous pressures and from [2] and [3] for continuous pressures. We are going to present it in its final general form given by [23]. It consists essentially in stabilizing an element by adding suitable bubble functions to the velocity field.

In order to do that, we first associate to every finite element discretization $Q_h \subset Q$ the space

$$B(\underline{\text{grad}}\, Q_h) = \left\{ \underline{\beta} \in V : \underline{\beta}|_K = b_K\, \underline{\text{grad}}\, q_h|_K \text{ for some } q_h \in Q_h \right\}, \tag{81}$$

where b_K is a bubble function defined in K. In particular, we can take as B_K the standard cubic bubble if K is a triangle, or a biquadratic bubble if K is a square or other obvious generalizations in 3D. In other words, the restriction of a $\underline{\beta} \in B(\underline{\text{grad}}\, Q_h)$ to an element K is the product of the bubble functions b_K times the gradient of a function of $Q_h|_K$.

Remark 4.8. Notice that the space $B(\underline{\text{grad}}\, Q_h)$ is not defined through a basic space $\hat{B}$ on the reference element. This could be easily done in the case of *affine* elements, for all the reasonable choices of Q_h. However, this is clearly *unnecessary*: if we know how to compute q_h on K we also know how to compute $\underline{\text{grad}}\, q_h$ and there is no need for a reference element.

We can now prove our basic results, concerning the two cases of continuous or discontinuous pressures.

Proposition 4.10. *(Stability of continuous pressure elements). Assume that there exists an operator $\Pi_1 \in \mathfrak{L}(V, V_h)$ satisfying the property of the Clément interpolant* (34). *If $Q_h \subset C^0(\Omega)$ and V_h contains the space $B(\underline{\text{grad}}\, Q_h)$ then the pair (V_h, Q_h) is a stable element, in the sense that it satisfies the inf–sup condition* (20).

Proof. We shall build a Fortin operator, like in Proposition 4.2. We only need to construct the operator Π_2. We define $\Pi_2 : V \to B(\underline{\text{grad}}\, Q_h)$, on each element, by requiring

$$\begin{aligned} &\Pi_2 \underline{v}|_K \in B(\underline{\text{grad}}\, Q_h)|_K = b_{3,K}\, \underline{\text{grad}}\, Q_h|_K, \\ &\int_K (\Pi_2 \underline{v} - \underline{v}) \cdot \underline{\text{grad}}\, q_h \, dx = 0, \quad \forall q_h \in Q_h. \end{aligned} \tag{82}$$

Problem (82) has obviously a unique solution and Π_2 satisfies (53). Finally (55) follows by a scaling argument. Hence Proposition 4.2 gives the desired result.

Corollary 4.1. *Assume that $Q_h \subset Q$ is a space of continuous piecewise smooth functions. If V_h contains $(\mathfrak{L}_1^1)^2 \oplus B(\underline{\text{grad}}\, Q_h)$ then the pair (V_h, Q_h) satisfies the inf–sup condition* (20).

Proof. Since V_h contains piecewise linear functions, there exists a Clément interpolant Π_1 satisfying (34). Hence we can apply Proposition (4.10).

We now consider the case of discontinuous pressure elements.

Proposition 4.11. *(Stability of discontinuous pressure elements). Assume that there exists an operator $\tilde{\Pi}_1 \in \mathfrak{L}(V, V_h)$ satisfying*

$$\begin{aligned} &||\tilde{\Pi}_1 \underline{v}||_V \le c||\underline{v}||_V, \quad \forall \underline{v} \in V, \\ &\int_K \text{div}(\underline{v} - \tilde{\Pi}_1 \underline{v})\, dx = 0, \quad \forall \underline{v} \in V\ \forall K \in \mathcal{T}_h. \end{aligned} \tag{83}$$

If V_h contains $B(\underline{\text{grad}}\, Q_h)$ then the pair (V_h, Q_h) is a stable element, in the sense that it satisfies the inf–sup condition (20).

Proof. We are going to use Proposition 4.10. We take $\tilde{\Pi}_1$ as operator Π_1. We are not defining Π_2 on the whole V, but only in the subspace

$$V^0 = \left\{ \underline{v} \in V : \int_K \operatorname{div} \underline{v} \, dx = 0, \quad \forall K \in \mathcal{T}_h \right\}. \tag{84}$$

This will be enough, since we need to apply Π_2 to the difference $\underline{v} - \tilde{\Pi}_1 \underline{v}$ which is in V^0 by (83).

For every $\underline{v} \in V^0$ we define $\Pi_2 \underline{v} \in B(\underline{\text{grad}}\, Q_h)$ by requiring that, in each element K,

$$\begin{aligned} &\Pi_2 \underline{v}|_K \in B(\underline{\text{grad}}\, Q_h)|_K = b_{3,K} \underline{\text{grad}}\, Q_h|_K, \\ &\int_K \operatorname{div}(\Pi_2 \underline{v} - \underline{v}) q_h \, dx = 0, \quad \forall q_h \in Q_h|_K. \end{aligned} \tag{85}$$

Note that (85) is uniquely solvable, if $\underline{v} \in V^0$, since the divergence of a bubble function has always zero mean value (hence the number of nontrivial equations is equal to $\dim(Q_h|_K) - 1$, which is equal to the number of unknowns; the nonsingularity then follows easily). It is obvious that Π_2, as given by (85), will satisfy (53) for all $\underline{v} \in V^0$. We have to check that

$$\|\Pi_2 \underline{v}\|_1 \le c \|v\|_V, \tag{86}$$

which actually follows by a scaling argument making use of the following bound

$$|\widehat{\Pi_2 \underline{v}}|_{0,\hat{K}} \le c(\theta_0) |\hat{\underline{v}}|_{1,\hat{K}}. \tag{87}$$

Corollary 4.2. *(Two-dimensional case). Assume that $Q_h \subset Q$ is a space of piecewise smooth functions. If V_h contains $(\mathcal{L}_2^1)^2 \oplus B(\underline{\text{grad}}\, Q_h)$ then the pair (V_h, Q_h) satisfies the inf–sup condition* (20).

Proof. The stability of the $P_2 - P_0$ element (see Sect. 3 implies the existence of $\tilde{\Pi}_1$ as in Proposition 4.11.

Propositions 4.10 and 4.11 are worth a few comments. They show that almost any element can be stabilized by using bubble functions. For continuous pressure elements this procedure is mainly useful in the case of triangular elements. For discontinuous pressure elements it is possible to stabilize elements which are already stable for piecewise constant pressure field. Examples of such a procedure can be found in [34]. Stability with respect to piecewise constant pressure implies that at least one degree of freedom on each side or face of the element is linked to the normal component of velocity (see [37] and Remark 3.1).

Example 4.6. The use of internal degrees of freedom can be used in the stability analysis of several methods. For instance, we use it for the analysis of the MINI element (see Sects. 6.1 and 7.1) in the case of continuous pressures and of the Crouzeix–Raviart element (see Remark 6.1 and Sect. 7.2) in the case of discontinuous pressures.

5 Spurious Pressure Modes

For a given choice of V_h and Q_h, the space S_h of spurious pressure modes is defined as follows

$$S_h = \ker B_h^t = \left\{ q_h \in Q_h : \int_\Omega q_h \operatorname{div} \underline{v}_h \, dx = 0 \ \forall \underline{v}_h \in V_h \right\}. \tag{88}$$

It is clear that a necessary condition for thc validity of the inf–sup condition (20) is the absence of spurios modes, that is,

$$S_h = \{0\}. \tag{89}$$

In particular, if S_h is nontrivial then the solution p_h to the discrete Stokes problem (15) is not unique, namely $p_h + s_h$ is still a solution when $s_h \in S_h$.

We shall illustrate how this situation may occur with the following example.

Example 5.1. The $Q_1 - P_0$ element

Among quadrilateral element, the $Q_1 - P_0$ element is the first that comes to mind. It is defined as (see Fig. 6):

$$V_h = (\mathfrak{L}^1_{[1]})^2 \cap V, \qquad Q_h = \mathfrak{L}^0_0 \cap Q. \tag{90}$$

This element is strongly related, for rectangular meshes, to some finite difference methods [38]. Its first appearence in a finite elment context seems to be in [46].

However simple it may look, the $Q_1 - P_0$ element is one of the hardest elements to analyze and many questions are still open about its properties. This element does not satisfy the inf–sup condition: it strongly depends on the mesh. For a regular mesh the kernel of the discrete gradients is one-dimensional. More precisely, $\underline{\text{grad}}_h \, q_h = 0$ implies that q_h is constant on the red and black cells if the mesh is viewed as a *checkerboard* (Fig. 7). This means that one singular value of the operator $B_h = \operatorname{div}_h$ is zero. Moreover, it has been checked by computation [49] that a large number of positive singular values converge to zero when h becomes small. In [48] indeed it has been proved that the second singular value is $O(h)$ and is not bounded below (see also [52]). The $Q_1 - P_0$ element has been the subject of a vast literature.

We shall now present a few more examples and distinguish between local and global spurious pressure modes.

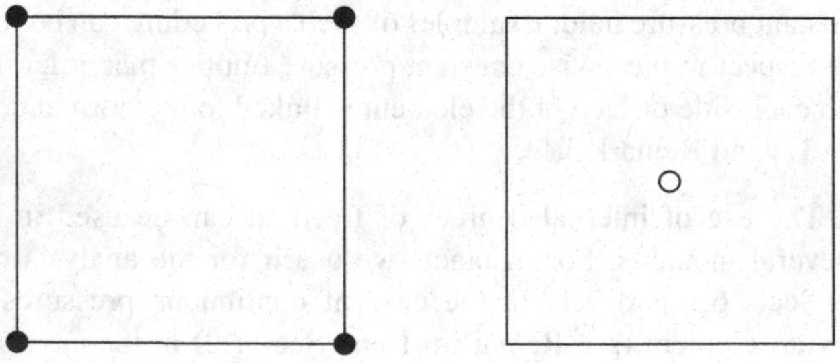

Fig. 6. The $Q_1 - P_0$ element

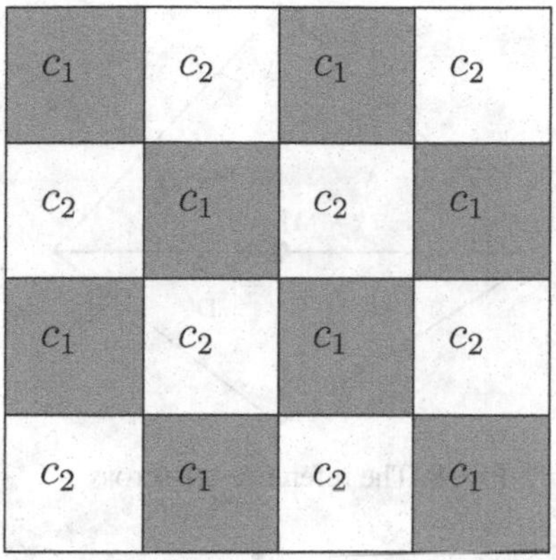

Fig. 7. Checkerboard spurious mode

Example 5.2. The crisscross $P_1 - P_0$ element

Let us consider a mesh of quadrilaterals divided into four triangles by their diagonals (Fig. 3). We observed in Example 3.2 that the $P_1 - P_0$ element, on *general meshes*, is affected by locking, that is, the computed velocity vanishes. On the mesh introduced above, however, it is easy to see that nonzero divergence free functions can be obtained. The divergence is constant on each triangle. This means four linear relations between the values of the partial derivatives. It is easily seen that one of them can be expressed as a combination of the others, this fact being caused by equality of tangential derivatives along the diagonals. To make things simpler, we consider the case where the diagonals are orthogonal (Fig. 8) and we label by A, B, C, D the four triangles. We then have, by taking locally the coordinates axes along the diagonals, and denoting by u^K the restriction of a function of V_h to the element K,

$$\frac{\partial u_1^K}{\partial x_1} + \frac{\partial u_2^K}{\partial x_2} = 0, \quad K = A, B, C, D. \tag{91}$$

On the other hand, one has at the point M

$$\frac{\partial u_1^A}{\partial x_2} = \frac{\partial u_2^B}{\partial x_2}, \quad \frac{\partial u_1^A}{\partial x_1} = \frac{\partial u_1^C}{\partial x_1}, \quad \frac{\partial u_2^C}{\partial x_2} = \frac{\partial u_2^D}{\partial x_2}, \quad \frac{\partial u_1^B}{\partial x_1} = \frac{\partial u_1^D}{\partial x_1}. \tag{92}$$

It is easy to check that this makes one of the four conditions (91) redundant. The reader may check the general case by writing the divergence operator in a nonorthogonal coordinate system.

The consequence of the above discussion is that on each composite quadrilateral one of the four constant pressure values will be undetermined. The dimension of $\ker B_h^t$ will be *at least* as large as the number of quadrilaterals minus one.

Thus, three constraints remain on each composite quadrilateral element. If we admit that two of them can be controlled, using the methods of Sect. 4.6, by the "internal" node M, we obtain an element that is very similar to the $Q_1 - P_0$ element

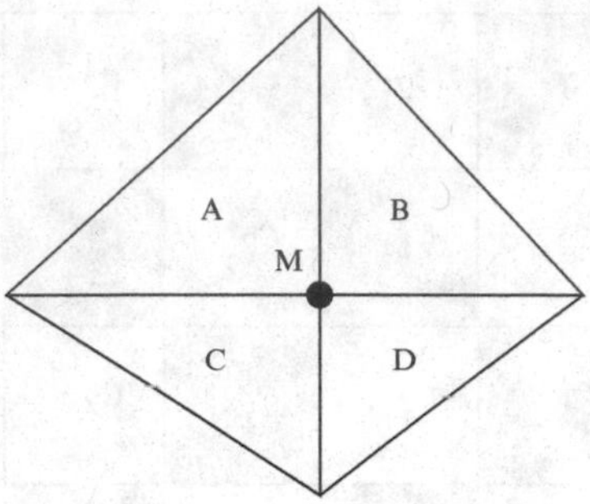

Fig. 8. The reference crisscross

with respect to the degrees of freedom. Indeed, it can be checked that on a regular mesh, an additional checkerboard mode occurs and that the behavior of this approximation is essentially the same as that of the $Q_1 - P_0$. These analogies have been pointed out, for instance, in [17].

The above example clearly shows the existence of two kinds of spurious pressure modes. Let us consider an element where S_h is nontrivial.

In the case of the crisscross $P_1 - P_0$ element presented in the previous example, dim S_h grows as h goes to 0 and there exists a basis of S_h with local support (that is, the support of each basis function can be restricted to one macroelement). We shall refer to these modes as *local spurious modes*. Such pressure modes can be eliminated by considering a composite mesh (in the previous example a mesh of quadrilaterals instead of triangles) and using a smaller space for the pressures by deleting some degrees of freedom from the composite elements.

If we now consider the $Q_1 - P_0$ example (see Example 5.1), the dimension of S_h does not grow when h goes to 0 and no basis can be found with a local support. We then have a *global spurious mode* which cannot be eliminated as easily as the local ones. Global modes usually appear on special (regular) meshes and are symptoms that the behavior of the element at hand is strongly mesh dependent and requires a special care. Some elements may generate both local and global modes as we have seen in the crisscross $P_1 - P_0$ method (see Example 5.2).

It must be emphasized that local spurious modes are source of troubles only when one prefers to work directly on the original mesh and not on the composite mesh on which they could easily be filtered out by a simple projection on each macroelement. We shall prove this in the next subsection in which a more precise framework will be given.

Example 5.3. The crisscross $P_2 - P_1$ element

Another simple example where a local mode occurs is the straightforward extension of the previous example to the case of a $P_2 - P_1$ approximation (Fig. 9). This means on each quadrilateral 12 discrete divergence-free constraints, and it is easily seen by the argument of Example 5.2, written at the point M, that one of them is

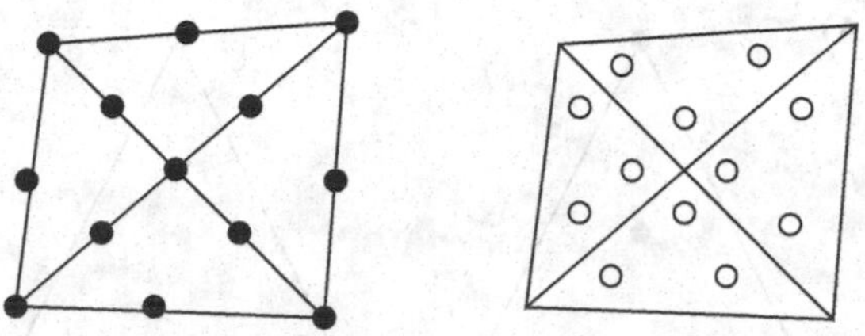

Fig. 9. The crisscross $P_2 - P_1$ element

redundant. Thus one spurious mode will appear for each composite quadrilateral. However, in this case, no global mode will appear. The analysis of this element is also related to the work of [29] by considering the stream function associated with a divergence-free function.

The presence of spurious modes can be interpreted as a signal that the pressure space used is in some sense too rich. We therefore can hope to find a cure by using a strict subspace $\hat{Q}_h$ of Q_h as the space of the discrete pressures, in order to obtain a stable approximation. The question arises whether or not this stability can be used to prove at least a partial result on the original approximation. One can effectively get some result in this direction.

6 Two-Dimensional Stable Elements

In this section we shall make use of the techniques presented in Sect. 4 to prove the stability for some of the most popular two-dimensional Stokes elements. The degrees of freedom corresponding to some of those are collected in Fig. 10.

We start with triangular elements and then we present schemes based on quadrilaterals.

The Hood–Taylor element (two- and three-dimensional) and its generalization will be presented in Sect. 8.

6.1 The MINI Element

This element, which is probably the cheapest one for the approximation of the Stokes equation, has been introduced in [3]. Given a mesh of triangles, the definition of the spaces is as follows

$$V_h = (\mathfrak{L}_1^1 \oplus B_3)^2 \cap V, \qquad Q_h = \mathfrak{L}_1^1 \cap Q, \tag{93}$$

where by B_3 we denotes the space of cubic bubbles.

The proof of the stability for the MINI element is an immediate consequence of Corollary 4.1

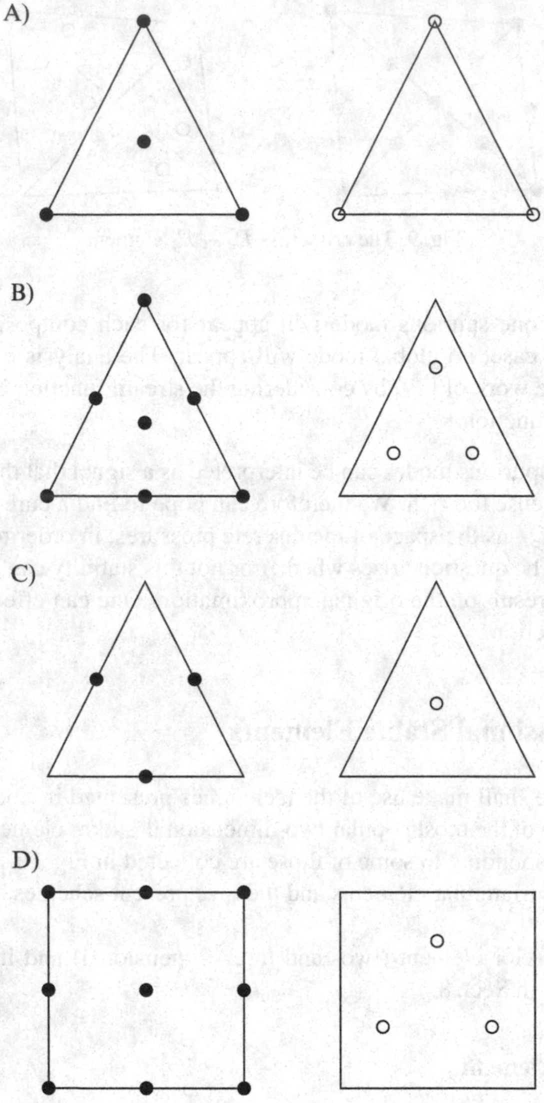

Fig. 10. Some stable two-dimensional Stokes elements: (a) the MINI element, (b) the Crouzeix–Raviart element, (c) the $P_1^{NC} - P_0$ element, (d) the $Q_2 - P_1$ element

6.2 The Crouzeix–Raviart Element

This element, presented in [31], is an enrichment to the $P_2 - P_0$ scheme which provides now well-balanced approximation properties. Given a mesh of triangles, the approximating spaces are

$$V_h = (\mathfrak{L}_2^1 \oplus B_3)^2 \cap V, \qquad Q_h = \mathfrak{L}_1^0 \cap Q. \tag{94}$$

The proof of the stability for this element can been carried out again with the help of Proposition 4.2. We use as operator Π_1 the Fortin operator of the $P_2 - P_0$ element (see also Proposition 4.11) and we take advantage of the internal degrees of freedom in V_h to define $\Pi_2 : V \to (B_3)^2$. Actually, we shall define $\Pi_2\underline{v}$ only in the case when $\operatorname{div}\underline{v}$ has zero mean value in each K. This will be sufficient, since we shall use in practice $\Pi_2(\underline{v} - \Pi_1\underline{v})$ and $\Pi_1\underline{v}$ satisfies (83). For all K and for all $\underline{v}$ with

$$\int_K \operatorname{div}\underline{v}\, dx = 0, \tag{95}$$

we then set $\Pi_2\underline{v}$ as the unique solution of

$$\Pi_2\underline{v} \in (B_3(K))^2, \tag{96}$$

$$\int_K \operatorname{div}(\Pi_2\underline{v} - \underline{v})q_h\, dx = 0, \quad \forall q_h \in P_1(K). \tag{97}$$

Note that (96), (97) is a linear system of three equations ($\dim P_1(K) = 3$) in two unknowns ($\dim(B_3(K))^2 = 2$) which is compatible since $\underline{v}$ is assumed to satisfy (95) and, on the other hand, for every $\underline{b} \in (B_3(K))^2$ we clearly have

$$\int_K \operatorname{div}\underline{b}\, dx = 0. \tag{98}$$

We have only to prove that

$$||\Pi_2\underline{v}||_{1,K} \leq c||\underline{v}||_{1,K} \tag{99}$$

for all $\underline{v} \in V$ satisfying (95). Indeed (97) can be written as

$$\int_K (\Pi_2\underline{v}) \cdot \underline{\operatorname{grad}}\, q_h\, dx = \int_K \operatorname{div}\underline{v}(q_h - \bar{q}_h)\, dx \tag{100}$$

where $\bar{q}_h$ is any piecewise constant approximation of q_h. A scaling argument yields

$$|\widehat{\Pi_2\underline{v}}|_{0,\hat{K}} \leq c(\theta_0)|\hat{\underline{v}}|_{1,\hat{K}} \tag{101}$$

that easily implies (99).

Remark 6.1. We could prove the same result also as a consequence of Proposition 4.11. The same proof applies quite directly to the $Q_2 - P_1$ rectangular element (see Sect. 6.4). It can also be used to create nonstandard elements. For instance, in [34], bubble functions were added to a $Q_1 - P_0$ element in order to use a P_1 pressure field. This element is not more, but neither less, stable than the standard $Q_1 - P_0$ and gives better results in some cases.

6.3 $P_1^{NC} - P_0$ Approximation

We consider the classical stable nonconforming triangular element introduced in [31], in which midside nodes are used as degrees of freedom for the velocities. This generates a piecewise linear nonconforming approximation; pressures are taken constant on each element (see Fig. 10). We do not present the stability analysis for this element, which does not fit within the framework of our general results, since V_h is not contained in V. However, we remark that this method is attractive for several reasons. In particular, the restriction to an element K of the solution $\underline{u}_h \in V_h$ is exactly divergence-free, since $\operatorname{div} V_h \subset Q_h$. Another important feature of this element is that it can be seen as a "mass conservation" scheme. The present element has been generalized to second order in [39]. It must also be said that coerciveness may be a problem for the $P_1^{NC} - P_0$ element, as it does not satisfy the discrete version of Korn's inequality. This issue has been deeply investigated and clearly illustrated in [5].

Remark 6.2. The generalization of nonconforming finite elements to quadrilaterals is not straightforward. In particular, approximation properties of the involved spaces are not obvious. More details can be found in [55].

6.4 $Q_k - P_{k-1}$ Elements

We now discuss the stability and convergence of a familiy of quadrilateral elements. The lowest order of this family, the $Q_2 - P_1$ element, is one of the most popular Stokes elements. Given $k \geq 2$, the discrete spaces are defined as follows:

$$V_h = (\mathfrak{L}^1_{[k]})^2 \cap V, \qquad Q_h = \mathfrak{L}^0_{[k-1]} \cap Q.$$

If the mesh is built of *rectangles*, the stability proof is an immediate consequence of Proposition 4.11, since (83) is satisfied for V_h (indeed, the $Q_2 - P_0$ is a stable Stokes element, see Remark 3.1). In the case of a general *quadrilateral* mesh things are not so easy; even the definition of the space Q_h is not so obvious and there have been different opinions, during the years, about two possible natural definitions. Following [15], we discuss in detail the case $k = 2$.

The $Q_2 - P_1$ Element

This element was apparently discovered around a blackboard at the Banff Conference on Finite Elements in Flow Problems (1979). Two different proofs of stability can be found in [41] and [59] for the rectangular case. This element is a relatively late comer in the field; the reason for this is that using a P_1 pressure on a quadrilateral is not a standard procedure. It appeared as a cure for the instability of the $Q_2 - Q_1$ element which appears quite naturally in the use of reduced integration penalty methods (see [9]). This last element is essentially related to the $Q_1 - P_0$ element and suffers the same problems altough to a lesser extent. Another cure can be obtained by adding internal nodes (see [34]).

On a general quadrilateral mesh, the space Q_h can be defined in two different ways: either Q_h consists of (discontinuous) piecewise linear functions, or it is built by considering three linear shape functions on the reference unit square and mapping them to the general elements like it is usually done for continuous finite elements. We point out that, since the mapping F_K from the reference element $\hat{K}$ to the general element K in this case is bilinear but not affine, the two constructions are not equivalent. We shall refer to the first possibility as *unmapped* pressure approach and to the second one as *mapped* pressure approach.

In order to analyze the stability of either scheme, we use the macroelement technique presented in section 4.5 with macroelements consisting of one single element. We start with the case of the *unmapped* pressure approach; this is the original proof presented in [59]. Let M be a macroelement and $q_h = a_0 + a_x x + a_y y \in Q_{0,M}$ an arbitrary function in K_M. If $b(x,y)$ denote the biquadratic bubble function on K, then $\underline{v}_h = (a_x b(x,y), 0)$ is an element of $V_{0,M}$ and

$$0 = \int_M q_h \operatorname{div} \underline{v}_h \, dx\, dy = -\int_M \underline{\operatorname{grad}}\, q_h \cdot \underline{v}_h \, dx\, dy = -a_x \int_M b(x,y)\, dx\, dy$$

implies $a_x = 0$. In a similar way, we get $a_y = 0$ and, since the average of q_h on M vanishes, we have the macroelement condition $q_h = 0$.

We now move to the *mapped* pressure approach, following the proof presented in [15]. There, it is recalled that the macroelement condition (79) can be related to an algebraic problem in which we are led to proof that a 2×2 matrix is nonsingular. Actually, it turns out that the determinant of such matrix is a multiple of the Jacobian determinant of the function mapping the reference square $\hat{K}$ onto M, evaluated at the barycenter of $\hat{K}$. Since this number must be nonzero for any element of a well-defined mesh, we can deduce that the macroelement condition is satisfied in this case also, and then conclude that the stability holds thanks to Proposition 4.8.

So far, we have shown that either the *unmapped* and the *mapped* pressure approach gives rise to a stable $Q_2 - P_1$ scheme. However, as a consequence of the results proved in [6], we have that the mapped pressure approach *cannot achieve optimal approximation order*. Namely, the unmapped pressure space provides a second-order convergence in L^2, while the mapped one achieves only $O(h)$ in the same norm. In [15] several numerical experiments have been reported, showing that on general quadrilateral meshes (with constant distortion) the unmapped pressure approach provides a second-order convergence (for both velocity in H^1 and pressure in L^2), while the mapped approach is only suboptimally first-order convergent. It is interesting to remark that in this case also the convergence of the velocities is suboptimal, according to the error estimate (21).

7 Three-Dimensional Elements

Most elements presented in Sect. 6 have a three-dimensional extension. Some of them are schematically plotted in Fig. 11.

7.1 The MINI Element

Consider a regular sequence of decompositions of Ω into tetrahedra. The spaces are defined as follows:

$$V_h = (\mathfrak{L}_1^1 + B_4)^3 \cap V, \qquad Q_h = \mathfrak{L}_1^1 \cap Q,$$

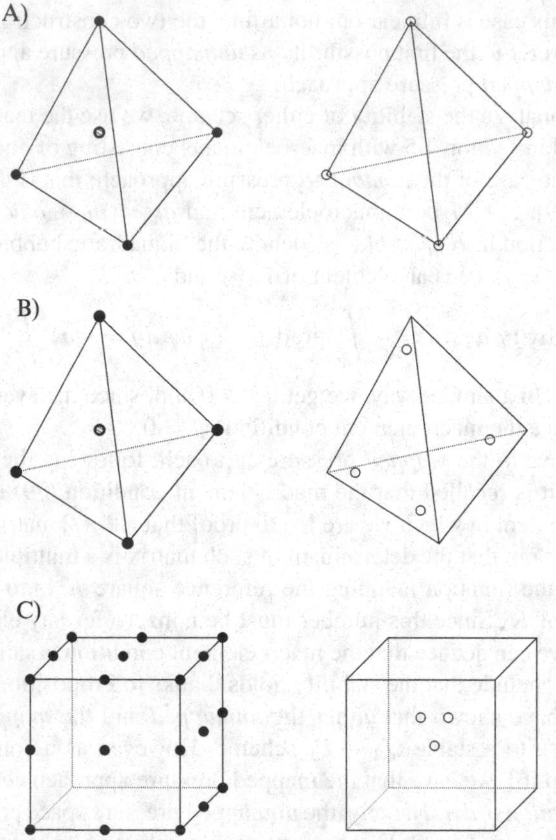

Fig. 11. Some stable three-dimensional Stokes elements: (a) the MINI element, (b) the Crouzeix–Raviart element, (c) the $Q_2 - P_1$ element

where B_4 denotes the space of quartic bubbles. Then the stability of this element follows easily, like in the 2D case, from Corollary 4.1.

7.2 The Crouseix–Raviart Element

The straightforward generalization of the Crouseix–Raviart element is given by

$$V_h = (\mathfrak{L}_1^1 + B_4)^3 \cap V, \qquad Q_h = \mathfrak{L}_1^0 \cap Q.$$

The stability is an easy consequence of Proposition 4.11.

7.3 $P_1^{NC} - P_0$ Approximation

The triangular $P_1^{NC} - P_0$ easily generalizes to tetrahedra in 3D. Also in this case, since div $V_h \subset Q_h$, the restriction of the discrete solution to every element is truly divergence free.

7.4 $Q_k - P_{k-1}$ Elements

Given a mesh of hexahedrons, we define

$$V_h = (\mathfrak{L}^1_{[k]})^3 \cap V, \qquad Q_h = \mathfrak{L}^0_{[k-1]} \cap Q,$$

for $k \geq 2$. We refer to the two-dimensional case (see Sect. 6.4) for the definition of the pressure space. In particular, we recall that Q_h on each element consists of true polynomials and is not defined via the reference element. With the correct definition of the pressure space, the proof of stability for this element is a simple generalization of the corresponding two-dimensional version.

8 $P_k - P_{k-1}$ Schemes and Generalized Hood–Taylor Elements

The main result of this section (see Theorems 8.1 and 8.2) consists in showing that a family of popular Stokes elements satisfies the inf–sup condition (20). The first element of this family has been introduced in [45] and for this reason the members of the whole family are usually referred to as *generalized* Hood–Taylor elements.

This section is organized in two subsections. In the first one we discuss discontinuous pressure approximations for the $P_k - P_{k-1}$ element in the two-dimensional triangular case; it turns out that this choice is not stable in the lower-order cases and requires suitable conditions on the mesh sequences for the stability of the higher-order elements.

The last subsection deals with the generalized Hood–Taylor elements, which provide a continuous pressure approximation in the plane (triangles and quadrilaterals) and in the three-dimensional space (tetrahedrons and hexahedra).

8.1 $P_k - P_{k-1}$ Elements

In this subsection we shall recall the statement of a basic result by Scott and Vogelius [58] which, roughly speaking, says: under suitable assumptions on the decomposition $\mathfrak{T}_h$ (in triangles) the pair $V_h = (\mathfrak{L}^1_2)^2$, $Q_h = \mathfrak{L}^1_{k-1}$ satisfies the inf–sup condition for $k \geq 4$.

On the other hand, the problem of finding stable lower-order approximations has been studied by Qin [54], where interesting remarks are made on this scheme and where the possibility of filtering out the spurious pressure modes is considered.

In order to state in a precise way the restrictions that have to be made on the triangulation for higher-order approximations, we assume that Ω is a polygon, and that its boundary $\partial\Omega$ has no double points. In other words, there exists two continuous piecewise linear maps $x(t)$, $y(t)$ from $[0,1[$ into $\mathbb{R}$ such that

$$\begin{cases} (x(t_1) = x(t_2) \text{ and } y(t_1) = y(t_2)) \text{ implies } t_1 = t_2, \\ \partial\Omega = \{(x,y) : x = x(t),\ y = y(t) \text{ for some } t \in [0,1[\}. \end{cases} \tag{102}$$

Clearly, we will have $\lim_{t\to 1} x(t) = x(0)$ and $\lim_{t\to 1} y(t) = y(0)$. We remark that we are considering a less general case than the one treated by [58]. We shall make further restrictions in what follows, so that we are actually going to present a particular case of their results.

Let now V be a vertex of a triangulation $\mathfrak{T}_h$ of Ω and let $\theta_1, \ldots, \theta_p$, be the angles, at V, of all the triangles meeting at V, ordered, for instance, in the counter-clockwise sense. If V is an internal vertex we also set $\theta_{p+1} := \theta_1$. Now we define $S(V)$ according to the following rules:

$$p = 1 \quad \Rightarrow \quad S(V) = 0 \tag{103}$$

$$p > 1,\ V \in \partial\Omega \quad \Rightarrow \quad S(V) = \max_{i=1,p-1} (\pi - \theta_1 - \theta_{i+1}) \tag{104}$$

$$V \notin \partial\Omega \quad \Rightarrow \quad S(V) = \max_{i=1,p} (\pi - \theta_i - \theta_{i+1}) \tag{105}$$

It is easy to check that $S(V) = 0$ if and only if all the edges of $\mathfrak{T}_h$ meeting at V fall on two straight lines. In this case V is said to be singular [58]. If $S(V)$ is positive but very small, then V will be "almost singular". Thus $S(V)$ measures how close V is to be singular.

We are now able to state the following result.

Proposition 8.1 ([58]). *Assume that there exists two positive constants c and* δ *such that*

$$ch \leq h_K,\ \forall K \in \mathfrak{T}_h, \tag{106}$$

and

$$S(V) \geq \delta,\ \forall V \text{ vertex of } \mathfrak{T}_h. \tag{107}$$

Then the choice $V_h = (\mathfrak{L}_1^1)^2$, $Q_h = \mathfrak{L}_{k-1}^0$, $k \leq 4$, *satisfies the inf–sup condition with a constant depending on c and* δ *but not on h.*

Condition (107) is worth a few comments. The trouble is that $S(V) = 0$ makes the linear constraints on u_h, arising from the divergence-free condition, linearly dependent (see, also, Examples 5.2 and 5.3). When this linear dependence appears, some part of the pressure becomes unstable. In the present case, this unstable part could be filtered out.

Remark 8.1. The P_k-P_{k-1} element can obviously be stabilized by adding bubbles to the velocity space in the spirit of Sect. 4.6 (see Proposition 4.11). For a less expensive stabilization, consisting in adding bubbles only in few elements, see [13].

8.2 Generalized Hood–Taylor Elements

In this subsection we recall the results proved in [12, 14] concerning the stability of the generalized Hood–Taylor schemes. On triangles or tetrahedra, velocities are approximated by a standard P_k element and pressures by a standard *continuous* P_{k-1}, that is $\underline{v}_h \in (\mathfrak{L}_k^1)^n$ $(n = 2, 3)$, $p \in \mathfrak{L}_{k-1}^1$. This choice has an analogue on rectangles or cubes using a Q_k element for velocities and a Q_{k-1} element for pressures. The lowest order element (i.e., $k = 2$) has been introduced by Hood and Taylor [45]. Several papers are devoted to the analysis of this popular element. The degrees of freedom of some elements belonging to this family are reported in Fig. 12.

The first proof of convergence was given for the two-dimensional case in [10] where a weaker form of the inf–sup condition was used. The analysis was subsequently improved in [66] who showed that the classical inf–sup condition is indeed satisfied (see Verfürth's trick in Sect. 4.3). The macroelement technique can easily be used for the stability proof of the rectangular and cubic element (of any order) as well as of the tetrahedral case when $k = 2$ (see [59]). In [24] an alternative technique of proof has been presented for the triangular and tetrahedral cases when $k = 2$. This proof generalizes to the triangular case when $k = 3$ (see [21]). Finally, a general proof of convergence can be found in [12] and [14] for the triangular and tetrahedral case, respectively.

We now state and prove the theorem concerning the two-dimensional triangular case (see [12]).

Theorem 8.1. *Let Ω be a polygonal domain and $\mathcal{T}_h$ a regular sequence of triangular decompositions of it. Then the choice $V_h = (\mathfrak{L}^1_k \cap H^1_0(\Omega))^2$ and $Q_h = \mathfrak{L}^1_{k-1} \cap L^2_0(\Omega)$ satisfies the inf–sup condition* (20) *for any $k \geq 2$ if and only if each triangulation contains at least three triangles.*

Proof. **Step 1: Necessary part.** Let us show first that the hypothesis on the mesh is necessary. If $\mathcal{T}_h$ only contains one element, then it is easy to see that the inf–sup constant is zero (otherwise it should be $\operatorname{div} V_h \subset Q_h$, which is not the case since the functions in Q_h are not zero at the vertices). We shall show that if $\mathcal{T}_h$ contains only two triangles T_1 and T_2, then there exists one spurious pressure mode. This implies that also in this case the inf–sup constant vanishes. We choose the coordinate system (x, y) in such a way that the common edge of T_1 and T_2 lies on the y-axis. Moreover, we suppose that T_2 is the reference triangle and T_1 the symmetric one with respect to the x-axis, see Fig. 13. The general case can then be handled by means of suitable affine mappings.

We denote by $\lambda_{i,a}$ and $\lambda_{i,b}$ the barycentric coordinates relative to the vertices a and b, respectively, belonging to the element T_i, $i = 1, 2$. It is easy to check that it holds: $\lambda_{1,a} = 1 + x - y$, $\lambda_{1,b} = y$, $\lambda_{2,a} = 1 - x - y$, and $\lambda_{2,b} = y$. We shall also make use of the function $\lambda_{2,c} = x$. Let $L(x)$ be the Legendre polynomial of degree $k - 2$ on the unit interval with respect to the weight $w(x) = x(1 - x)^3$ and consider the function $p(x) \in Q_h$ defined as follows:

$$p'(x) = \begin{cases} -L(-x) & \text{for } x < 0, \\ L(x) & \text{for } x > 0. \end{cases} \tag{108}$$

We shall show that $\operatorname{grad} p$ is orthogonal to any velocity $\underline{v} \in V_h$. Since p does not depend on y, we can consider the first component v_1 of $\underline{v}$ only, which, by virtue of the continuity at $x = 0$ and of the boundary conditions, has the following general form:

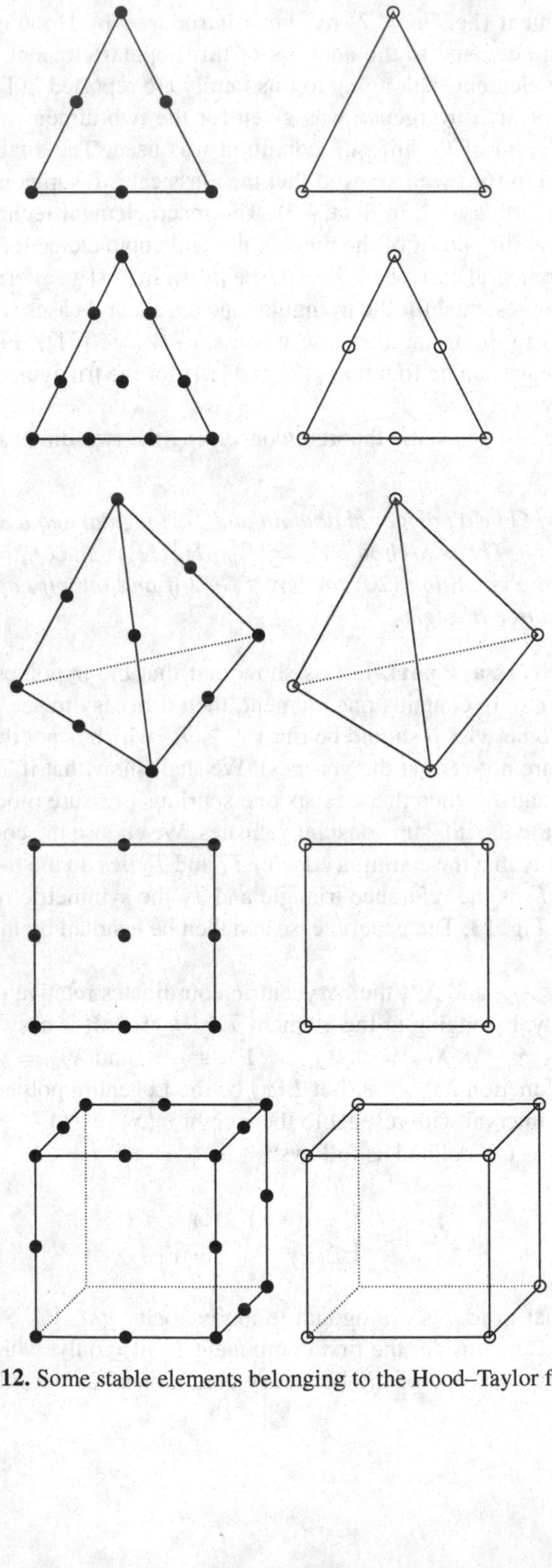

Fig. 12. Some stable elements belonging to the Hood–Taylor family

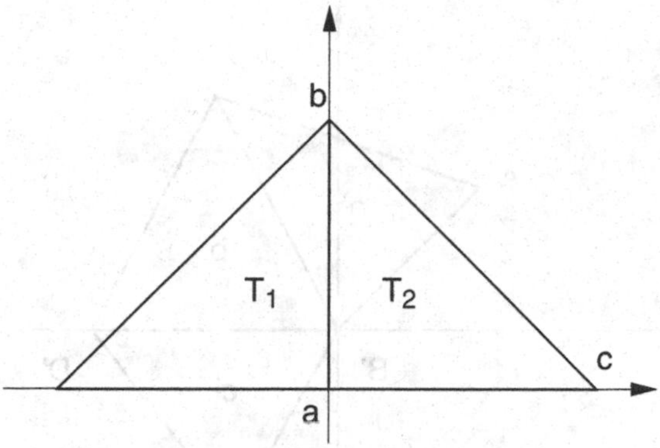

Fig. 13. The reference triangle and its symmetric

$$v_1 = \begin{cases} \lambda_{1,a}\lambda_{1,b}(C_{k-2}(y) + xA_{k-3}(x,y)) & \text{in } T_1, \\ \lambda_{2,a}\lambda_{2,b}(C_{k-2}(y) + xB_{k-3}(x,y)) & \text{in } T_2, \end{cases} \tag{109}$$

where the subscripts denote the degrees of the polynomials A, B and C. We then have

$$\begin{aligned} \int_{T_1 \cup T_2} \underline{v} \cdot \operatorname{grad} p \, dx \, dy &= \int_{T_1} v_1 p' \, dx \, dy + \int_{T_2} v_1 p' \, dx \, dy \\ &= \int_{T_2} \lambda_{2,a}\lambda_{2,b} L(x) x (B_{k-3}(x,y) - A_{k-3}(-x,y)) \, dx \, dy \\ &= \int_{T_2} \lambda_{2,a}\lambda_{2,b}\lambda_{2,c} L(x) q(x,y) \, dx \, dy \end{aligned} \tag{110}$$

where $q(x,y)$ is a polynomial of degree $k-3$ and where the term involving C disappears by virtue of the symmetries. The last integral reads

$$\int_{T_2} xy(1-x-y)L(x)q(x) \, dx \, dy = \int_0^1 xL(x)Q(x) \, dx \tag{111}$$

and an explicit calculation shows that $Q(x)$ is of the form

$$Q(x) = (1-x)^3 p_{k-3}(x), \tag{112}$$

where p_{k-3} is a polynomial of degree $k-3$. We can now conclude with the final computation

$$\int_{T_1 \cup T_2} \underline{v} \cdot \operatorname{grad} p \, dx \, dy = \int_0^1 x(1-x)^3 L(x) p_{k-3}(x) \, dx = 0. \tag{113}$$

Step 2: Sufficient part. The idea of the proof consists in considering, for each h, a partition of the domain Ω in subdomains containing exactly three adjacent triangles.

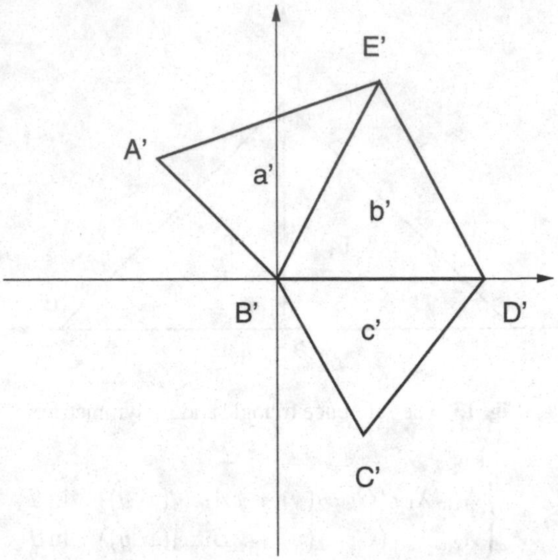

Fig. 14. A generic triplet of triangles

By making use of Proposition 4.6 and the technique presented in Sect. 4.2, it will be enough to prove the inf–sup condition for a single macroelement, provided we are able to bound the number of intersections between different subdomains (basically, everytimes two subdomains intersect each other, a factor $1/\sqrt{2}$ shows up in front of the final inf–sup constant). Indeed, it is possible to prove that, given a generic triangulation of a polygon, it can be presented as the disjoint union of triplets of triangles and of polygons that can be obtained as unions of triplets with at most three intersections.

Given a generic macroelement $a' \cup b' \cup c'$, consider the (x, y) coordinate system shown in Fig. 14, so that the vertices are $B' = (0,0)$, $D' = (1,0)$, $E' = (\alpha, \beta)$. By means of the affine mapping $x' = x + \alpha y$, $y' = \beta y$, the Jacobian of which is β, we can consider the macroelement $a \cup b \cup c$ shown in Fig. 15, so that b is the unit triangle. Since $\beta \neq 0$, the considered affine mapping is invertible. With an abuse in the notation, we shall now denote by Ω the triplet $a \cup b \cup c$ and by V_h and Q_h the finite element spaces built on it.

We denote by λ^a_{AB} the barycentric coordinate of the triangle a vanishing on the edge AB (analogous notation holds for the other cases). Moreover, we denote by $L^a_{i,x}(x)$ the i-th Legendre polynomial in $[x_A, 0]$, with respect to the measure $\mu_{a,x}$ defined by

$$\int_{x_A}^{0} f(x)\, d\mu_{a,x} = \int_a \lambda^a_{AB}\lambda^a_{AE} f(x)\, dx\, dy, \qquad \forall f(x) : [x_A, 0] \to \mathbb{R}, \tag{114}$$

where x_A is the x-coordinate of the vertex A. We shall make use of the following Legendre polynomials, which are defined in a similar way: $L^b_{i,x}$ (its definition involves λ^b_{ED} and λ^b_{BD}), $L^b_{i,y}$ (using λ^b_{BE} and λ^b_{BD}), and $L^c_{i,y}$ (using λ^c_{BC} and λ^c_{CD}).

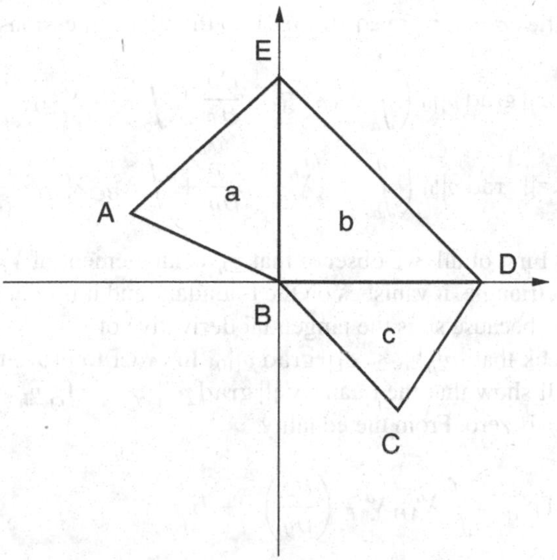

Fig. 15. A macroelement where b is the reference triangle

Standard properties of the Legendre polynomials ensures that we can normalize them, for instance, by requiring that they assume the same value (say 1) at the origin. We now prove by induction with respect to the degree k that a modified inf–sup condition holds true (see Verfürth's trick in Sect. 4.3). Namely, for any $q_h \in Q_h$, we shall construct $\underline{v}_h \in V_h$ such that

$$\begin{aligned} -\int_{a\cup b\cup c} \underline{v}_h \cdot \operatorname{grad} q_h \, dx\, dy &\geq c_1 || \operatorname{grad} q_h ||_0^2, \\ ||\underline{v}_h||_0 &\leq c_2 || \operatorname{grad} q_h ||_0. \end{aligned} \tag{115}$$

The case $k = 2$. This is the original Hood–Taylor method. Given $p \in Q_h$, we define $\underline{v}_h = (v_1(x, y), v_2(x, y))$ triangle by triangle as follows:

$$v_1(x,y)|_a = -\lambda^a_{AB}\lambda^a_{AE} || \operatorname{grad} p ||_0 \cdot \sigma, \tag{116}$$

$$v_2(x,y)|_a = -\lambda^a_{AB}\lambda^a_{AE} \frac{\partial p}{\partial y}, \tag{117}$$

$$v_1(x,y)|_b = -\lambda^b_{ED}\lambda^b_{BD} || \operatorname{grad} p ||_0 \cdot \sigma - \lambda^b_{ED}\lambda^b_{EB} \frac{\partial p}{\partial x}, \tag{118}$$

$$v_2(x,y)|_b = -\lambda^b_{ED}\lambda^b_{BD} \frac{\partial p}{\partial y} - \lambda^b_{ED}\lambda^b_{EB} || \operatorname{grad} p ||_0 \cdot \tau, \tag{119}$$

$$v_1(x,y)|_c = -\lambda^c_{BC}\lambda^c_{CD} \frac{\partial p}{\partial x}, \tag{120}$$

$$v_2(x,y)|_c = -\lambda^c_{BC}\lambda^c_{CD} || \operatorname{grad} p ||_0 \cdot \tau, \tag{121}$$

where the quantities σ and τ are equal to ± 1 so that the expressions

$$H = \sigma \|\operatorname{grad} p\|_0 \left(\int_a \lambda^a_{AB} \lambda^a_{AE} \cdot \frac{\partial p}{\partial x} + \int_b \lambda^b_{ED} \lambda^b_{BD} \cdot \frac{\partial p}{\partial x} \right), \tag{122}$$

$$K = \tau \|\operatorname{grad} p\|_0 \left(\int_b \lambda^b_{EB} \lambda^b_{ED} \cdot \frac{\partial p}{\partial y} + \int_c \lambda^c_{BC} \lambda^c_{CD} \cdot \frac{\partial p}{\partial y} \right) \tag{123}$$

are nonnegative. First of all, we observe that $\underline{v}_h$ is an element of V_h: its degree is at most two in each triangle, it vanishes on the boundary and it is continuous across the the internal edges because so is the tangential derivative of p.

It easy to check that $\|\underline{v}_h\|_0 \leq c_1 \|\operatorname{grad} p\|_0$. In order to prove the first equation in (115), we shall show that the quantity $|||\operatorname{grad} p||| := -\int_\Omega \underline{v}_h \cdot \operatorname{grad} p$ vanishes only when $\operatorname{grad} p$ is zero. From the equality

$$\begin{aligned} 0 = |||\operatorname{grad} p||| = \int_a \lambda^a_{AB} \lambda^a_{AE} \left(\frac{\partial p}{\partial y} \right)^2 + H \\ + \int_b \left(\lambda^b_{ED} \lambda^b_{EB} \left(\frac{\partial p}{\partial x} \right)^2 + \lambda^b_{ED} \lambda^b_{BD} \left(\frac{\partial p}{\partial y} \right)^2 \right) \\ + K + \int_c \lambda^c_{BC} \lambda^c_{CD} \left(\frac{\partial p}{\partial x} \right)^2 \end{aligned} \tag{124}$$

it follows that

$$\frac{\partial p}{\partial y} = 0 \quad \text{in } a, \tag{125}$$

$$\frac{\partial p}{\partial x} = \frac{\partial p}{\partial y} = 0 \quad \text{in } b, \tag{126}$$

$$\frac{\partial p}{\partial x} = 0 \quad \text{in } c, \tag{127}$$

$$H = K = 0. \tag{128}$$

These last equations, together with the fact that each component of $\operatorname{grad} p$ is constant if $p \in Q_h$, easily imply that

$$\operatorname{grad} p = (0,0) \quad \text{in } \Omega. \tag{129}$$

The case $k > 2$. Given p in Q_h, if p is locally of degree $k-2$, then the result follows from the induction hypothesis. Otherwise, there exists at least one triangle of Ω in which p is exactly of degree $k-1$. Like in the previous case, we define $\underline{v}_h = (v_1(x,y), v_2(x,y))$ as follows:

$$v_1(x,y)|_a = -\lambda^a_{AB}\lambda^a_{AE}||\operatorname{grad} p||_0 L^a_{k-2,x}\cdot\sigma, \tag{130}$$

$$v_2(x,y)|_a = -\lambda^a_{AB}\lambda^a_{AE}\frac{\partial p}{\partial y}, \tag{131}$$

$$v_1(x,y)|_b = -\lambda^b_{ED}\lambda^b_{BD}||\operatorname{grad} p||_0 L^b_{k-2,x}\cdot\sigma - \lambda^b_{ED}\lambda^b_{EB}\frac{\partial p}{\partial x}, \tag{132}$$

$$v_2(x,y)|_b = -\lambda^b_{ED}\lambda^b_{BD}\frac{\partial p}{\partial y} - \lambda^b_{ED}\lambda^b_{EB}||\operatorname{grad} p||_0 L^b_{k-2,y}\cdot\tau, \tag{133}$$

$$v_1(x,y)|_c = -\lambda^c_{BC}\lambda^c_{CD}\frac{\partial p}{\partial x}, \tag{134}$$

$$v_2(x,y)|_c = -\lambda^c_{BC}\lambda^c_{CD}||\operatorname{grad} p||_0 L^c_{k-2,y}\cdot\tau, \tag{135}$$

with the same assumption on σ and τ, so that the terms

$$H = \sigma||\operatorname{grad} p||_0\left(\int_a \lambda^a_{AB}\lambda^a_{AE}L^a_{k-2,x}\cdot\frac{\partial p}{\partial x} + \int_b \lambda^b_{ED}\lambda^b_{BD}L^b_{k-2,x}\cdot\frac{\partial p}{\partial x}\right), \tag{136}$$

$$K = \tau||\operatorname{grad} p||_0\left(\int_b \lambda^b_{EB}\lambda^b_{ED}L^b_{k-2,y}\cdot\frac{\partial p}{\partial y} + \int_c \lambda^c_{BC}\lambda^c_{CD}L^c_{k-2,y}\cdot\frac{\partial p}{\partial y}\right) \tag{137}$$

are nonnegative. The same arguments as for $k=2$, together with the described normalization of the Legendre polynomials, show that $\underline{v}_h$ belongs to V_h.

In order to conclude the proof, we need show that if $|||\operatorname{grad} p||| = 0$ then the degree of $\operatorname{grad} p$ is strictly less than $k-2$. As before, $|||\operatorname{grad} p||| = 0$ implies

$$\frac{\partial p}{\partial y} = 0 \quad \text{in } a, \tag{138}$$

$$\operatorname{grad} p = 0 \quad \text{in } b, \tag{139}$$

$$\frac{\partial p}{\partial x} = 0 \quad \text{in } c, \tag{140}$$

$$H = K = 0. \tag{141}$$

The last equalities imply

$$\int_a \lambda^a_{AB}\lambda^a_{AE}L^a_{k-2,x}\cdot\frac{\partial p}{\partial x} = 0 \tag{142}$$

and

$$\int_c \lambda^c_{BC}\lambda^c_{CD}L^c_{k-2,y}\cdot\frac{\partial p}{\partial y} = 0 \tag{143}$$

It follows that the degree of $\operatorname{grad} p$ is strictly less than $k-2$ in contrast to our assumption.

Remark 8.2. The proof of the theorem shows that the continuity hypothesis on the pressure space Q_h can be weakened up to require that q_h is only continuous on triplets of elements.

We conclude this subsection by stating the three-dimensional analogous to the previous theorem and by recalling the main argument of the proof presented in [14].

Theorem 8.2. *Let Ω be a polyhedral domain and $\mathcal{T}_h$ a regular sequence of decompositions of it into tetrahedra. Assume that every tetrahedron has at least one internal vertex. Then the choice $V_h = (\mathfrak{L}^1_k \cap H^1_0(\Omega))^3$ and $Q_h = \mathfrak{L}^1_{k-1}$ satisfies the inf–sup condition* (20) *for any $k \geq 2$.*

Proof. We shall make use of the macroelement technique presented in Sect. 4.5. In particular, we shall use Proposition 4.9 and the comments included in Remark 4.7.

We consider an overlapping macroelement partition of $\mathcal{T}_h$ as follows: for each internal vertex x_0 we define a corresponding macroelement M_{x_0} by collecting all elements which touch x_0. Thanks to the regularity assumptions on the mesh, we only have to show that the macroelement condition (79) holds true (see, in particular, Remark 4.7).

Let us consider an element $K \in M = M_{x_0}$ and an edge e of K which touches x_0. With a suitable choice of the coordinate system, we can suppose that the direction of e coincides with that of the x axis. With the notation of Sect. 4.5 we shall show that a function in K_M cannot contain functions which depend on x in K. Namely, given a function $p \in Q_{0,M}$, we can define a function $\underline{v} \in V_{0,M}$ as follows.

$$\underline{v} = \left(-\lambda_{1,i}\lambda_{2,i}\frac{\partial p}{\partial x}, 0, 0 \right) \qquad \text{in } K_i,$$

where K_i is a generic element of M sharing the edge e with K and $\lambda_{j,i}$, $j = 1, 2$, are the barycentric coordinates of K_i associated with the two faces of K_i which do not touch e. On the remaining elements, each component of $\underline{v}$ is set equal to zero. It is clear that $\underline{v}$ is a kth-order polynomial in K_i and, since p is continuous in M, $\partial p/\partial x$ is continuous across the faces which meet at e and the function $\underline{v}$ is continuous as well. Hence, $\underline{v}$ belongs to $V_{0,M}$.

From the definition of $Q_{0,M}$ it turns out that

$$0 = \int_M p \operatorname{div} \underline{v} = -\int_M \operatorname{grad} p \cdot \underline{v} = \sum_i \int_{K_i} \lambda_{1,i}\lambda_{2,i} \left| \frac{\partial p}{\partial x} \right|^2$$

The last relation implies that p does not depend on x in K_i for any i and, in particular, in K. On the other hand, we can repeat the same argument using as e the other two edges of K meeting at x_0 and, since the directions of the three used edges are independent, we obtain that p is constant in K.

Remark 8.3. From the previous proof we can deduce that the hypotheses on the triangulation can be weakened, by assuming that each tetrahedron has at least three edges which do not lie on the boundary of Ω and which are not in the same plane. On the other hand, given a generic mesh of tetrahedra, it is not difficult to add suitable elements in order to meet the requirements of the previous theorem.

Remark 8.4. The main argument in the proof of the previous theorem is the straightforward generalization of the two-dimensional case. Indeed, the proof of Theorem 8.1 could be carried out using the macroelement technique as well.

9 Nearly Incompressible Elasticity, Reduced Integration Methods and Relation with Penalty Methods

9.1 Variational Formulations and Admissible Discretizations

Let us now turn our attention on problems associated with approximations of nearly incompressible materials. Considering, to make things simpler, a problem with homogeneous Dirichlet conditions and the standard variational principle

$$\inf_{\underline{v}\in(H_0^1(\Omega))^2} \mu \int_\Omega |\underline{\underline{\varepsilon}}(\underline{v})|^2\, dx + \frac{\lambda}{2}\int_\Omega |\operatorname{div}\, \underline{v}|^2\, dx - \int_\Omega \underline{f}\cdot\underline{v}\, dx, \tag{144}$$

it can be noticed that this problem is closely related to a penalty method to solve the Stokes problem.

It was soon recognized in practice that a brute force use of (144) could lead, for large values of λ, to bad results, the limiting case being the locking phenomenon that is an identically zero solution. A cure was found in using a reduced (that is inexact) numerical quadrature when evaluating the term $\lambda\int_\Omega |\operatorname{div}\, \underline{v}|^2\, dx$ associated with compressibility effects. We refer the reader to the papers of [50] and [9] for a discussion of the long history of this idea. We shall rather develop in details on this example the relations of reduced integrations and mixed methods and try to make clear to what extent they may be claimed to be equivalent. For this we first recall that problem (144) can be transformed by a straightforward application of duality techniques into a saddle point problem

$$\inf_{\underline{v}}\sup_{q} \mu \int_\Omega |\underline{\underline{\varepsilon}}(\underline{v})|^2\, dx - \frac{2}{2\lambda}\int_\Omega |q|^2\, dx + \int_\Omega q \operatorname{div}\, \underline{v}\, dx - \int_\Omega \underline{f}\cdot\underline{v}\, dx \tag{145}$$

for which optimality conditions are, denoting $(\underline{u}, p)$ the saddle point,

$$\mu\int_\Omega \underline{\underline{\varepsilon}}(\underline{u}) : \underline{\underline{\varepsilon}}(\underline{v})\, dx + \int_\Omega p \operatorname{div}\, \underline{v}\, dx = \int_\Omega \underline{f}\cdot\underline{v}\, dx, \forall \underline{v}\in (H_0^1(\Omega))^2 \tag{146}$$

$$\int_\Omega \operatorname{div}\, \underline{u}\, q\, dx = \frac{1}{\lambda}\int_\Omega pq\, dx, \forall q\in L^2(\Omega). \tag{147}$$

This is obviously very close to a Stokes problem and is also an example of a problem of the following nature

$$a(u,v) + b(v,p) = (f,v), \forall v\in V,\ u\in V, \tag{148}$$

$$b(u,q) - c(p,q) = (g,q), \forall q\in Q,\ p\in Q. \tag{149}$$

We can then derive that an approximation of (146) and (147) (that is a choice of an approximation for both $\underline{u}$ and p) which leads to error estimates independent of λ must be a good approximation for Stokes problem. The preceding sections of this chapter therefore give us a good idea of what should (or should not) be used as an approximation. What we shall now see is that reduced integration methods correspond to *an implicit choice* of a mixed approximation. The success of the reduced integration method will thus rely on the qualities of this underlying mixed method.

9.2 Reduced Integration Methods

Let us consider a (more or less) standard approximation of the original problem (144). An exact evaluation of the "penalty term" $\lambda \int_{\Omega} |\operatorname{div}\ \underline{v}|^2 dx$ means that for λ large one tries to get an approximation of $\underline{u}$ which is *exactly* divergence-free. But as we have already seen few finite elements can stand such a condition that will in most cases lead to locking phenomenon due to overconstraining. In a mixed formulation one relaxes the incompressibility condition by the choice of the approximation for p. Let us now see how this will be translated as a reduced integration method at least in some cases. Let us then consider $V_h \subset V = (H_0^1(\Omega))^2$, $Q_h \subset Q = L^2(\Omega)$, these approximation spaces being built from finite elements defined on a partition of Ω. On each element K, let there be given a set of k points x_i and weights ω_i defining a numerical quadrature formula

$$\int_K f(x)\, dx = \sum_{i=1}^{k} \omega_i\, f(x_i). \tag{150}$$

Remark 9.1. It will be convenient to define the numerical quadrature on a reference element K and to evaluate integrals by a change of variables.

$$\int_K f(x)\, dx = \int_{\hat{K}} f(\hat{x})\, J(\hat{x})\, d\hat{x} = \sum_{i=1}^{k} \omega_1\, f(\hat{x}_i)\, J(\hat{x}_i). \tag{151}$$

The presence of the Jacobian $J(x)$ should be taken into account when discussing the precision of the quadrature rule on K.

Let us now make the hypothesis that for $\underline{v}_h \in V_h$ and $p_h,\ q_h \in Q_h$, one has exactly

$$\int_K q_h\ \operatorname{div}\ \underline{v}_h\, dx = \sum_{i=1}^{k} \omega_i\, \hat{q}_h(\hat{x}_i) \widehat{\operatorname{div}\ \underline{v}_h}(\hat{x}_i)\, J(\hat{x}_i) \tag{152}$$

and

$$\int_K p_h\, q_h\, dx = \sum_{k=1}^{k} \omega_i\, \hat{p}_h(\hat{x}_i)\, \hat{q}_h(\hat{x}_i)\, J(\hat{x}_i). \tag{153}$$

Let us now consider the discrete form of (147)

$$\int_{\Omega} \operatorname{div}\ \underline{u}_h\, q_h\, dx = \frac{1}{\lambda} \int_{\Omega} p_h\, q_h\, dx,\ \forall q_h \in Q_h. \tag{154}$$

When the space Q_h is built from discontinuous functions, this can be read element by element

$$\int_K q_h\ \operatorname{div}\ \underline{u}_h\, dx = \frac{1}{\lambda} \int_K p_h\, q_h\, dx,\ \forall q_h \in Q_h, \tag{155}$$

so that using (152) and (153) one gets

$$\hat{p}_h(\hat{x}_i) = \lambda \widehat{\operatorname{div}\ \underline{u}_h}(\hat{x}_i) \text{ or } p_h(x_i) = \lambda \operatorname{div}\ u_h(x_i). \tag{156}$$

Formula (151) can in turn be used in the discrete form of (146) which now gives

$$\begin{cases} 2\mu \displaystyle\int_\Omega \underline{\underline{\varepsilon}}(\underline{u}_h) : \underline{\underline{\varepsilon}}(\underline{v}_h)\, dx + \lambda \sum_K \Big(\sum_{i=1}^{k} \omega_i\, J(\hat{x}_i) (\widehat{\text{div } \underline{u}_h}(\hat{x}_i) (\widehat{\text{div } \underline{v}_h}(\hat{x}_i)) \Big) \\ \qquad\qquad\qquad\qquad\qquad\qquad = \displaystyle\int_\Omega \underline{f} \cdot \underline{v}_h \, dx. \end{cases} \tag{157}$$

In general the term $\sum_K \Big(\sum_{i=1}^{k} \omega_i\, J(\hat{x}_i) (\widehat{\text{div } \underline{u}_h}(\hat{x}_i) (\widehat{\text{div } \underline{v}_h}(\hat{x}_i)) \Big)$ *is not* an exact evaluation of $\int_\Omega \text{div } \underline{u}_h \text{ div } \underline{v}_h \, dx$ and reduced integration is effectively introduced. In the case where (152) and (153) hold there is a perfect equivalence between the mixed method and the use of reduced integration. Whatever will come from one can be reduced to the other one. It will however not be in general possible to get equalities (152) and (153) so that a further analysis will be needed. But we shall first consider some examples of this complete equivalence case.

Example 9.1. Let us consider the $Q_1 - P_0$ approximation on a *rectangle* and a one-point quadrature rule. It is clear that div $\underline{u}_h \in P_1(K)$ and is integrated exactly. In the same way a one-point rule is exact for $\int_\Omega p_h\, q_h \, dx$ whenever $p_h,\ q_h \in P_0(K)$. There is thus a perfect equivalence between reduced integration and the exact penalty method defined by (154).

Example 9.2. We now consider again the same $Q_1 - P_0$ element on a general quadrilateral. To show that we still have equivalence requires a somewhat more delicate analysis. Indeed at first sight the quadrature rule is not exact for $\int_{\hat{K}} \widehat{\text{div } \underline{u}_h}\ J_K(\hat{x})\ d\hat{x}$. Let us however consider in detail the term $\widehat{\text{div } \underline{u}_h} = \widehat{\frac{\partial u_1}{\partial x_1}} + \widehat{\frac{\partial u_2}{\partial x_2}}$. Let $B = DF$ be the Jacobian matrix of the transformation F from $\hat{K}$ into K. Writing explicitly

$$F = \begin{cases} a_0 + a_1\hat{x} + a_2\hat{y} + a_3\hat{x}\hat{y} \\ b_0 + b_1\hat{x} + b_2\hat{y} + b_3\hat{x}\hat{y} \end{cases} \tag{158}$$

one has

$$B = \begin{pmatrix} a_1 + a_3\hat{y} & a_2 + a_3\hat{x} \\ b_1 + b_3\hat{y} & b_2 + b_3\hat{x} \end{pmatrix} \tag{159}$$

so that we get

$$B^{-1} = \frac{1}{J(\hat{x})} \begin{pmatrix} b_2 + b_3\hat{x} & -a_2 - a_3\hat{x} \\ -b_1 - b_3\hat{y} & a_1 + a_3\hat{y} \end{pmatrix}. \tag{160}$$

But

$$\widehat{\frac{\partial u_1}{\partial x_1}} = \Big(\frac{\partial \hat{u}_1}{\partial \hat{x}_1}(b_2 + b_3\hat{x}) - \frac{\partial \hat{u}_1}{\partial \hat{x}_2}(b_1 + b_3\hat{y}) \Big) \frac{1}{J(\hat{x})}, \tag{161}$$

$$\widehat{\frac{\partial u_2}{\partial x_2}} = \Big(\frac{\partial \hat{u}_2}{\partial \hat{x}_1}(-a_2 - a_3\hat{x}) + \frac{\partial \hat{u}_2}{\partial \hat{x}_2}(a_1 + a_3\hat{y}) \Big) \frac{1}{J(\hat{x})}. \tag{162}$$

When computing $\int_{\hat{K}} \widehat{\text{div } \underline{u}_h} \; J(\hat{x}) \; d\hat{x}$, Jacobians cancel and one is left with the integral of a function which is linear in each variable and which can be computed exactly by a one-point formula.

Example 9.3. Using a four-point integration formula on a straight-sided quadrilateral can be seen as in the previous example to be exactly equivalent to a $Q_2 - Q_1$ approximation [8, 9].

The above equivalence is however not the general rule. Consider the following examples.

Example 9.4. We want to use a reduced integration procedure to emulate the Crouzeix–Raviart $P_2 - P_1$ element. To define a P_1 pressure, we need three integration points which can generate a formula that will be exact for second degree polynomials (but not more). The bubble function included in velocity however makes div $\underline{u}_h \in P_2(K)$ and $\int_K$ div $\underline{u}_h \; q_h dx$ will not be evaluated exactly.

Example 9.5. A full isoparametric $Q_2 - Q_1$ element is not equivalent to its four-point reduced integration analogue.

Example 9.6. A $Q_2 - P_0$ approximation is not, even on rectangles, equivalent to a one-point reduced integration method for div $\underline{u}_h$ contains second-order term which are not taken into account by a one-point quadrature.

9.3 Effects of Inexact Integration

If we now consider into more details the cases where a perfect equivalence does not hold between the mixed method and some reduced integration procedure we find ourselves in the setting of nonconforming approximation. In particular $b(\underline{v}_h, q_h)$ is replaced by an approximate bilinear form $b_h(\underline{v}_h, q_h)$. We shall suppose to simplify that the scalar product on Q_h is exactly evaluated. Two questions must then be answered.

— Does $b_h(.,.)$ satisfy the inf–sup condition?
— Do error estimates still hold without loss of accuracy?

Example 9.7. We in fact come back to Example 9.6 and study on a rectangular mesh, the $Q_2 - P_0$ approximation (see Sect. 6.4) with a one-point quadrature rule. This is not, as we have said, equivalent to the standard $Q_2 - P_0$ approximation. We now want to check, using Proposition 4.1, that it satisfies the inf–sup condition. We thus have to build a continuous operator (in $H^1(\Omega)$-norm) such that

$$\int_\Omega \text{div } \underline{u}_h \; q_h \; dx = \sum_K [(\text{div } \Pi_h \underline{u}_h)(M_{0,K}) q_K] \; \text{area}(K) \tag{163}$$

where $M_{0,K}$ is the barycenter of K and q_K the restriction of q_h to K. We can restrict our analysis to one element as q_h is discontinuous and we study both sides of equality (163). We have of course, taking $q_K = 1$,

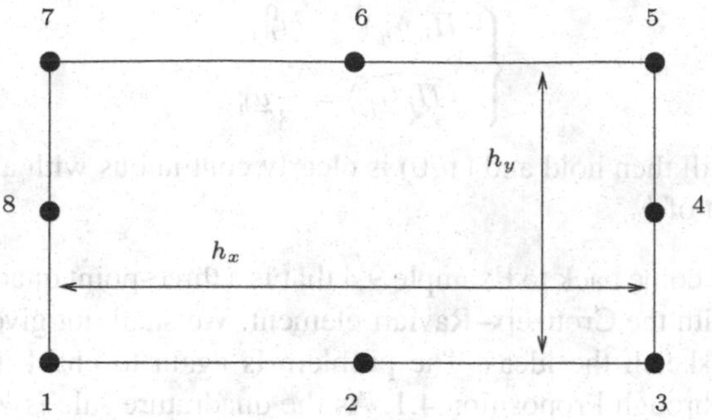

Fig. 16. A rectangle

$$\int_K \operatorname{div} \underline{u}_h \, dx = \int_{\partial K} \underline{u}_h \cdot \underline{n} \, d\sigma. \tag{164}$$

Using the numbering of Fig. 16 and denoting by u_i, v_i the horizontal and vertical components of velocity at node i, we can write (164) by Simpson's quadrature rule in the form

$$\left\{ \begin{aligned} \int_K \operatorname{div} \underline{u}_h \, dx = {} & \frac{h_y}{6}[u_5 + 4u_4 + u_3] - \frac{h_y}{6}[u_1 + 4u_8 + u_7] \\ & + \frac{h_x}{6}[v_7 + 4v_6 + v_5] - \frac{h_x}{6}[v_1 + 4v_2 + v_3]. \end{aligned} \right. \tag{165}$$

If we write

$$u_4 = \frac{u_5 + u_3}{2} + \hat{u}_4, \; u_8 = \frac{u_1 + u_7}{2} + \hat{u}_8 \tag{166}$$

$$v_6 = \frac{v_5 + v_7}{2} + \hat{v}_6, \; v_2 = \frac{v_1 + v_3}{2} + \hat{v}_2 \tag{167}$$

where $\hat{u}_4, \hat{u}_6, \hat{v}_6$ and $\hat{v}_2$ are corrections with respect to a bilinear interpolation we may rewrite (165) as

$$\left\{ \begin{aligned} \int_K \operatorname{div} \underline{u}_h \, dx = {} & \frac{h_y}{2}[u_5 + u_3 + \frac{4}{3}\hat{u}_4] - \frac{h_y}{2}[u_1 + u_7 + \frac{4}{3}\hat{u}_8] \\ & + \frac{h_y}{2}[v_7 + v_5 + \frac{4}{3}\hat{v}_6] - \frac{h_x}{2}[v_1 + v_3 + \frac{4}{3}\hat{v}_2]. \end{aligned} \right. \tag{168}$$

On the other hand area (K) div $\underline{u}_h(M_{0,K})$ can be seen to be equal to

$$\left\{ \begin{aligned} & \frac{h_y}{2}[u_5 + u_3 + 2\hat{u}_4] - \frac{h_y}{2}[u_1 + u_7 + 2\hat{u}_8] \\ & - \frac{h_x}{2}[u_7 + v_5 + 2\hat{v}_6] - \frac{h_x}{2}[v_1 + v_3 + 2\hat{u}_2]. \end{aligned} \right. \tag{169}$$

If we thus split $\underline{u}_h$ into a bilinear part $\underline{u}_h^0$ and a mid-point correction part $\hat{\underline{u}}_h$, one can define $\Pi_h \underline{u}_h$ by setting

$$\begin{cases} (\Pi_h \underline{u}_h)^0 = \underline{u}_h^0, \\ (\widehat{\Pi_h \underline{u}_h}) = \dfrac{2}{3}\hat{\underline{u}}_h. \end{cases} \tag{170}$$

Equality (164) will then hold and (170) is clearly continuous with a continuity constant independent of h.

Example 9.8. We come back to Example 9.4 that is a three-point quadrature rule used in conjunction with the Crouzeix–Raviart element. We shall not give the analysis in details but only sketch the ideas. The problem is again to check that the inf–sup condition holds through Proposition 4.1. As the quadrature rule *is exact* when q_h is *piecewise constant*, the obvious idea is to build $\Pi_h \underline{u}_h$ by keeping the trace of $\underline{u}_h$ on ∂K and *only modifying the coefficients of the bubble functions*. This can clearly be done. Continuity is now to be checked and the proof is essentially the same as the standard proof of the inf–sup condition (Sect. 7.2).

Example 9.9 (A modified $Q_1 - P_0$ element). We now present a puzzling example of an element which is stable but for which convergence is tricky due to a consistency error term. We have here a case where using a one-point quadrature rule will change the situation with respect to the inf–sup condition. In fact it will make a stable element from an unstable one but will also introduce an essential change in the problem. The departure point is thus the standard $Q_1 - P_0$ element which, as we know, does not satisfy the inf–sup condition. We now make it richer by adding to velocity $\underline{u}_h|_K = \{u_1, u_2\}$ what we shall call wave functions. On the reference element $\hat{K} =]-1,1[\times]-1,1[$, those functions are defined by

$$\begin{cases} w_1 = \hat{x}\, b_2(\hat{x},\hat{y}), \\ w_2 = \hat{y}\, b_2(\hat{x},\hat{y}), \end{cases} \tag{171}$$

where $b_2(\hat{x},\hat{y}) = (1-\hat{x}^2)(1-\hat{y}^2)$ is the Q_2 bubble function. If we now consider

$$\hat{\underline{u}}_h|_K = \{u_1 + \alpha_K w_1, u_2 + \alpha_K w_2\} = \underline{u}_h|_K + \alpha_K \underline{w}_K, \tag{172}$$

we obtain a new element with an internal degree of freedom. The wave functions that we added vanish on the boundary and nothing is changed for the stability of the mixed method with exact integration. If we rather use a one-point quadrature rule, things become different. We shall indeed check that the modified bilinear form $b_h(\hat{\underline{v}}_h, q_h)$ satisfies the inf–sup condition. We thus have to show that

$$\sup_{\hat{\underline{u}}_h} \frac{\sum_K \operatorname{div}\, \hat{\underline{u}}_h(M_{0,K}) p_K\, h_K^2}{\|\hat{\underline{u}}_h\|_1} \geq k_0\, |p_h|_0. \tag{173}$$

This is easily checked by posing on K (we suppose a rectangular mesh to simplify)

$$\hat{\underline{u}}_h|_K = h_K\, p_K \underline{w}_K. \tag{174}$$

We then have div $\hat{\underline{u}}_h = p_h$ and

$$\|\underline{\hat{u}}_h\|_{1,K} = h\, p_K\, \|\underline{w}_K\|_{1,K}, \tag{175}$$

which implies

$$\|\underline{\hat{u}}_h\|_1 \leq c\, |p_h|_0, \tag{176}$$

and (173) follows. A remarkable point is that even the hydrostatic mode has disappeared. This is an indication that something incorrect has been introduced in the approximation. An analysis of *consistency error* indeed shows that usual error estimates fail and that we are actually approximating a continous problem in which the incompressibility condition has been replaced by $\operatorname{div} u + kp = 0$ where $k = 1575/416$. We then see that if in general for Stokes problem, making the space of velocities richer improves (at least does not reduce) the quality of the method, this fact can become false when numerical integration is used.

Let us now turn our attention to the problem of error estimation. From [24] all we have to do is to estimate the consistency terms

$$\sup_{v_h} \frac{|b_h(v_h, p) - b_h(v_h, p)|}{\|v_h\|_V} \tag{177}$$

and

$$\sup_{q_h} \frac{|b(u, q_h) - b_h(u, q_h)|}{\|q_h\|_0}. \tag{178}$$

We thus have to estimate quadrature errors. It would be out of purpose to enter into details and we refer the reader to [27, 28] where examples of such analysis are presented exhaustively. The first step is to transform (177) into a form which is sometimes more tractable. We may indeed write

$$\begin{cases} b(v_h, p) - b_h(v_h, p) = (b(v_h, p - q_h) - b_h(v_h, p - q_h)) \\ \qquad\qquad\qquad\qquad\quad + (b(v_h, q_h) - b_h(v_h, q_h)) \end{cases} \tag{179}$$

and

$$\begin{cases} b(u, q_h) - b_h(u, q_h) = (b(u - v_h, q_h) - b_h(u - v_h, q_h)) \\ \qquad\qquad\qquad\qquad\quad + (b(v_h, q_h) - b_h(v_h, q_h)). \end{cases} \tag{180}$$

The first parenthesis in the right-hand side of (179) and (180) can be reduced to an approximation error. The second parenthesis implies only polynomials.

Let us therefore consider (180) for the three approximations introduced above. (Coming back to the notations of the present section). For the Crouzeix–Raviart triangle taking $\underline{v}_h$ the standard interpolate of $\underline{u}$ makes the second parenthesis vanish while the first yields an $O(h)$ estimate. For the two other approximations taking $\underline{v}_h$ to be a standard bilinear approximation of $\underline{u}$ makes the second parenthesis vanish while the first yields on $O(h)$ estimate, which is the best that we can hope anyway. The real trouble is therefore with (177) with or without (179). In the case of the Crouzeix–Raviart triangle, we can use directly (177) and the following result of [27, 28], (Theorem IV.1.5).

Proposition 9.1. *Let $f \in W_{k,q}(\Omega)$, $p_k \in P_k(K)$ and denote $E_k(fp_k)$ the quadrature error on element K when numerical integration is applied to fp_k. Let us suppose that $E_K(\hat{\phi}) = 0$, $\forall \hat{\phi} \in P_{2k-2}(K)$ then one has*

$$|E_K(fp_k)| \leq ch_K^k\,(meas(K))^{\frac{1}{2}-\frac{1}{q}}|f|_{k,q,K}|p_k|_1. \tag{181}$$

Taking $k = 2, q = \infty$ and using the inverse inequality to go from $|p_k|_1$ to $|p_k|_0$ one gets an $O(h^2)$ estimate for (177).

The two other approximations cannot be reduced to Proposition 9.1 and must be studied through (179). We must study a term like

$$\sup_{v_h} \frac{|b(v_h, q_h) - b_h(v_h, q_h)|}{\|v_h\|_1}. \tag{182}$$

This can at best be *bounded*. For instance in the case of the $Q_2 - P_0$ approximation we can check by hand that the quadrature error on K reduces to $h_K^3 \mid \text{div } \underline{v}_h|_{2,K}\, p_k$.

10 Divergence-Free Basis, Discrete Stream Functions

We have dealt in this note with the mixed formulation of the Stokes problem and we have built finite element approximations in which discrete divergence-free functions approximate the continuous ones. It is sometimes useful to consider directly the constrained minimization problem

$$\inf_{\underline{v}_0 \in V_0} \frac{1}{2} \int_\Omega |\underline{\underline{\varepsilon}}(\underline{v}_0)|^2\, dx - \int_\Omega \underline{f} \cdot \underline{v}_0\, dx, \tag{183}$$

where V_0 is the subspace of divergence-free functions. In this subspace we have a standard minimization problem and the discrete form would lead to a positive definite linear system. Indeed the solution of problem (183) satisfies the variational equation,

$$\int_\Omega \underline{\underline{\varepsilon}}(\underline{u}_0) : \underline{\underline{\varepsilon}}(\underline{v}_0)\, dx = \int_\Omega \underline{f} \cdot \underline{v}_0\, dx,\ \forall \underline{v}_0 \in V_0, \underline{u}_0 \in V_0. \tag{184}$$

In the discrete problem, if one knows a basis $\{\underline{w}_0, \ldots, \underline{w}_m\}$ of V_{0h} the solution is reduced to the solution of the linear system

$$A_0\, U_0 = F_0, \tag{185}$$

where

$$a_{ij}^0 = \int_\Omega \underline{\underline{\varepsilon}}(\underline{w}_i) : \underline{\underline{\varepsilon}}(\underline{w}_j)\, dx,\ f_i^0 = \int_\Omega \underline{f} \cdot \underline{w}_i\, dx, \tag{186}$$

and

$$A_0 = \{a_{ij}^0\},\ F_0 = \{f_i^0\}. \tag{187}$$

Building a basis for the divergence-free subspace could therefore lead to a neat reduction of computational costs: pressure is eliminated, along with a certain amount of velocity degrees of freedom. System (185) is smaller than the original one. It must however be noted that with respect to the condition number, (185) is behaving like a fourth-order problem, which makes its practical usefulness often dubious. As to pressure, it can be recovered a posteriori (see [25, 26]).

The construction of such a basis is not however a very popular method and is considered as a hard task although it has been numerically implemented (see [43, 65, 44]).

As we shall see the two-dimensional case is quite readily handled in many cases. The degrees of freedom can be associated with those of a discrete stream function. The three-dimensional problem is harder to handle: a generating system can often easily be found but the construction of a basis requires the elimination of some degrees of freedom in a not so obvious way.

We shall also consider rapidly a numerical procedure, related to static condensation that will require a partly divergence-free basis.

Finally we want to emphasize that the construction which we describe will make sense only if the finite element approximation is good so that the previous analysis is still necessary even if it might seem to be bypassed.

We first consider a simple example of a divergence-free basis.

Example 10.1 (The nonconforming $P_1 - P_0$ element). We consider the classical nonconforming element introduced in [31] (cf. Sect. 6.3) in which mid-side nodes are used as degrees of freedom for velocity. This generates a piecewise linear nonconforming approximation; pressure is taken constant on each element (Fig. 17). The restriction to an element K of $\underline{u}_h \in V_h$ is then exactly divergence-free and is therefore locally the curl of a quadratic polynomial. This discrete stream function cannot be continuous on interfaces but must have continuous derivatives at mid-side points: it can be built from Morley's triangle. The degrees of freedom of the divergence-free basis can be associated to the degrees of freedom of this nonconforming stream function (Fig. 18). This assigns a basis function to each vertex and to each mid-side node. They are depicted schematically in Fig. 19. One observes a general pattern: divergence-free functions are made from small vortices.

Remark 10.1. The kind of basis obtained in the previous example is typical of a domain without holes with homogeneous Dirichlet boundary conditions. Whenever a hole is present, an extra basis function must be added in order to ensure circulation around the hole (Fig. 20). This function is not local.

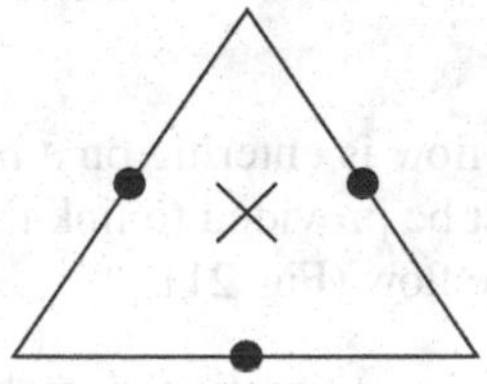

Fig. 17. Nonconforming element

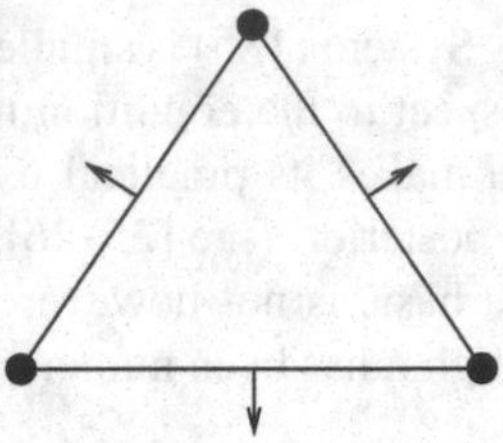

Fig. 18. Nonconforming stream function

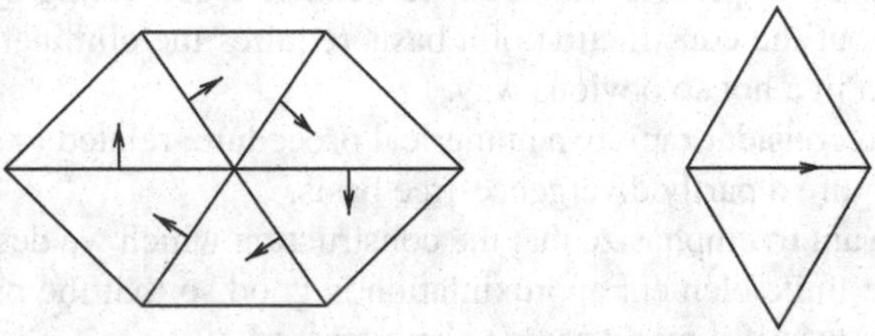

Fig. 19. Basis functions for a divergence free P_1–P_0 nonconforming element

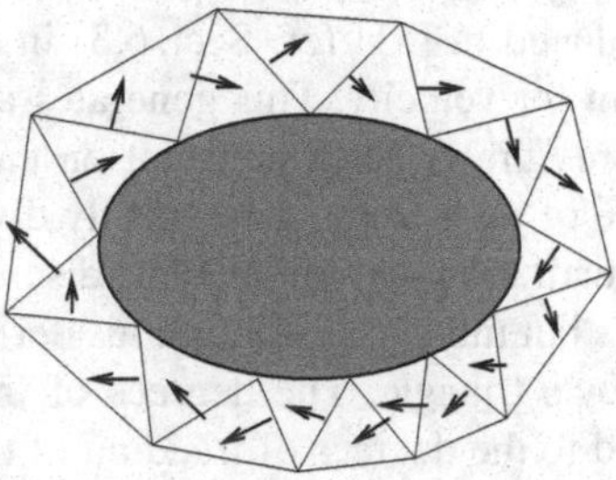

Fig. 20. Divergence free function around a hole

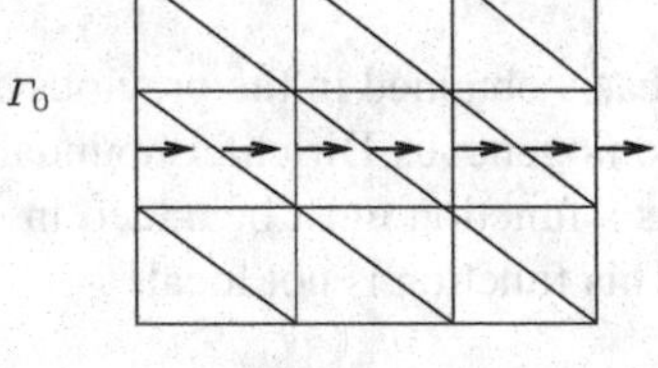

Fig. 21. Divergence free function with different boundary conditions

In the same way when the flow is entering on a part Γ_0 of $\partial\Omega$ and outgoing on a part Γ_1, a basis function must be provided to link those parts and to thus take into account the potential part of the flow (Fig. 21).

We now consider a conforming approximation, namely the popular $Q_2 - P_1$ element (see Sect. 6.4).

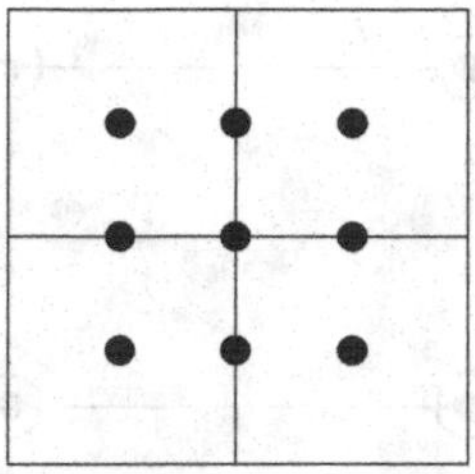

Fig. 22. A 2 × 2 macroelement

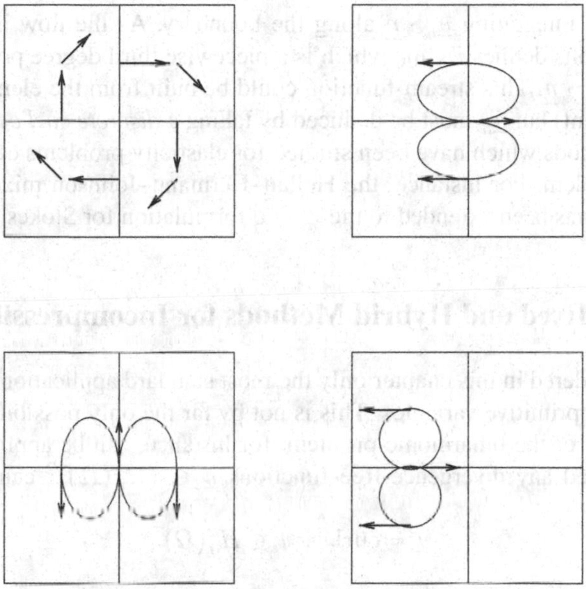

Fig. 23. Divergence-free functions

Example 10.2 (The conforming $Q_2 - P_1$ element). We shall sketch in this example the construction of a divergence-free basis for the $Q_2 - P_1$ element. To make things simple we shall assume that the mesh is formed of 2 × 2 macroelements. The general case can easily be deduced. Let us first look for divergence-free (in the discrete sense of course) functions with their support on a macroelement. We have 18 degrees of freedom for velocity (Fig. 22) linked by $(12 - 1) = 11$ linear constraints. This leaves seven linearly independent functions which can be described by the diagrams of Fig. 23.

Three of them are associated with each vertex and one to each mid-side node. It must be noted that internal nodes are no longer degrees of freedom.

Remark 10.2. The "divergence-free" functions described above cannot be taken as the curl of a stream-function as they are not exactly divergence-free. However a discrete stream-function can nevertheless be built. Its trace on ∂K can be totally

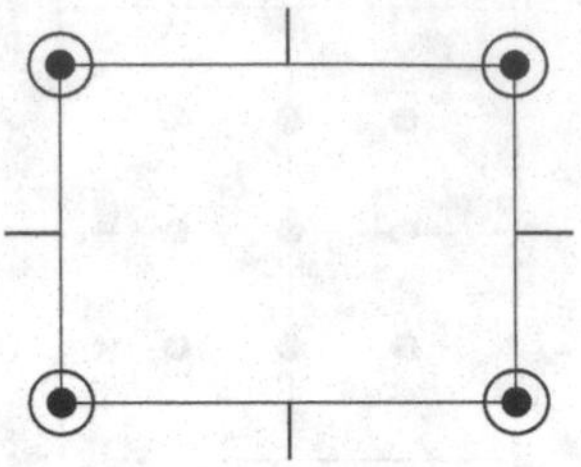

Fig. 24. Adini's element

determined by integrating $\underline{u}_h \cdot \underline{n}$ along the boundary. As the flow is conserved at element level this defines $\psi_h|_{\partial K}$ which is a piecewise third degree polynomial such that $\frac{\partial \psi_h}{\partial \tau} = \underline{u}_h \cdot \underline{n}$. This stream-function could be built from the element of Fig. 24 (Adini's element) but $\underline{u}_h$ must be deduced by taking a *discrete curl operation*.

Other methods which have been studied for elasticity problems can be extended to Stokes problem. For instance, the Hellan–Hermann–Johnson mixed method for plate bending has been extended to the $\psi - \omega$ formulation for Stokes by [22].

11 Other Mixed and Hybrid Methods for Incompressible Flows

We have considered in this chapter only the most standard applications to the Stokes problem using primitive variables. This is not by far the only possibility; the $\psi - \omega$ decomposition of the biharmonic problem, for instance, can be applied to a Stokes problem. Indeed any divergence-free functions $\underline{u} \in (H_0^1(H))^2$ can be written in the form

$$\underline{u} = \operatorname{curl} \psi, \ \psi \in H_0^2(\Omega). \tag{188}$$

From (188) we get

$$\operatorname{curl} \underline{u} = \omega = -\Delta \psi. \tag{189}$$

On the other hand, taking the curl of equation (1) gives

$$-\Delta \omega = \operatorname{curl} \underline{f} = f_1. \tag{190}$$

This procedure can be extended to the Navier–Stokes equation (indeed in many ways) including, if wanted, some upwinding procedure for the nonlinear terms (see [40, 47]). The reader will find a fairly complete study of such procedures in [42], [53]. It must be noted that the simplest case of such a procedure, using for ψ_h a bilinear approximation yields as an approximation of $\underline{u}$ the famous MAC cells (Fig. 25).

Indeed this is nothing but the space $RT_{[0]}$ for which the subspace of divergence-free functions can be obtained from a bilinear stream-function. The Hellan–Hermann–Johnson mixed method for elasticity can also be applied to the Stokes problem with $\underline{u}_h$ chosen in some approximation of $H(\operatorname{div}, \Omega)$. A direct approach precludes to use a symmetric tensor and forces to use $\underline{\underline{\operatorname{grad}}}\, \underline{u}$ instead of

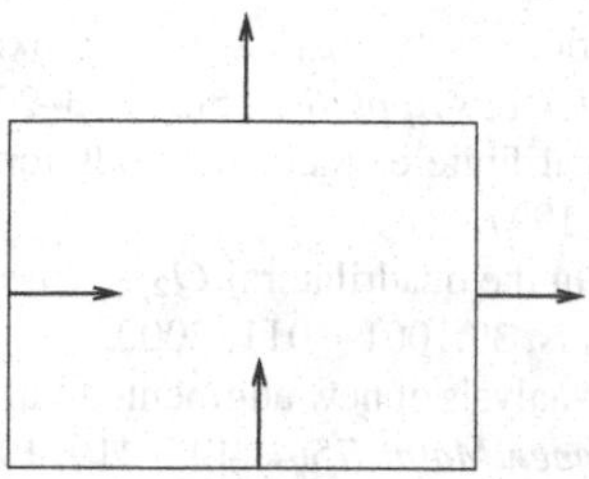

Fig. 25. MAC cell

$\underline{\underline{\varepsilon}}(\underline{u})$ as dual variable [4]. This difficulty has been circumvented by [51] by enriching the spaces by the trick of [1] or [2] or [20].

Finally it must be said that dual hybrid methods have been applied by [7] to the Stokes problem. This generates elements which are defined only by the traces at the boundaries and for which internal values can be chosen arbitrarily. This can be seen as the ultimate case of enrichment by bubble functions: enriching by a (potentially infinite dimensional) space enables to use exactly divergence-free function, provided the inf–sup condition is satisfied for *piecewise constant pressure*.

References

1. M. Amara and J.M. Thomas. Equilibrium finite elements for the linear elastic problem. *Numer. Math.*, 33:367–383, 1979.
2. D.N. Arnold, F. Brezzi, and J. Douglas. PEERS: a new mixed finite element for plane elasticity. *Jpn. J. Appl. Math.*, 1:347–367, 1984.
3. D.N. Arnold, F. Brezzi, and M. Fortin. A stable finite element for the Stokes equations. *Calcolo*, 21:337–344, 1984.
4. D.N. Arnold and J.S. Falk. A new mixed formulation for elasticity. *Numer. Math.*, 53:13–30, 1988.
5. D.N. Arnold. On nonconforming linear-constant elements for some variants of the Stokes equations. *Istit. Lombardo Accad. Sci. Lett. Rend. A*, 127(1):83–93 (1994), 1993.
6. D.N. Arnold, D. Boffi, and R.S. Falk. Approximation by quadrilateral finite elements. *Math. Comp.*, 71(239):909–922, 2002.
7. S.N. Atluri and C. Yang. A hybrid finite element for Stokes flow II. *Int. J. Numer. Methods Fluids*, 4:43–69, 1984.
8. M. Bercovier. *Régularisation duale des problèmes variationnels mixtes.* PhD thesis, Université de Rouen, 1976.
9. M. Bercovier. Perturbation of a mixed variational problem, applications to mixed finite element methods. *R.A.I.R.O. Anal. Numer.*, 12:211–236, 1978.
10. M. Bercovier and O.A. Pironneau. Error estimates for finite element method solution of the Stokes problem in the primitive variables. *Numer. Math.*, 33:211–224, 1977.
11. C. Bernardi and G. Raugel. Méthodes d'éléments finis mixtes pour les équations de Stokes et de Navier–Stokes dans un polygone non convexe. *Calcolo*, 18:255–291, 1981.
12. D. Boffi. Stability of higher order triangular Hood–Taylor methods for stationary Stokes equations. *Math. Models Methods Appl. Sci.*, 2(4):223–235, 1994.

13. D. Boffi. Minimal stabilizations of the $P_{k+1} - P_k$ approximation of the stationary Stokes equations. *Math. Models Methods Appl. Sci.*, 5(2):213–224, 1995.
14. D. Boffi. Three-dimensional finite element methods for the Stokes problem. *SIAM J. Numer. Anal.*, 34:664–670, 1997.
15. D. Boffi and L. Gastaldi. On the quadrilateral $Q_2 - P_1$ element for the Stokes problem. *Int. J. Numer. Methods Fluids*, 39:1001–1011, 2002.
16. D. Boffi and C. Lovadina. Analysis of new augmented Lagrangian formulations for mixed finite element schemes. *Numer. Math.*, 75(4):405–419, 1997.
17. D. Boffi, F. Brezzi, and L. Gastaldi. On the problem of spurious eigenvalues in the approximation of linear elliptic problems in mixed form. *Math. Comput.*, 69(229):121–140, 2000.
18. J.M. Boland and R.A. Nicolaides. Stability of finite elements under divergence constraints. *SIAM J. Numer. Anal.*, 20(4):722–731, 1983.
19. F. Brezzi and K.J. Bathe. A discourse on the stability conditions for mixed finite element formulations. *CMAME*, 82:27–57, 1990.
20. F. Brezzi, J. Douglas, Jr., and L.D. Marini. Recent results on mixed finite element methods for second order elliptic problems. *Vistas in applied mathematics,* 25–43, Transl. Ser. Math. Engrg., Optimization Software, New York, 1986.
21. F. Brezzi and R.S. Falk. Stability of higher-order Hood-Taylor methods. *SIAM J. Numer. Anal.*, 28(3):581–590, 1991.
22. F. Brezzi, J. Le Tellier, and T. Olier. Mixed finite element approximation for the stationary Navier–Stokes equations (in Russian). In *Viceslitelnia Metodii V. Prikladnoi Mathematiceskie*, NAUKA, Novosibirsk, 1982, pp. 96–108. Meeting INRIA, Novosibirsk.
23. F. Brezzi and J. Pitkäranta. On the stabilization of finite element approximations of the Stokes equations. In W. Hackbush, editor, *Efficient Solutions of Elliptic Systems, Notes on Numerical Fluid Mechanics*, vol. 10, Braunschweig, Wiesbaden, Vieweg, 1984.
24. F. Brezzi and M. Fortin. *Mixed and Hybrid Finite Element Methods*. Springer, Berlin Heidelberg New York, 1991.
25. P. Caussignac. Explicit basis functions of quadratic and improved quadratic finite element spaces for the Stokes problem. *Commun. Appl. Numer. Methods*, 2:205–211, 1986.
26. P. Caussignac. Computation of pressure from the finite element vorticity stream-function approximation of the Stokes problem. *Commum. Appl. Numer. Methods*, 3:287–295, 1987.
27. P.G. Ciarlet. *Mathematical elasticity, vol. I. Three-Dimensional Elasticity*. North-Holland, Amsterdam, 1988.
28. P.G. Ciarlet. *Mathematical elasticity, vol. II. Theory of Plates*. North-Holland, Amsterdam, 1997.
29. J.F. Ciavaldini and J.C. Nédélec. Sur l'élément de Fraeijs de Veubeke et Sander. *R.A.I.R.O. Anal. Numer.*, 8:29–45, 1974.
30. P. Clément. Approximation by finite element functions using local regularization. *R.A.I.R.O. Anal. Numer.*, 9:77–84, 1975.
31. M. Crouzeix and P.A. Raviart. Conforming and nonconforming finite element methods for solving the stationary Stokes equations. *R.A.I.R.O. Anal. Numer.*, 7:33–76, 1973.
32. G. Duvaut and J.L. Lions. *Les inéquations en mécanique et en physique*. Dunod, Paris, 1972.
33. A. Fortin. *Méthodes d'éléments finis pour les équations de Navier–Stokes*. PhD thesis, Université Laval, 1984.

34. A. Fortin and M. Fortin. Newer and newer elements for incompressible flow. In R.H. Gallagher, G.F. Carey, J.T. Oden, and O.C. Zienkiewicz, editors, *Finite Elements in Fluids*, Volume 6. Chichester, England and New York, Wiley-Interscience, 1985, p. 171–187.
35. M. Fortin. Utilisation de la méthode des éléments finis en mécanique des fluides. *Calcolo*, 12:405–441, 1975.
36. M. Fortin. An analysis of the convergence of mixed finite element methods. *R.A.I.R.O. Anal. Numer.*, 11:341–354, 1977.
37. M. Fortin. Old and new finite elements for incompressible flows. *Int. J. Numer. Methods Fluids*, 1:347–364, 1981.
38. M. Fortin, R. Peyret, and R. Temam. Résolution numérique des équations de Navier–Stokes pour un fluide visqueux incompressible. *J. Mécanique*, 10, 3:357–390, 1971.
39. M. Fortin and M. Soulie. A nonconforming piecewise quadratic finite element on triangles. *Int. J. Numer. Methods Eng.*, 19:505–520, 1983.
40. M. Fortin and F. Thomasset. Mixed finite element methods for incompressible flow problems. *J. Comput. Physics*, 37:173–215, 1979.
41. V. Girault and P.A. Raviart. *Finite Element Methods for Navier–Stokes Equations, Theory and Algorithms*. Springer, Berlin Heidelberg New York, 1986.
42. R. Glowinski and O. Pironneau. Numerical methods for the first biharmonic equation and for the two-dimensional Stokes problem. *SIAM Rev.*, 17:167–212, 1979.
43. D. Griffiths. Finite elements for incompressible flow. *Math. Methods Appl. Sci.*, 1:16–31, 1979.
44. F. Hecht. Construction d'une base de fontions P1 non-conformes à divergence nulle dans $\mathbb{R}^3$. *R.A.I.R.O. Anal. Numer.*, 15:119–150, 1981.
45. P. Hood and C. Taylor. Numerical solution of the Navier–Stokes equations using the finite element technique. *Comput. Fluids*, 1:1–28, 1973.
46. T.J.R. Hughes and H. Allik. Finite elements for compressible and incompressible continua. In *Proceedings of the Symposium on Civil Engineering*, Nashville, TN. Vanderbilt University, 1969, pp. 27–62.
47. C. Johnson. On the convergence of a mixed finite element method for plate bending problems. *Numer. Math.*, 21:43–62, 1973.
48. C. Johnson and J. Pitkäranta. Analysis of some mixed finite element methods related to reduced integration. *Math. Comput.*, 38:375–400, 1982.
49. D.S. Malkus. Eigenproblems associated with the discrete LBB-condition for incompressible finite elements. *Int. J. Eng. Sci.*, 19:1299–1310, 1981.
50. D.S Malkus and T.J.R. Hughes. Mixed finite element methods. reduced and selective integration techniques: a unification of concepts. *Comput. Methods Appl. Mech. Eng.*, 15:63–81, 1978.
51. Z. Mghazli. *Une méthode mixte pour les équations de l'hydrodynamique*. PhD thesis, Université de Montréal, 1987.
52. J.T. Oden and O. Jacquotte. Stability of some mixed finite element methods for Stokesian flows. *Comput. Methods Appl. Mech. Eng.*, 43:231–247, 1984.
53. O. Pironneau. *Finite Element Methods for Fluids*. John Wiley, Chichester, 1989. Translated from the French.
54. J. Qin. *On the convergence of some simple finite elements for incompressible flows*. PhD thesis, Penn State University, 1994.
55. R. Rannacher and S. Turek. Simple nonconforming quadrilateral Stokes element. *Numer. Methods Partial Differ. Equations*, 8(2):97–111, 1992.

56. R.L. Sani, P.M. Gresho, R.L. Lee, and D.F. Griffiths. The cause and cure (?) of the spurious pressures generated by certain FEM solutions of the incompressible Navier–Stokes equations. I. *Int. J. Numer. Methods Fluids*, 1(1):17–43, 1981.
57. R. L. Sani, P. M. Gresho, R. L. Lee, D. F. Griffiths, and M. Engelman. The cause and cure (!) of the spurious pressures generated by certain FEM solutions of the incompressible Navier–Stokes equations. II. *Int. J. Numer. Methods Fluids*, 1(2):171–204, 1981.
58. L.R. Scott and M. Vogelius. Norm estimates for a maximal right inverse of the divergence operator in spaces of piecewise polynomials. *Math. Model. Numer. Anal.*, 9:11–43, 1985.
59. R. Stenberg. Analysis of mixed finite element methods for the Stokes problem: a unified approach. *Math. Comput.*, 42:9–23, 1984.
60. R. Stenberg. On the construction of optimal mixed finite element methods for the linear elasticity problem. *Numer. Math.*, 48:447–462, 1986.
61. R. Stenberg. On some three-dimensional finite elements for incompressible media. *Comput. Methods Appl. Mech. Eng.*, 63:261–269, 1987.
62. R. Stenberg. On the postprocessing of mixed equilibrium finite element methods. In W. Hackbusch and K. Witsch, editors, *Numerical Tehchniques in Continuum Mechanics*. Veiweg, Braunschweig, 1987. Proceedings of the Second GAMM-Seminar, Kiel, 1986.
63. R. Stenberg. Error analysis of some finite element methods for the Stokes problem. *Math. Comput.*, 54(190):495–508, 1990. Chesnay, France, 1988.
64. R. Temam. *Navier–Stokes Equations*. North-Holland, Amsterdam, 1977.
65. F. Thomasset. *Implementation of Finite Element Methods for Navier–Stokes Equations. Springer Series in Computational Physics*. Springer, Berlin Heidelberg New York, 1981.
66. R. Verfürth. Error estimates for a mixed finite element approximation of the Stokes equation. *R.A.I.R.O. Anal. Numer.*, 18:175–182, 1984.
67. O.C. Zienkiewicz, S. Qu, R.L. Taylor, and S. Nakazawa. The patch text for mixed formulations. *Int. J. Numer. Methods Eng.*, 23:1873–1883, 1986.

Polynomial Exact Sequences and Projection-Based Interpolation with Application to Maxwell Equations

Leszek Demkowicz*

Institute for Computational Engineering and Sciences, The University of Texas at Austin, ACES 6.322, 105, Austin, TX 78712, USA
leszek@ices.utexas.edu

1 Introduction

The presented notes review the concept and main results concerning commuting projections and projection-based interpolation operators defined for one-, two- and three-dimensional exact sequences involving the gradient, curl, and divergence operators, and Sobolev spaces. The discrete sequences correspond to polynomial spaces defining the classical, continuous finite elements, the "edge elements" of Nédélec, and "face elements" of Raviart–Thomas. All discussed results extend to the elements of variable order as well as parametric elements. The presentation reproduces results for 2D from [19] and 3D from [17, 25, 20, 13] and attempts to present them in a unified manner for all types of finite elements forming the exact sequences. The idea of the projection-based interpolation for elliptic problems was introduced in [37] and generalized to the exact sequence in [21]. The presented results build on the existence of polynomial preserving extension operators [15, 22].

Sobolev Spaces

I am assuming that the reader is familiar with essentials of Sobolev spaces. For those who seek a complete and compact presentation on the subject, I highly recommend the book of McLean [32] to which I will refer for most of technical details relevant to this paper. We will use the Hörmander's definition for spaces $H^s(\Omega)$ that remains valid for the whole range of $s \in \mathbb{R}$. The H^s spaces are isomorphic with duals of spaces $\widetilde{H}^s$, the closure of $\mathcal{D}(\Omega)$ in $H^s(\mathbb{R}^n)$. For $s \geq -\frac{1}{2}$, the restriction operator from $\widetilde{H}^s$ into $H^s(\Omega)$ is injective and, for this range of s, space $\widetilde{H}^s$ can be identified with a subspace of $H^s(\Omega)$. For values s different from half-integers, space $\widetilde{H}^s(\Omega)$ coincides with the space $H^s_0(\Omega)$, the closure of test functions in the $H^s(\Omega)$-norm, with the equivalence constants blowing up with s approaching the half-integers. The energy spaces $H^1(\Omega)$, $\boldsymbol{H}(\mathbf{curl},\Omega)$, $\boldsymbol{H}(\mathrm{div},\Omega)$ are imposed by physics, and so are the corresponding spaces of boundary traces:

* The work has been supported by Air Force under Contract F49620-98-1-0255.

$H^{\frac{1}{2}}(\partial\Omega), H^{-\frac{1}{2}}(\partial\Omega), \boldsymbol{H}^{-\frac{1}{2}}(\mathrm{curl}, \partial\Omega)$. The nonlocality of norms and the breakdown of Trace Theorem and so-called localization results for the half-integers, are the source of notorious technical difficulties. The use of fractional spaces H^s and a careful monitoring of equivalence constants allows for alleviating most of these difficulties as we shall present it in the text.

Throughout the paper, Ω will denote a single element, interval in 1D, a polygon in 2D, or a polyhedron in 3D. The domains fall into the general category of Lipshitz domains covered by McLean for scalar-valued functions. For details concerning the vector-valued spaces, we refer to the work of Buffa and Ciarlet [12].

We use the higher-order Sobolev spaces to express regularity of projected and interpolated functions. It has been well established that this is a wrong choice for elliptic or Maxwell problems formulated in polyhedral domains or/and material interfaces. Most of the research on *hp* methods and exponential convergence is based on the notion of countably normed Besov spaces introduced by Babuška and Guo, see e.g. [40]. It is for that reason that we always try to estimate the interpolation errors with the corresponding best approximation errors. The last step of the interpolation error estimation resulting in optimal *p*- or *hp*-convergence rates reduces then to the best approximation results using more sophisticated means to access the regularity of approximated functions.

Scope of the Presentation

The following four sections correspond to four lectures. In the first lecture we discuss the grad–curl–div exact sequence and review the known polynomial exact sequences corresponding to various finite elements, and the concept of parametric elements. This part is mostly algebraic, although some of the details and even the notation may be a little overwhelming for a first time reader of the material. The second lecture focuses on a seemingly trivial one-dimensional sequence. We proceed with an attention to details, and invoke already at this level the main arguments on Sobolev spaces. The third lecture covers the two-dimensional case. Finally the fourth lecture proceeds at a faster pace zooming through the three-dimensional case covered in Sect. 5, and discussing applications of the presented techniques to the analysis and approximation of time-harmonic Maxwell equations. We conclude with a short discussion of open problems. Contrary to the original contribution [20], this presentation "marches" from 1D to 3D problems.

2 Exact Polynomial Sequences

2.1 One-Dimensional Sequences

We begin our discussion with the simplest one-dimensional exact sequence.

$$\mathbb{R} \to H^s(I) \xrightarrow{\partial} H^{s-1}(I) \to \{0\}. \tag{1}$$

Here $s \geq 0$, both Sobolev spaces are defined on the unit interval $I = (0,1)$, and ∂ denotes the derivative operator. The space of real numbers $\mathbb{R}$ symbolizes the one-dimensional space of constant functions, and $\{0\}$ denotes the trivial space consisting of the zero function only. The first operator (not shown) is identity, and the last one is the trivial map setting all arguments to the zero vector. The notion of the exact sequence conveys in this case the non-so-trivial (in context of real s) message that the derivative operator is well-defined, it is a surjection, and that its null space consists of constants. Let H^s_{avg} denote the subspace of functions of zero average,

$$H^s_{avg}(I) = \{u \in H^s : \int_I u = 0\}. \tag{2}$$

The exact sequence property is a consequence of the following result.

Proposition 2.1. *The derivative operator is an isomorphism from $H^s_{avg}(I)$ onto $H^{s-1}(I)$.*

Proof. The proof is carried out in three steps.

- We first demonstrate that ∂ is well-defined. Recall first [32, p. 309] that there exists a continuous extension operator,

$$H^s(I) \ni u \to U \in H^s(\mathbb{R}). \tag{3}$$

Take an arbitrary $\phi \in \mathcal{D}(I)$. We have,

$$\begin{aligned} |< u', \phi >| &= |- < U, \phi' >| \\ &\leq \|U\|_{H^s(\mathbb{R})} \|\phi'\|_{H^{-s}(\mathbb{R})} \\ &\leq C\|u\|_{H^s(I)} \|\phi\|_{H^{1-s}(\mathbb{R})} \quad \text{(Exercise 2.1)} \\ &= C\|u\|_{H^s(I)} \|\phi\|_{\widetilde{H}^{1-s}(I)} . \end{aligned} \tag{4}$$

Recall the density of test functions in $\widetilde{H}^{1-s}(I)$ and the fact that $H^{s-1}(I)$ is isomorphic with the dual of $\widetilde{H}^{1-s}(I)$.
- Next we show injectivity. Let $u' = 0$. It is sufficient to show that,

$$< u, \phi >= \int_I u\phi \tag{5}$$

vanishes for all test functions with zero average. Indeed, an arbitrary test function can always be decomposed into a constant and a function with zero average,

$$\phi = c + \phi_0, \quad c = \int_I \phi, \quad \int_I \phi_0 = 0\,. \tag{6}$$

Then,

$$< u, \phi >= \int_I u(c + \phi_0) = c\int_I u + < u, \phi_0 >=< u, \phi_0 >, \tag{7}$$

since we have restricted the derivative operator to functions of zero average. Next,

$$\psi(x) = \int_0^x \phi_0(t)\,dt\,, \tag{8}$$

is also a test function and $\psi' = \phi_0$. Thus,

$$< u, \phi_0 >=< u, \psi' >= 0\,, \tag{9}$$

since $u' = 0$.

- We show surjectivity by constructing a continuous right inverse. For $s = 1$ we need to integrate simply the derivative. Let $v \in L^2(I)$. Define,

$$u(x) = \int_0^x v(t)\,dt, \quad u_0 = u - \int_I u\,. \tag{10}$$

Obviously, $u_0' = v$ and $\|u_0\|_{H^1(I)} \leq C\|v\|_{L^2(I)}$. For $s = 0$ we utilize the following characterization of space $H^{-1}(I)$ [32, 74].

$$H^{-1}(I) = \{v = u_1' + v_1 \;:\; u_1, v_1 \in L^2(I)\}\,, \tag{11}$$

with the norm defined by taking the infimum over all possible (nonunique) decompositions of v,

$$\|v\|_{H^{-1}(I)} = \inf_{u_1,v_1} \left(\|u_1\|_{L^2(I)} + \|v_1\|_{L^2(I)}\right)\,. \tag{12}$$

We can define then the right-inverse by setting,

$$u = u_1 + \int_0^x v_1, \quad u_0 = u - \int_I u\,. \tag{13}$$

Again, u_0 depends continuously upon v in the right norms. By the interpolation argument, the right-inverse can be extended to $H^{s-1}(I)$, for an arbitrary $0 \leq s \leq 1$.

Exercise 2.1. Show that,

$$\|u'\|_{H^{s-1}(\mathbb{R})} \leq \|u\|_{H^s(\mathbb{R})} \quad \forall s \in \mathbb{R}. \tag{14}$$

We introduce now the corresponding polynomial exact sequence,

$$\mathbb{R} \rightarrow \mathcal{P}^p(I) \xrightarrow{\partial} \mathcal{P}^{p-1}(I) \rightarrow \{0\}\,, \tag{15}$$

where $\mathcal{P}^p(I)$ denotes the space of polynomials of order less or equal p, defined on unit interval I. In the next section, we shall study various projection operators P_i and projection-based interpolation operators Π_i that make the following diagram commute,

$$\begin{array}{ccccccc} \mathbb{R} \to & H^s & \xrightarrow{\partial} & H^{s-1} & \to \{0\} \\ \downarrow & P_1 \big\downarrow \varPi_1 & & P_2 \big\downarrow \varPi_2 & \downarrow \\ \mathbb{R} \to & \mathcal{P}^p(I) & \xrightarrow{\partial} & \mathcal{P}^{p-1}(I) & \to \{0\}. \end{array} \tag{16}$$

The projections operators P_i will always be defined on the whole spaces but interpolation operators $\varPi_i$ may be defined only on a subspace due to increased regularity requirements necessary to define e.g. function values at vertices, or average of a function over the integral. We shall also abbreviate the notation by dropping the constants and the trivial spaces, with the understanding however that all properties resulting from the presence of these spaces (surjectivity of ∂, $\mathcal{N}(\partial) = \mathbb{R}$, preservation of constants by $P_1, \varPi_1$) are satisfied.

$$\begin{array}{ccc} H^s & \xrightarrow{\partial} & H^{s-1} \\ P_1 \big\downarrow \varPi_1 & & P_2 \big\downarrow \varPi_2 \\ \mathcal{P}^p(I) & \xrightarrow{\partial} & \mathcal{P}^{p-1}(I). \end{array} \tag{17}$$

2.2 Two Dimensional Sequences

Let $\varOmega$ be a master triangle,

$$\varOmega = \{(x_1, x_2) \,:\, x_1 > 0,\ x_2 > 0,\ x_1 + x_2 < 1\} \tag{18}$$

or master square $\varOmega = (0,1)^2$. We shall study the following exact sequence,

$$\mathbb{R} \to H^s(\varOmega) \xrightarrow{\nabla} \boldsymbol{H}^{s-1}(\mathrm{curl}, \varOmega) \xrightarrow{\mathrm{curl}} H^{s-1}(\varOmega) \to \{0\} \tag{19}$$

where ∇ is the gradient operator, curl denotes the scalar-valued curl operator,

$$\mathrm{curl}\, \boldsymbol{E} = E_{1,2} - E_{2,1}\,, \tag{20}$$

and $\boldsymbol{H}^{s-1}(\mathrm{curl}, \varOmega)$ denotes the subspace of vector fields with both components in $H^{s-1}(\varOmega)$ such that the curl is in $H^{s-1}(\varOmega)$,

$$\boldsymbol{H}^{s-1}(\mathrm{curl}, \varOmega) = \{\boldsymbol{E} \in \boldsymbol{H}^{s-1}(\varOmega) \,:\, \mathrm{curl}\, \boldsymbol{E} \in H^{s-1}(\varOmega)\}\,. \tag{21}$$

We shall restrict ourselves to the range $s \geq \frac{1}{2}$. We will introduce in Sect. 4.3 a right-inverse of the curl operator demonstrating that the curl operator is a surjection.

Exercise 2.2. Follow the first step in the proof of Proposition 2.1 to prove that the gradient operator is well defined.

The two-dimensional exact sequence can be reproduced with several families of polynomials.

Nédélec's Triangle of the Second Type [35]

We have an obvious exact sequence,

$$\mathbb{R} \rightarrow \mathcal{P}^p \xrightarrow{\nabla} \boldsymbol{P}^{p-1} \xrightarrow{\text{curl}} \mathcal{P}^{p-2} \rightarrow \{0\}. \tag{22}$$

Here $\mathcal{P}^p$ denotes the space of polynomials of (group) order less or equal p, e.g. $x_1^2 x_2^3 \in \mathcal{P}^5$, and $\boldsymbol{P}^p = \mathcal{P}^p \times \mathcal{P}^p$. Obviously, the construction starts with $p \geq 2$, i.e. the $\boldsymbol{H}(\text{curl})$-conforming elements are at least of first order.

The construction can be generalized to triangles of *variable order*. With each triangle's edge we associate the corresponding edge order p_e. We assume that,

$$p_e \leq p \quad \text{for every edge } e.$$

We introduce now the following polynomial spaces:

- The space of scalar-valued polynomials u of order less or equal p, whose traces on edges e reduce to polynomials of (possibly smaller) order p_e,

$$\mathcal{P}^p_{p_e} = \{u \in \mathcal{P}^p \;:\; u|_e \in \mathcal{P}^{p_e}(e)\}.$$

- The space of vector-valued polynomials $\boldsymbol{E}$ of order less or equal p, whose *tangential traces* $E_t|_e$ on edges e reduce to polynomials of order p_e,

$$\boldsymbol{P}^p_{p_e} = \{\boldsymbol{E} \in \boldsymbol{P}^p \;:\; E_t|_e \in \mathcal{P}^{p_e}(e)\}.$$

- The space of scalar-valued polynomials of order less or equal p, with zero average

$$\mathcal{P}^p_{avg} = \{u \in \mathcal{P}^p \;:\; \int_T u = 0\}.$$

We have then the exact sequence,

$$\mathcal{P}^p_{p_e} \xrightarrow{\nabla} \boldsymbol{P}^{p-1}_{p_e-1} \xrightarrow{\nabla\times} \mathcal{P}^{p-2}(\mathcal{P}^{p-2}_{avg}) \,. \tag{23}$$

The case $p_e = -1$ corresponds to the homogeneous Dirichlet boundary condition. In the case of homogeneous Dirichlet boundary conditions imposed on *all* edges e, the last space in the sequence, corresponding to polynomials of order $p-2$, must be replaced with the space of polynomials with zero average.

Exercise 2.3. Prove that (23) is an exact sequence.

Nédélec's Rectangle of the First Type [34]

All spaces are defined on the unit square. We introduce the following polynomial spaces.

$$\begin{aligned} W_p &= \mathcal{Q}^{(p,q)}, \\ \boldsymbol{Q}_p &= \mathcal{Q}^{(p-1,q)} \times \mathcal{Q}^{(p,q-1)}, \\ Y_p &= \mathcal{Q}^{(p-1,q-1)}. \end{aligned} \tag{24}$$

Here, $\mathcal{Q}^{p,q} = \mathcal{P}^p \otimes \mathcal{P}^q$ denotes the space of polynomials of order less or equal p, q with respect to x, y, respectively. For instance, $2x^2y^3 \in \mathcal{Q}^{(2,3)}$. The polynomial spaces form again an exact sequence,

$$W_p \xrightarrow{\nabla} \boldsymbol{Q}_p \xrightarrow{\nabla\times} Y_p. \tag{25}$$

The generalization to variable order elements is a little less straightforward than for the triangles. For each horizontal edge e, we introduce order p_e, and with each vertical edge e, we associate order q_e. We assume again that the minimum rule holds, i.e.

$$p_e \le p, \quad q_e \le q. \tag{26}$$

By $\mathcal{Q}^{(p,q)}_{p_e,q_e}$ we understand the space of polynomials of order less or equal p with respect to x and order less or equal q with respect to y, such that their traces to horizontal edges e reduce to polynomials of (possibly smaller than p) degree p_e, and restrictions to vertical edges e reduce to polynomials of (possibly smaller than q) order q_e,

$$\begin{aligned} \mathcal{Q}^{(p,q)}_{p_e,q_e} = \{u \in \mathcal{Q}^{(p,q)} \;:\; & u(\cdot,0) \in \mathcal{P}^{p_1}(0,1), u(\cdot,1) \in \mathcal{P}^{p_2}(0,1), \\ & u(0,\cdot) \in \mathcal{P}^{q_1}(0,1), u(1,\cdot) \in \mathcal{P}^{q_2}(0,1)\}. \end{aligned} \tag{27}$$

With spaces

$$\begin{aligned} W_p &= \mathcal{Q}^{(p,q)}_{p_e,q_e}, \\ \boldsymbol{Q}_p &= \mathcal{Q}^{(p-1,q)}_{p_e-1} \times \mathcal{Q}^{(p,q-1)}_{q_e-1}, \\ Y_p &= \mathcal{Q}^{(p-1,q-1)}, \end{aligned} \tag{28}$$

we have the exact sequence,

$$W_p \xrightarrow{\nabla} \boldsymbol{Q}_p \xrightarrow{\nabla\times} Y_p. \tag{29}$$

Notice that space $\boldsymbol{Q}_p$ cannot be obtained by merely differentiating polynomials from $\mathcal{Q}^{(p,q)}_{p_e,q_e}$. For the derivative in x, this would lead to space $\mathcal{Q}^{(p-1,q)}_{p_e-1,q_e}$ for the first component, whereas in our definition above q_e has been increased to q. This is motivated by the fact that the traces of E_1 along the *vertical edges* are interpreted as *normal* components of the $\boldsymbol{E}$ field. The $\boldsymbol{H}(\text{curl})$-conforming fields "connect" only through tangential components and, therefore, shape functions corresponding to the normal components on the boundary are classified as interior modes, and they should depend only on the order of the element and not on the order of neighboring elements.

Exercise 2.4. Prove that (29) is an exact sequence.

Nédélec's Triangle of the First Type [34]

There is a significant difference between the triangular and square elements presented so far. For the triangle, the order p drops upon differentiation from p to $p-2$, see the exact sequence (22). This merely reflects the fact that differentiation always lowers the polynomial order by 1. In the case of the rectangular element and the Q-spaces, however, the order in the diagram has dropped only by 1, from (p, q) to $(p-1, q-1)$, cf. exact sequence (25). A similar effect can be obtained for triangles. We shall discuss the concept within the general context of the variable order element.

The goal is to switch from $p-2$ to $p-1$ in the last space in sequence (23) *without* increasing the order p in the first space in the sequence. We begin by rewriting (23) with p increased by 1.

$$\mathcal{P}^{p+1}_{p_e} \xrightarrow{\nabla} \mathbf{P}^{p}_{p_e-1} \xrightarrow{\nabla\times} \mathcal{P}^{p-1} . \tag{30}$$

Notice that we have not increased the order along the edges. This is motivated with the fact that the edge orders do not affect the very last space in the diagram.[2] Next, we decompose the space of potentials into the previous space of polynomials $\mathcal{P}^p_{p_e}$ and *an algebraic complement* $\widetilde{\mathcal{P}}^{p+1}_{p_e}$,

$$\mathcal{P}^{p+1}_{p_e} = \mathcal{P}^{p}_{p_e} \oplus \widetilde{\mathcal{P}}^{p+1}_{p_e} . \tag{31}$$

The algebraic complement is *not unique*, it may be constructed in (infinitely) many different ways. The decomposition in the space of potentials implies a corresponding decomposition in the $\boldsymbol{H}$(curl)-conforming space,

$$\boldsymbol{P}^{p}_{p_e-1} = \boldsymbol{P}^{p-1}_{p_e-1} \oplus \nabla(\widetilde{\mathcal{P}}^{p+1}_{p_e}) \oplus \widetilde{\boldsymbol{P}}^{p}_{p_e-1} . \tag{32}$$

The algebraic complement $\widetilde{\boldsymbol{P}}^{p}_{p_e-1}$ is again *not unique*. The desired extension of the original sequence can now be constructed by removing the gradients of order $p+1$,

$$\mathcal{P}^{p}_{p_e} \xrightarrow{\nabla} \mathbf{P}^{p-1}_{p_e-1} \oplus \widetilde{\boldsymbol{P}}^{p}_{p_e-1} \xrightarrow{\nabla\times} \mathcal{P}^{p-1} . \tag{33}$$

Exercise 2.5. Prove that (33) is an exact sequence.

Note the following facts:

- The modified sequence (33) enables the $\boldsymbol{H}$(curl)-conforming discretization of lowest order on triangles. For $p = p_e = 1$,

[2] Except for the case of the homogeneous Dirichlet boundary condition imposed on the whole boundary which forces the use of polynomials of zero average for the last space in the diagram.

$$\mathbf{P}_0^0 \oplus \widetilde{\boldsymbol{P}}_0^1 = \boldsymbol{P}_0^1, \quad \dim \boldsymbol{P}_0^1 = 3\,. \tag{34}$$

The complement $\widetilde{\mathcal{P}}_1^2$ *is empty* and, therefore, in this case, the resulting space $\boldsymbol{P}_0^1 = \mathbf{P}_0^0 \oplus \widetilde{\boldsymbol{P}}_0^1$, corresponding to the famous construction of Whitney [42], *is unique*. This is the smallest space to enforce the continuity of the (constant) tangential component of $\boldsymbol{E}$ across the interelement boundaries.

- It is not necessary but natural to construct the complements using spans of scalar and vector *bubble functions*. In this case the notation $\widetilde{\mathcal{P}}_{-1}^{p+1}$ and $\widetilde{\boldsymbol{P}}_{-1}^{p}$ is more appropriate. The concept is especially natural if one uses *hierarchical shape functions*. We can always enforce the zero trace condition by augmenting original shape functions with functions of lower order. In other words, we change the complement but *do not alter* the ultimate polynomial space.

The choice of the complements may be made unique by imposing additional conditions. Nédélec's original construction for elements of uniform order p employs skewsymmetric polynomials,

$$\boldsymbol{R}^p = \{\boldsymbol{E} \in \boldsymbol{P}^p \,:\, \epsilon^p(\boldsymbol{E}) = 0\}\,, \tag{35}$$

where ϵ^p is the Nédélec *symmetrization operator*,

$$\begin{aligned}(\epsilon^p(\boldsymbol{E}))_{i_1,\ldots,i_{p+1}} = \frac{1}{p+1}\Bigg(& \frac{\partial^p E_{i_1}}{\partial x_{i_2}\ldots\partial x_{i_p}\partial x_{i_{p+1}}} \\ & + \frac{\partial^p E_{i_2}}{\partial x_{i_3}\ldots\partial x_{i_{p+1}}\partial x_{i_1}} + \cdots + \frac{\partial^p E_{i_{p+1}}}{\partial x_{i_1}\ldots\partial x_{i_{p-1}}\partial x_{i_p}}\Bigg)\,.\end{aligned} \tag{36}$$

The algebraic complement can then be selected as the subspace of *homogeneous*[3] symmetric polynomials $\boldsymbol{D}^p$,

$$\boldsymbol{R}^p = \boldsymbol{P}^p \oplus \boldsymbol{D}^p\,. \tag{37}$$

There are many equivalent conditions characterizing the space $\boldsymbol{D}^p$. The most popular one reads as follows

$$\boldsymbol{E} \in \boldsymbol{D}^p \Leftrightarrow \boldsymbol{E} \text{ is homogeneous and } \boldsymbol{x}\cdot\boldsymbol{E}(\boldsymbol{x}) = 0 \;\; \forall \boldsymbol{x}\,. \tag{38}$$

The space $\boldsymbol{D}^p$ can also nicely be characterized as the image of homogeneous polynomials of order $p-1$ under the Poincaré map, see [29, 30],

$$\begin{aligned} E_1(\boldsymbol{x}) &= -x_2 \int_0^1 t\psi(t\boldsymbol{x})\,dt \\ E_2(\boldsymbol{x}) &= x_1 \int_0^1 t\psi(t\boldsymbol{x})\,dt\,. \end{aligned} \tag{39}$$

[3] A polynomial of order p is homogeneous if it can be represented as a sum of monomials of order p. Equivalently, $u(\xi x_1,\ldots,\xi x_n) = \xi^p u(x_1,\ldots,x_n)$.

The Poincaré map is a right inverse of the curl map, $\nabla \times \boldsymbol{E} = \psi$, for the $\boldsymbol{E}$ defined above. Consistently with our discussion, it can be shown that the tangential trace of a symmetric polynomial of order p is always a polynomial of order less or equal $p-1$. For other characterizations of the space $\boldsymbol{D}^p$, see [26]. An important property of the Nédélec space $\boldsymbol{R}^p$ is that it is invariant under affine transformations, cf. Exercise 2.6. Consequently, the polynomial space is independent of the way in which the vertices of the triangle are enumerated.

Exercise 2.6. Prove that Nédélec's space is affine invariant. More precisely, let $\boldsymbol{x} \rightarrow \boldsymbol{y} = \boldsymbol{B}\boldsymbol{x} + \boldsymbol{b}$ denote a nonsingular affine map from $\mathbb{R}^n$ into itself. Let $\hat{\boldsymbol{E}} = \hat{\boldsymbol{E}}(\boldsymbol{x})$ be a symmetric polynomial of order p, i.e.,

$$\boldsymbol{x} \cdot \hat{\boldsymbol{E}}(\boldsymbol{x}) = 0, \quad \forall \boldsymbol{x}\,. \tag{40}$$

Define,

$$E_i(\boldsymbol{y}) = \sum_j \hat{E}_j(\boldsymbol{x}) \frac{\partial x_j}{\partial y_i}, \quad \boldsymbol{y} = \boldsymbol{B}\boldsymbol{x} + \boldsymbol{b}\,. \tag{41}$$

Show that

$$\boldsymbol{y} \cdot \boldsymbol{E}(\boldsymbol{y}) = 0, \quad \forall \boldsymbol{y}\,. \tag{42}$$

Uniqueness of the spaces could also be naturally enforced by requesting *orthogonality* of algebraic complements [41, 16],

$$\begin{aligned} &\mathcal{P}^{p+1}_{p_e} = \mathcal{P}^p_{p_e} \oplus \tilde{\mathcal{P}}^{p+1}_{-1}\,, \quad \mathcal{P}^{p+1}_{-1} = \mathcal{P}^p_{-1} \overset{\perp}{\oplus} \tilde{\mathcal{P}}^{p+1}_{-1} \\ &\boldsymbol{P}^p_{p_e-1} = \boldsymbol{P}^{p-1}_{p_e-1} \oplus \nabla(\tilde{\mathcal{P}}^{p+1}_{p_e}) \oplus \tilde{\boldsymbol{P}}^p_{-1}, \quad \boldsymbol{P}^p_{-1} = \boldsymbol{P}^{p-1}_{-1} \overset{\perp}{\oplus} \nabla(\tilde{\mathcal{P}}^{p+1}_{-1}) \overset{\perp}{\oplus} \tilde{\boldsymbol{P}}^p_{-1}\,. \end{aligned} \tag{43}$$

The orthogonality for the scalar-valued and the vector-valued polynomial spaces is usually understood in the sense of H^1_0 and $\boldsymbol{H}(\mathrm{curl})$ scalar products, respectively.

Parametric Elements

The concept of an exact sequence of discrete (finite-dimensional) spaces goes beyond polynomial spaces. Study of the construction of the parametric element and the corresponding exact sequence is not only necessary for dealing with curved elements but it enhances essentially the understanding of the polynomial spaces, e.g. the concept of affine and "rotational" invariance. We will discuss the notion of parametric elements after we present the 3D exact polynomial sequences.

Nédélec Tetrahedron of the Second Type [35]

All polynomial spaces are defined on the master tetrahedron,

$$\Omega = \{(x_1, x_2, x_3) \, : \, x_1 > 0, x_2 > 0, x_3 > 0, x_1 + x_2 + x_3 < 1\}\,. \tag{44}$$

We have the following exact sequence,

$$\mathcal{P}^p \xrightarrow{\nabla} \boldsymbol{P}^{p-1} \xrightarrow{\nabla\times} \boldsymbol{P}^{p-2} \xrightarrow{\nabla\circ} \mathcal{P}^{p-3}. \tag{45}$$

Here $\mathcal{P}^p$ denotes the space of polynomials of (group) order less or equal p, e.g. $x_1^2x_2^3x_3^2 \in \mathcal{P}^7$, and $\boldsymbol{P}^p = \mathcal{P}^p \times \mathcal{P}^p \times \mathcal{P}^p$. Obviously, the construction starts with $p \geq 3$, i.e. the $\boldsymbol{H}(\text{curl})$-conforming elements are at least of second order.

The construction can be generalized to tetrahedra of *variable order*. With each tetrahedron's face we associate the corresponding face order p_f, and with each tetrahedron's edge, we associate the corresponding edge order p_e. We assume that,

$$p_f \leq p \,\forall \text{ face } f, \quad p_e \leq p_f \forall \text{ face } f \text{ adjacent to edge } e, \;\forall \text{ edge } e\,. \tag{46}$$

The assumption is satisfied in practice by enforcing the *minimum rule*, i.e. setting the face and edge orders to the minimum of the orders of the adjacent elements. We introduce now the following polynomial spaces.

- The space of scalar-valued polynomials of order less or equal p, whose traces on faces f reduce to polynomials of (possibly smaller) order p_f, and whose traces on edges e reduce to polynomials of (possibly smaller) order p_e,

$$\mathcal{P}^p_{p_f,p_e} = \{u \in \mathcal{P}^p \;:\; u|_f \in \mathcal{P}^{p_f}(f),\, u|_e \in \mathcal{P}^{p_e}(e)\}. \tag{47}$$

- The space of vector-valued polynomials of order less or equal p, whose *tangential traces* on faces f reduce to polynomials of order p_f, and whose *tangential traces* on edges e reduce to polynomials of order p_e,

$$\boldsymbol{P}^p_{p_f,p_e} = \{\boldsymbol{E} \in \boldsymbol{P}^p \;:\; \boldsymbol{E}_t|_f \in \boldsymbol{P}^{p_f}(f),\, E_t|_e \in \mathcal{P}^{p_e}(e)\}. \tag{48}$$

- The space of vector-valued polynomials of order less or equal p, whose *normal traces* on faces f reduce to polynomials of order p_f

$$\boldsymbol{P}^p_{p_f} = \{\boldsymbol{E} \in \boldsymbol{P}^p \;:\; E_n|_f \in \mathcal{P}^{p_f}(f)\}. \tag{49}$$

We have then the exact sequence,

$$\mathcal{P}^p_{p_f,p_e} \xrightarrow{\nabla} \boldsymbol{P}^{p-1}_{p_f-1,p_e-1} \xrightarrow{\nabla\times} \boldsymbol{P}^{p-2}_{p_f-2} \xrightarrow{\nabla\circ} \mathcal{P}^{p-3}\,, \tag{50}$$

The case $p_f, p_e = -1$ corresponds to the homogeneous Dirichlet boundary condition.

Nédélec's Hexahedron of the First Type [34]

All polynomial spaces are defined on a unit cube. We introduce the following polynomial spaces.

$$
\begin{aligned}
W_p &= Q^{(p,q,r)} \\
\boldsymbol{Q}_p &= Q^{(p-1,q,r)} \times Q^{(p,q-1,r)} \times Q^{(p,q,r-1)} \\
\boldsymbol{V}_p &= Q^{(p,q-1,r-1)} \times Q^{(p-1,q,r-1)} \times Q^{(p-1,q-1,r)} \\
Y_p &= Q^{(p-1,q-1,r-1)} .
\end{aligned}
\tag{51}
$$

Here $Q^{p,q,r} = \mathcal{P}^p \otimes \mathcal{P}^q \otimes \mathcal{P}^r$ denotes the space of polynomials of order less or equal p, q, r with respect to x, y, z, respectively. For instance, $2x^2y^3 + 3x^3z^8 \in Q^{(3,3,8)}$. The polynomial spaces form again the exact sequence,

$$
W_p \xrightarrow{\nabla} \boldsymbol{Q}_p \xrightarrow{\nabla\times} \boldsymbol{V}_p \xrightarrow{\nabla\circ} Y_p . \tag{52}
$$

The generalization to variable order elements is a little less straightforward than for the tetrahedra. Review the 2D construction first. In three dimensions, spaces get more complicated and notation more cumbersome. We start with the space,

$$
Q^{(p,q,r)}_{(p_f,q_f),(p_f,r_f),(q_f,r_f),p_e,q_e,r_e} , \tag{53}
$$

that consists of polynomials in $Q^{(p,q,r)}$ such that:

- their restrictions to faces f parallel to axes x, y reduce to polynomials in $Q^{(p_f,q_f)}$,
- their restrictions to faces f parallel to axes x, z reduce to polynomials in $Q^{(p_f,r_f)}$,
- their restrictions to faces f parallel to axes y, z reduce to polynomials in $Q^{(q_f,r_f)}$,
- their restriction to edges parallel to axis x, y, z reduce to polynomials of order p_e, q_e, r_e respectively,

with the minimum rule restrictions:

$$
p_f \le p, q_f \le q, r_f \le r, \quad p_e \le p_f, q_e \le q_f, r_e \le r_f, \text{ for adjacent faces } f . \tag{54}
$$

The 3D polynomial spaces forming the de Rham diagram, are now introduced as follows,

$$
\begin{aligned}
W_p &= Q^{(p,q,r)}_{(p_f,q_f),(p_f,r_f),(q_f,r_f),p_e,q_e,r_e} \\
\boldsymbol{Q}_p &= Q^{(p-1,q,r)}_{(p_f-1,q_f),(p_f-1,r_f),p_e-1,q_f,r_f} \times Q^{(p,q-1,r)}_{(p_f,q_f-1),(q_f-1,r_f),p_f,q_e-1,r_f} \\
&\quad \times Q^{(p,q,r-1)}_{(p_f,r_f-1),(q_f,r_f-1),p_f,q_f,r_e-1} \\
\boldsymbol{V}^p &= Q^{(p,q-1,r-1)}_{(q_f-1,r_f-1)} \times Q^{(p-1,q,r-1)}_{(p_f-1,r_f-1)} \times Q^{(p-1,q-1,r)}_{(p_f-1,q_f-1)} \\
Y_p &= Q^{(p-1,q-1,r-1)} .
\end{aligned}
\tag{55}
$$

Note the following points:

- There is no restriction on edge order in the $\boldsymbol{H}(\mathrm{div})$-conforming space. The only order restriction is placed on faces normal to the particular component, e.g. for the first component H_1, the order restriction is imposed only on faces parallel to y, z faces.

- For the $\boldsymbol{H}(\mathrm{curl})$-conforming space, there is no restriction on face order for faces perpendicular to the particular component. For instance, for E_1, there is no order restriction on faces parallel to y, z axes. The edge orders for edges perpendicular to x are inherited from faces *parallel* to the x axis. This is related to the fact that elements connecting through the first component E_1, connect only through faces and edges parallel to the first axis only.

Exercise 2.7. Prove that the spaces defined above form an exact sequence.

Nédélec Tetrahedron of the First Type [34]

Review the construction of the corresponding triangular element first. The 3D construction goes along the same lines but it becomes more technical. We discuss the element of variable order. The following decompositions are relevant.

$$\begin{aligned}
\mathcal{P}^{p+1}_{p_e,p_f+1} &= \mathcal{P}^{p}_{p_e,p_f} \oplus \widetilde{\mathcal{P}}^{p+1}_{-1,p_f+1} \\
\boldsymbol{P}^{p}_{p_e-1,p_f} &= \boldsymbol{P}^{p-1}_{p_e-1,p_f-1} \oplus \nabla(\widetilde{\mathcal{P}}^{p+1}_{-1,p_f+1}) \oplus \widetilde{\boldsymbol{P}}^{p}_{-1,p_f} \\
\boldsymbol{P}^{p}_{p_f} &= \boldsymbol{P}^{p-1}_{p_f-1} \oplus \nabla(\widetilde{\boldsymbol{P}}^{p+1}_{-1,p_f+1}) \oplus \widetilde{\boldsymbol{P}}^{p}_{-1}
\end{aligned} \tag{56}$$

The ultimate sequence looks as follows:

$$\mathcal{P}^{p}_{p_e,p_f} \xrightarrow{\nabla} \boldsymbol{P}^{p-1}_{p_e-1,p_f-1} \oplus \widetilde{\boldsymbol{P}}^{p}_{-1,p_f} \xrightarrow{\nabla\times} \boldsymbol{P}^{p-1}_{p_f-1} \oplus \widetilde{\boldsymbol{P}}^{p}_{-1} \xrightarrow{\nabla\circ} \mathcal{P}^{p-1} \,. \tag{57}$$

Referring to [41, 16] for details, we emphasize only that switching to the tetrahedra of the first type in 3D, requires adding not only extra interior bubbles but face bubbles as well. The actual construction of Nédélec involves the choice of a special complement $\widetilde{\boldsymbol{P}}^{p}_{-1,p_f}$ consisting of skewsymmetric polynomials; all remarks on the 2D element, including a characterization using Poincaré maps, remain valid.

Exercise 2.8. Prove that the spaces defined above form an exact sequence.

Prismatic Elements

We shall not discuss here the construction of the exact sequences for the prismatic elements. The prismatic element shape functions are constructed as tensor products of triangular element and 1D element shape functions. We can use both Nédélec's triangles for the construction and, consequently, we can also produce two corresponding exact sequences.

Parametric Elements

Given a bijective map $\boldsymbol{x} = \boldsymbol{x}_{\Omega}(\boldsymbol{\xi})$ transforming master element $\hat{\Omega}$ onto a physical element Ω, and master element shape functions $\hat{\phi}(\boldsymbol{\xi})$, we define the H^1-conforming shape functions on the physical element in terms of master element coordinates,

$$\phi(\boldsymbol{x}) = \hat{\phi}(\boldsymbol{\xi}) = \hat{\phi}(\boldsymbol{x}_{\Omega}^{-1}(\boldsymbol{x})) = (\hat{\phi} \circ \boldsymbol{x}_{\Omega}^{-1})(\boldsymbol{x}) \,. \tag{58}$$

The definition reflects the fact that the integration of master element matrices is always done in terms of master element coordinates and, therefore, it is simply convenient to define the shape functions in terms of master coordinates $\boldsymbol{\xi}$. This implies that the parametric element shape functions are compositions of the inverse $\boldsymbol{x}_\Omega^{-1}$ and the master element polynomial shape functions. In general, we do not deal with polynomials anymore. In order to keep the exact sequence property, we have to define the $\boldsymbol{H}(\text{curl})$-, $\boldsymbol{H}(\text{div})$-, and L^2-conforming elements consistently with the way the differential operators transform. For gradients we have,

$$\frac{\partial u}{\partial x_i} = \frac{\partial \hat{u}}{\partial \xi_k}\frac{\partial \xi_k}{\partial x_i} \tag{59}$$

and, therefore,

$$E_i = \hat{E}_k \frac{\partial \xi_k}{\partial x_i}\,. \tag{60}$$

For the curl operator we have,

$$\begin{aligned}\epsilon_{ijk}\frac{\partial E_k}{\partial x_j} &= \epsilon_{ijk}\frac{\partial}{\partial x_j}\left(\hat{E}_l \frac{\partial \xi_l}{\partial x_k}\right)\\ &= \epsilon_{ijk}\frac{\partial \hat{E}_l}{\partial x_j}\frac{\partial \xi_l}{\partial x_k} + \hat{E}_l\,\underbrace{\epsilon_{ijk}\frac{\partial^2 \xi_l}{\partial x_k \partial x_j}}_{=0} = \epsilon_{ijk}\frac{\partial \hat{E}_l}{\partial \xi_m}\frac{\partial \xi_m}{\partial x_j}\frac{\partial \xi_l}{\partial x_k}\,.\end{aligned} \tag{61}$$

But,

$$\epsilon_{ijk}\frac{\partial \xi_m}{\partial x_j}\frac{\partial \xi_l}{\partial x_k} = J^{-1}\epsilon_{nml}\frac{\partial x_i}{\partial \xi_n}\,, \tag{62}$$

where J^{-1} is the inverse jacobian. Consequently,

$$\epsilon_{ijk}\frac{\partial E_k}{\partial x_j} = J^{-1}\frac{\partial x_i}{\partial \xi_n}\left(\epsilon_{nml}\frac{\partial \hat{E}_l}{\partial \xi_m}\right). \tag{63}$$

This leads to the definition of the $\boldsymbol{H}(\text{div})$-conforming parametric element,

$$H_i = J^{-1}\frac{\partial x_i}{\partial \xi_n}\hat{H}_n. \tag{64}$$

Finally,

$$\frac{\partial H_i}{\partial x_i} = \underbrace{\frac{\partial}{\partial x_i}\left(J^{-1}\frac{\partial x_i}{\partial \xi_k}\right)}_{=0}\hat{H}_k + J^{-1}\frac{\partial x_i}{\partial \xi_k}\frac{\partial \hat{H}_k}{\partial \xi_l}\frac{\partial \xi_l}{\partial x_i} = J^{-1}\frac{\partial \hat{H}_k}{\partial \xi_k}, \tag{65}$$

which establishes the transformation rule for the L^2-conforming elements,

$$f = J^{-1}\hat{f}\,. \tag{66}$$

Defining the parametric element spaces $W_p, \boldsymbol{Q}_p, \boldsymbol{V}_p, Y_p$ using the transformation rules listed above, we preserve for the parametric element the exact sequence (52).

In the case of the *isoparametric element*, the components of the transformation map $\boldsymbol{x}_\Omega$ come from the space of the H^1-conforming master element,

$$x_j = \sum_k x_{j,k}\hat{\phi}_k(\boldsymbol{\xi}) = \sum_k \boldsymbol{x}_k \phi_k(\boldsymbol{x}).$$

Here $x_{j,k}$ denote the (vector-valued) geometry degrees-of-freedom corresponding to element shape functions $\phi_k(\boldsymbol{x})$. By construction, therefore, the parametric element shape functions can reproduce any linear function $a_j x_j$. As they also can reproduce constants, the isoparametric element space of shape functions contains the space of all linear polynomials in $\boldsymbol{x}$ - $a_j x_j + b$, in mechanical terms - the space of linearized rigid body motions. The exact sequence property implies that the $\boldsymbol{H}(\mathbf{curl})$-conforming element can reproduce only constant fields, but the $\boldsymbol{H}(\mathrm{div})$-conforming element, in general, cannot reproduce even constants. This indicates in particular that, in context of general parametric (nonaffine) elements[4] unstructured mesh generators should be used with caution, cf. [2]. The critique does not apply to (algebraic) mesh generators based on a consistent representation of the domain as a manifold, with underlying global maps parametrizing portions of the domain. Upon a change of variables, the original problem can then be redefined in the reference domain discretized with affine elements, see [18] for more details.

3 Commuting Projections and Projection-Based Interpolation Operators in One Space Dimension

3.1 Commuting Projections: Projection Error Estimates

Let $I = (0,1)$. We consider the following diagram.

$$\begin{array}{ccccccc}
\mathbb{R} \longrightarrow & H^s(I) & \xrightarrow{\partial} & H^{s-1}(I) & \longrightarrow & \{0\} \\
\downarrow & \downarrow P_s^{\partial} & & \downarrow P_{s-1} & & \downarrow \\
\mathbb{R} \longrightarrow & \mathcal{P}^p & \xrightarrow{\partial} & \mathcal{P}^{p-1} & \longrightarrow & \{0\}.
\end{array} \tag{67}$$

Here P_{s-1} is the standard orthogonal projection in $H^{s-1}(I)$-norm, and the operator P_s^{∂} is defined as follows.

$$\begin{cases} P_s^{\partial} u =: u_p \in \mathcal{P}^p(I) \\ \|u_p' - u'\|_{H^{s-1}(I)} \to \min \\ (u_p - u, 1)_{H^s(I)} = 0. \end{cases} \tag{68}$$

We are interested in the range $0 \le s \le 1$.

[4] Note that general quadrilaterals or hexahedra with straight edges are not affine elements.

Exercise 3.1. Show that the diagram above commutes.

Finding the projection $P_s^\partial u$ can be interpreted as the solution of a constrained minimization problem leading to the following mixed formulation.

$$\begin{cases} P_s^\partial u =: u_p \in \mathcal{P}^p(I),\ \lambda \in \mathbb{R} \\ (u_p' - u', v')_{H^{s-1}(I)} \ +(\lambda, v)_{H^s(I)} = 0 \quad \forall v \in \mathcal{P}^p(I) \\ (u_p - u, \mu)_{H^s(I)} \qquad\qquad\qquad\quad = 0 \quad \forall \mu \in \mathbb{R}. \end{cases} \tag{69}$$

Here λ is a constant Lagrange multiplier. Substituting $v = \lambda$ in the first equation, we learn that the Lagrange multiplier must be zero. By the Brezzi's theory [11], estimation of the projection error involves the satisfaction of two inf–sup conditions:

- the inf–sup condition relating the space of solutions $\mathcal{P}^p$ and the multiplier space $\mathbb{R}$,

$$\sup_{v \in \mathcal{P}^p} \frac{|(\lambda, v)_{H^s(I)}|}{\|v\|_{H^s(I)}} \geq \beta |\lambda|, \quad \forall \lambda \in \mathbb{R}\,, \tag{70}$$

- the inf–sup in kernel condition,

$$\sup_{v \in \mathcal{P}^p_{avg}} \frac{|(u', v')_{H^{s-1}(I)}|}{\|v\|_{H^s(I)}} \geq \alpha \|u\|_{H^s(I)}, \quad \forall u \in \mathcal{P}^p_{avg}\,. \tag{71}$$

Notice that (Exercise 3.2),

$$(u, 1)_{H^s(I)} = (u, 1)_{L^2(I)} = \int_I u\,. \tag{72}$$

The first inf–sup condition is a direct consequence of the discrete exact sequence property, i.e. the fact that constants are reproduced by the polynomials, and that (cf. (72)),

$$\|1\|_{H^s(I)} = 1\,. \tag{73}$$

The choice of $v = \lambda$ then gives $\beta = 1$. The second inf–sup condition is implied by a Poincaré-like inequality,

$$\|u\|_{H^s(I)} \leq C \|u'\|_{H^{s-1}(I)}, \quad \forall u \in H^s(I)\ :\ \int_I u = 0\,. \tag{74}$$

This follows immediately from Proposition 2.1. Notice that we actually need only a discrete version of the inequality but with a constant independent of p.

Exercise 3.2. Prove (72).

We can formulate now our projections errors estimate.

Theorem 3.1. *There exist constants C, independent[5] of p such that,*

[5] and s as well

$$\|u - P_s^\partial u\|_{H^s(I)} \le C \inf_{w\in\mathcal{P}^p} \|u-w\|_{H^s(I)} \le Cp^{-(r-s)}\|u\|_{H^r(I)},$$
$$\forall u \in H^r(I),$$
$$\|E - P_{s-1}E\|_{H^{s-1}(I)} = \inf_{F\in\mathcal{P}^{p-1}} \|E-F\|_{H^{s-1}(I)} \le C(p-1)^{-(r-s)}\|E\|_{H^{r-1}(I)},$$
$$\forall E \in H^{r-1}(I), \tag{75}$$

for $s < r$.

Proof. The proof of the first estimate follows immediately from Brezzi's theory, standard best approximation error estimates for polynomials [40, p. 75],

$$\begin{aligned}\inf_{w\in\mathcal{P}^p} \|u-w\|_{L^2(I)} &\le Cp^{-r}\|u\|_{H^r(I)}, \quad r \ge 1\\ \inf_{w\in\mathcal{P}^p} \|u-w\|_{H^1(I)} &\le Cp^{-(r-1)}\|u\|_{H^r(I)}, r \ge 1\end{aligned} \tag{76}$$

and an interpolation argument. We first interpolate with $0 \le s \le 1$ to obtain,

$$\inf_{w\in\mathcal{P}^p} \|u-w\|_{H^s(I)} \le Cp^{-(r-s)}\|u\|_{H^r(I)}, \quad r \ge 1 \tag{77}$$

and next with r in between s and the original r (possibly < 1), to get the final estimate. The second estimate follows e.g. from the first one and Proposition 2.1.

An Alternative Characterization of P_s^∂

Let K be the inverse of the derivative operator studied in Proposition 2.1. Operator K is continuous and polynomial preserving. Let P_{s-1} be the orthogonal projection in $H^{s-1}(I)$-norm onto polynomials $\mathcal{P}^{p-1}$, and let P_s^0 be the orthogonal projection in $H^s(I)$-norm[6] onto the null space of the derivative operator, i.e. the constants. Then,

$$P_s^\partial = KP_{s-1}\partial + P_s^0(I - KP_{s-1}\partial). \tag{78}$$

Consequently,

$$I - P_s^\partial = (I - P_s^0)(I - KP_{s-1}\partial), \tag{79}$$

and the corresponding error estimate follows simply from the continuity of the operator K. The characterization illuminates the role of the inverse operator K.

3.2 Commuting Interpolation Operators: Interpolation Error Estimates

The Range $\frac{1}{2} < s \le 1$

We consider the following diagram.

[6] Projection onto constants in H^s norm is equivalent to the L^2-projection.

$$
\begin{array}{ccccccc}
\mathbb{R} & \longrightarrow & H^s & \xrightarrow{\partial} & H^{s-1} & \longrightarrow & \{0\} \\
\Big\downarrow id & & \Big\downarrow \Pi_s^{\partial} & & \Big\downarrow \Pi_{s-1} & & \\
\mathbb{R} & \longrightarrow & \mathcal{P}^p & \xrightarrow{\nabla} & \mathcal{P}^{p-1} & \longrightarrow & \{0\}.
\end{array}
\tag{80}
$$

Here Π_s^{∂} and Π_{s-1} are the projection-based interpolation operators defined as follows.

$$
\begin{cases}
\Pi_s^{\partial} u =: u_p \in \mathcal{P}^p(I) \\
u_p = u \text{ at } 0, 1 \\
\|u_p' - u'\|_{H^{s-1}(I)} \to \min
\end{cases}
\tag{81}
$$

and,

$$
\begin{cases}
\Pi_{s-1} E =: E_{p-1} \in \mathcal{P}^{p-1}(I) \\
< E_{p-1} - E, 1 >= 0 \\
\|E_{p-1} - E\|_{H^{s-1}(I)} \to \min\,.
\end{cases}
\tag{82}
$$

We are restricting ourselves first to $\frac{1}{2} < s \leq 1$. The problem of finding the $\Pi^{\partial} u$-interpolant can again be interpreted as a constrained minimization problems that leads to the following variational characterization,

$$
\begin{cases}
\Pi_s^{\partial} u =: u_p \in \mathcal{P}^p(I) \\
u_p(0) = u(0), \quad u_p(1) = u(1), \\
(u_p' - u', v')_{H^{s-1}(I)}, \quad \forall v \in \mathcal{P}^p : v(0) = v(1) = 0.
\end{cases}
\tag{83}
$$

Thus, finding the value of the commuting projection operator reduces to the solution of a local Neumann problem, and finding the interpolant is equivalent to a local Dirichlet problem. Similarly, determining $\Pi_{s-1} E$ is equivalent to the variational problem,

$$
\begin{cases}
\Pi_{s-1} E =: E_{p-1} \in \mathcal{P}^{p-1}(I) \\
< E_{p-1} - E, 1 >= 0, \\
(E_{p-1} - E, v)_{H^{s-1}(I)}, \quad \forall v \in \mathcal{P}^{p-1} : \displaystyle\int_I v = 0.
\end{cases}
\tag{84}
$$

Proposition 3.1. *Diagram above commutes.*

Proof. First notice that the interpolation operator Π_s^{∂} is well defined and that it preserves constants, i.e. the first part of the diagram commutes. The operator Π_{s-1} is defined on distributions from H^{s-1} with range $-\frac{1}{2} < s - 1 \leq 0$. Constant function 1 belongs to the dual $\widetilde{H}_{1-s}$, $0 \leq 1 - s < \frac{1}{2}$, so the average value is well-defined, cf. Exercise 3.3. We need to show that for $u \in H^s(I), u(0) = u(1) = 0$,

$$< u', 1 >= 0\,. \tag{85}$$

Let $\phi_n \in \mathcal{D}(I)$ be a sequence of test functions converging to 1 in $\widetilde{H}_{1-s}$-norm. By Proposition 2.1 and the duality argument, the derivative operator ∂ is a continuous map from $\widetilde{H}^s \to \widetilde{H}^{s-1}$. Consequently $\phi_n' \to 0$ in $\widetilde{H}^{s-1}$-norm. Next, let $\psi_m \in \mathcal{D}(I)$ be a sequence of test functions converging to u in the H^s-norm. Integration by parts yields,

$$< \psi_m', \phi_n >= \int_I \psi_m' \phi_n = -\int_I \psi_m \phi_n' = - < \psi_m, \phi_n' > . \tag{86}$$

Passing to the limit with n and m, we get the required result. Finally, the orthogonality condition for the derivative implies that,

$$(u_p' - u', v)_{H^{s-1}(I)} = 0, \quad \forall v \in \mathcal{P}^{p-1} : \int_I v = 0. \tag{87}$$

This is a consequence of the fact that the range of the derivative operator restricted to polynomials of order p that vanish at the endpoints, coincides with polynomials of order $p-1$ with zero average.

Exercise 3.3. Let $u \in H^{-r}(I)$, $0 \le r < \frac{1}{2}$. Let $\phi_n \in \mathcal{D}(I)$ be a sequence of test functions converging to 1 in $\widetilde{H}_r$-norm, cf. [32, p. 77]. Prove that the limit,

$$\lim_{n\to\infty} < u, \phi_n > \tag{88}$$

exists, and it is independent of the choice of the sequence.

Lemma 3.1. *Linear extension,*

$$u(x) = u_0(1-x) + u_1 x \tag{89}$$

defines a continuous extension operator $Ext : \mathbb{R}^2 \ni (u_0, u_1) \to u \in H^s(I)$, *with a norm independent of s.*

Proof. The result follows from obvious cases for $s = 0$ and $s = 1$, and the interpolation argument.

Theorem 3.2. *There exist constants* C_1, C_2, *independent of p such that,*

$$\|u - \Pi_s^\partial u\|_{H^s(I)} \le C_1 \inf_{w \in \mathcal{P}^p} \|u - w\|_{H^s(I)} \le C_1 p^{-(r-s)} \|u\|_{H^r(I)}, \qquad \forall u \in H^r(I),$$

$$\|E - \Pi_{s-1} E\|_{H^{s-1}(I)} \le C_2 \inf_{F \in \mathcal{P}^{p-1}} \|E - F\|_{H^{s-1}(I)} \le C_2 (p-1)^{-(r-s)} \|E\|_{H^{r-1}(I)}, \qquad \forall E \in H^{r-1}(I), \tag{90}$$

for $\frac{1}{2} < s < r$. *With* $s = \frac{1}{2} + \epsilon$, *and* $\epsilon \to 0$, *constant* $C_1 = O(\epsilon^{-1})$ *and constant* $C_2 = O(\epsilon^{-\frac{1}{2}})$.

Proof. The main strategy to derive the interpolation error estimates is now to compare the interpolation errors with the projection errors. We begin with the triangle inequality,

$$\|u - \Pi^{\partial}_s u\|_{H^s(I)} \le \|u - P^{\partial}_s u\|_{H^s(I)} + \|P^{\partial}_s u - \Pi^{\partial}_s u\|_{H^s(I)}. \tag{91}$$

Polynomial $\psi = P^{\partial}_s u - \Pi^{\partial}_s u$ satisfies,

$$(\psi', \phi')_{H^{s-1}(I)} = 0, \quad \forall \phi \in \mathcal{P}^p \; : \; \phi(0) = \phi(1) = 0 \tag{92}$$

and, therefore, it is the discrete polynomial minimum energy extension with the energy defined by the $H^{s-1}(I)$-norm of derivative ψ'.

Moreover, for $s > \frac{1}{2}$, the derivative operator ∂ is an isomorphism mapping $H^s_0(I)$ onto the subspace $H^{s-1}_{avg}(I)$ of $H^{s-1}(I)$ consisting of distributions with zero average. Consequently, its inverse is continuous and we have,

$$\|u\|_{H^s(I)} \le C\|u'\|_{H^{s-1}(I)}, \quad \forall u \in H^s_0(I). \tag{93}$$

Exercise 3.4. Prove that the constant C in (93) is $C = O(\epsilon^{-\frac{1}{2}})$ for $s = \frac{1}{2} + \epsilon$.

Denote now the trace of ψ at the end-points of interval I by $tr\psi$. We have,

$$\begin{aligned}
&\|\psi'\|_{H^{s-1}(I)} \\
&\le \|(Ext\, tr\psi)'\|_{H^{s-1}(I)} && (\psi \text{ is the polynomial minimum seminorm extension}) \\
&\le \|Ext\, tr\psi\|_{H^s(I)} && (\text{def. of } H^s\text{-norm}) \\
&\le \|Ext\|\, |\Pi^{\partial}_s u - P^{\partial}_s u|_{\partial I} && (\text{continuity of the extension operator}) \\
&= \|Ext\|\, |u - P^{\partial}_s u|_{\partial I} && (\text{def. of } \Pi^{\partial}_s) \\
&\le C_{tr}\|Ext\|\, \|u - P^{\partial}_s u\|_{H^s(I)} && (\text{Trace Theorem})
\end{aligned} \tag{94}$$

Here C_{tr} is the norm of the trace operator of order $O(\epsilon^{-\frac{1}{2}})$ for $s = \frac{1}{2} + \epsilon$, cf. [32, p. 100].

Combining the estimate above with the 1D Poincaré inequality (93), and the triangle inequality (91), we see that the interpolation error is bounded by the projection error, and the result follows from Theorem 3.1. Notice that the $O(\epsilon^{-\frac{1}{2}})$ blow up comes from the trace constant and the other $O(\epsilon^{-\frac{1}{2}})$ from the constant in the Poincaré inequality. If we content ourselves with the estimate in terms of the seminorm only, i.e. measure the interpolation error,

$$\|(u - \Pi^{\partial}_s u)'\|_{H^{s-1}(I)} \le C_2\|u - P^{\partial}_s u\|_{H^s(I)}, \tag{95}$$

the blow up of constant C_2 is of order $O(\epsilon^{-\frac{1}{2}})$ only.

The estimate for operator Π_{s-1} follows now from the estimate above, and the commutativity argument. Let $E \in H^{r-1}(I)$ and let $u \in H^r(I)$ be the value of the inverse of the derivative operator defined in Proposition 2.1. Then,

$$
\begin{aligned}
\|E - \Pi_{s-1}E\|_{H^{s-1}(I)} &= \|(u - \Pi_s^\partial u)'\|_{H^{s-1}(I)} && \text{(commutativity argument)} \\
&\le C_2 \|u - P_s^\partial\|_{H^s(I)} && \text{(estimate (95))} \\
&\le C_2 \|(u - P_s^\partial)'\|_{H^{s-1}(I)} && \text{(Proposition 2.1)} \\
&\le C_2 \|E - P_{s-1}\|_{H^{s-1}(I)}. && \text{(commutativity argument)}
\end{aligned} \tag{96}
$$

Remark 3.1. An alternative proof of the projection estimates for the whole range $0 \le s \le 1$, and the interpolation estimates for $\frac{1}{2} < s \le 1$, follows from the standard argument for continuous, polynomial preserving operators. We have, e.g.

$$
\begin{aligned}
\|u - \Pi_s^\partial u\|_{H^s(I)} &= \|(u - \phi) - \Pi_s^\partial (u - \phi)\|_{H^s(I)} \qquad (\forall \phi \in \mathcal{P}^p(I)) \\
&\le \|I - \Pi_s^\partial\|_{\mathcal{L}(H^s, H^s)} \inf_{\phi \in \mathcal{P}^p(I)} \|u - \phi\|_{H^s(I)}.
\end{aligned} \tag{97}
$$

Our presentation reflects the strategy that we will use for the two- and the three-dimensional case, and it allows for the estimates of the blow up constants.

The Case $s = \frac{1}{2}$

Contrary to the commuting projection operators that exhibit the best approximation property for the whole range of $s \in [0, 1]$, the minimum regularity assumption for the functions being interpolated is $r > \frac{1}{2}$. This does not prohibit defining the projection-based interpolation operators for $s = \frac{1}{2}$. The corresponding interpolation errors, measured in $H^{\frac{1}{2}}$ and $H^{-\frac{1}{2}}$ norms are no longer bounded by the best approximation errors in the same norms. We do get, however, almost optimal p-estimates "polluted" with logarithmic terms only. Repeating argument from the proof of Theorem 3.2,

$$
\begin{aligned}
\|\psi'\|_{H^{-\frac{1}{2}}(I)} &\le \|(Ext\, tr\psi)'\|_{H^{-\frac{1}{2}}(I)} \\
&\le C\|Ext\| \, |\Pi_{\frac{1}{2}}^\partial u - P_{\frac{1}{2}}^\partial u|_{\partial I} \\
&= C\|Ext\| \, |u - P_{\frac{1}{2}}^\partial u|_{\partial I} \\
&\le C\epsilon^{-\frac{1}{2}} \|Ext\| \, \|u - P_{\frac{1}{2}}^\partial u\|_{H^{\frac{1}{2}+\epsilon}(I)}
\end{aligned} \tag{98}
$$

where $\epsilon > 0$. On the other side, it follows from inequality (93) and the inverse inequality for polynomials,

$$
|\psi|_{H^{s+\epsilon}} \le C p^{2\epsilon} |\psi|_{H^s} \tag{99}
$$

that,

$$
\|\psi\|_{H^{\frac{1}{2}}(I)} \le \|\psi\|_{H^{\frac{1}{2}+\epsilon}(I)} \le C\epsilon^{-\frac{1}{2}} \|\psi'\|_{H^{-\frac{1}{2}+\epsilon}(I)} \le C\epsilon^{-\frac{1}{2}} p^{2\epsilon} \|\psi'\|_{H^{-\frac{1}{2}}(I)}. \tag{100}
$$

Combining (98) with (100), triangle inequality (91), and the projection error estimates, we get,

$$\|u - \Pi^{\partial}_{\frac{1}{2}} u\|_{H^{\frac{1}{2}}(I)} \le C\epsilon^{-1} p^{3\epsilon} p^{-(r-\frac{1}{2})} \|u\|_{H^r(I)} \tag{101}$$

for $\frac{1}{2} + \epsilon \le r$. Choosing $\epsilon = 1/\ln p$, we have,

$$\ln(p^\epsilon) = \epsilon \ln p = 1, \quad \text{so } p^\epsilon = e \tag{102}$$

and

$$\epsilon^{-1} = \ln p. \tag{103}$$

An analogous argument holds for operator Π_{s-1}. We obtain the following result.

Theorem 3.3. *There exist constants C_1, C_2, independent of p such that,*

$$\begin{aligned}
\|u - \Pi^{\partial}_{\frac{1}{2}} u\|_{H^{\frac{1}{2}}(I)} &\le C_1 \ln p \, p^{-(r-\frac{1}{2})} \|u\|_{H^r(I)}, & \forall u \in H^r(I), \\
\|E - \Pi_{-\frac{1}{2}} E\|_{H^{-\frac{1}{2}}(I)} &\le C_2 (\ln(p-1))^{\frac{1}{2}} (p-1)^{-(r-\frac{1}{2})} \|E\|_{H^{r-1}(I)}, & \forall E \in H^{r-1}(I),
\end{aligned} \tag{104}$$

and $\frac{1}{2} < r$.

The Case $0 \le s < \frac{1}{2}$

The interpolation operators are defined on spaces $H^r(I), H^{r-1}(I)$ with $r > \frac{1}{2}$ but the projections are done in the weaker norms. Specifically, we will be interested later in the case $s = 0$, corresponding to interpolation on edges for the 3D case

$$\begin{array}{ccccccc}
\mathbb{R} & \longrightarrow & H^r(I) & \xrightarrow{\partial} & H^{r-1}(I) & \longrightarrow & \{0\} \\
\downarrow & & \downarrow \Pi^{\partial}_s & & \downarrow \Pi_{s-1} & & \downarrow \\
\mathbb{R} & \longrightarrow & \mathcal{P}^p & \xrightarrow{\nabla} & \mathcal{P}^{p-1} & \longrightarrow & \{0\}.
\end{array} \tag{105}$$

The commuting interpolation operators need to be redefined[7]

$$\begin{cases}
\Pi^{\partial}_s u =: u_p \in \mathcal{P}^p(I) \\
u_p = u \text{ at } 0, 1 \\
\|u_p - u\|_{H^s(I)} \to \min
\end{cases} \tag{106}$$

and

$$\begin{cases}
\Pi_{s-1} E =: E_p = E_1 + E_{2,p} \in \mathcal{P}^{p-1}(I), \; E_1 = const, \; E_2 \in \mathcal{P}^{p-1}_{avg}(0,1), \\
< E_1 - E, 1 >= 0, \\
\| \int_0^x (E - E_1) - \int_0^x E_{2,p} \|_{H^s(I)} \to \min .
\end{cases} \tag{107}$$

[7] Notice the use of projection in H^s-norm in the modified definition of operator Π^{∂}_s as opposed to the use of the H^s-seminorm in the original definition.

In other words, given a distribution E, we first compute its average, then introduce potential,

$$u(x) = \int_0^x (E - E_1) := < E - E_1, 1_{[0,x]} > , \tag{108}$$

project it in the H^s-norm onto polynomials vanishing at the endpoints to get u_p, and differentiate back the projection u_p to get contribution $E_{2,p}$. Notice that, for $s > \frac{1}{2}$ and $u \in H^s(I)$, the projections,

$$\|u - u_p\|_{H^s(I)} \text{ and } \|u' - u_p'\|_{H^{s-1}(I)} \tag{109}$$

are equivalent, but with the equivalence constant blowing up for $s = \frac{1}{2}$, due to the breakdown of inequality (93).

Exercise 3.5. Prove that the diagram commutes.

Theorem 3.4. *Let $0 \le s < \frac{1}{2} < r$. There exist constants C, independent of p such that,*

$$\begin{aligned} \|u - \Pi_s^{\partial} u\|_{H^s(I)} &\le Cp^{-(r-s)}\|u\|_{H^r(I)}, & \forall u \in H^r(I), \\ \|E - \Pi_{s-1} E\|_{H^{s-1}(I)} &\le C(p-1)^{-(r-s)}\|E\|_{H^{r-1}(I)}, & \forall E \in H^{r-1}(I). \end{aligned} \tag{110}$$

Proof. Let $\frac{1}{2} < \mu < r$. We start with the best approximation estimate,

$$\|u - u_p\|_{H^\mu(I)} = \inf_{w_p \in \mathcal{P}^p} \|u - w_p\|_{H^\mu(I)} \le Cp^{-(r-\mu)}\|u\|_{H^r(I)}. \tag{111}$$

Let w be the solution of the dual problem,

$$\begin{cases} w \in H^\mu(I) \\ (\delta u, w)_{H^\mu(I)} = (\delta u, g)_{L^2(I)}, \quad \forall \delta u \in H^\mu(I) \end{cases} \tag{112}$$

with $g = u - u_p$. We can show, cf. Exercise 3.6, that,

$$\|w\|_{H^{2\mu}(I)} \le C\|g\|_{L^2(I)}. \tag{113}$$

Setting $\delta u = u - u_p$, we introduce the best approximation $w_p \in \mathcal{P}^p(I)$ of w in the $H^\mu(I)$-norm, and apply the standard duality argument,

$$\begin{aligned} \|u - u_p\|^2_{L^2(I)} &= (u - u_p, w)_{H^\mu(I)} \\ &= (u - u_p, w - w_p)_{H^\mu(I)} \\ &\le \|u - u_p\|_{H^\mu(I)}\|w - w_p\|_{H^\mu(I)} \\ &\le Cp^{-(r-\mu)}\|u\|_{H^r(I)}\, p^{-\mu}\|w\|_{H^{2\mu}(I)} \\ &\le Cp^{-r}\|u\|_{H^r(I)}\|u - u_p\|_{L^2(I)} \end{aligned} \tag{114}$$

to conclude that,

$$\|u - u_p\|_{L^2(I)} \le Cp^{-r}\|u\|_{H^r(I)}. \tag{115}$$

Next we define a correction

$$v(x) = (u(0) - u_p(0))\phi_0(x) + (u(1) - u_p(1))\phi_1(x), \tag{116}$$

where

$$\begin{cases} \phi_0 \in \mathcal{P}^p, \quad \phi_0(0) = 1 \\ \|\phi_0\|_{L^2(I)} \to \min \end{cases} \tag{117}$$

with ϕ_1 defined analogously. It has been proved in [38, Lemma 4.1] that

$$\|v\|_{L^2(I)} \le Cp^{-1}\max\{|u(0) - u_p(0)|, |u(1) - u_p(1)|\}. \tag{118}$$

It follows from the Trace Theorem that

$$\begin{aligned} \|v\|_{L^2(I)} &\le C(\mu)p^{-1}\|u - u_p\|_{H^\mu(I)} \\ &\le C(\mu)p^{-1}p^{-(r-\mu)}\|u\|_{H^r(I)} \le Cp^{-r}\|u\|_{H^r(I)}. \end{aligned} \tag{119}$$

Applying the triangle inequality finishes the argument for $s = 0$,

$$\begin{aligned} \|u - \Pi_0^\partial u\|_{L^2(I)} &\le \|u - (u_p + v)\|_{L^2(I)} \\ &\le \|u - u_p\|_{L^2(I)} + \|v\|_{L^2(I)} \le Cp^{-r}\|u\|_{H^r(I)}. \end{aligned} \tag{120}$$

For $s > \frac{1}{2}$, the projections in H^s norm and seminorms are equivalent, and a corresponding estimate for $s > \frac{1}{2}$ follows from Theorem 3.2. We can interpolate now with s between 0 and any $s > \frac{1}{2}$ to conclude that,

$$\|u - \Pi_s^\partial u\|_{H^s(I)} \le Cp^{-(r-s)}\|u\|_{H^r(I)}. \tag{121}$$

The corresponding estimate for the Π_{s-1} follows now from the commutativity argument, see proof of Theorem 3.2.

Remark 3.2. An alternative strategy would be to keep the $H^{\frac{1}{2}}$ interpolation operators for all values of s. The duality argument implies then the optimal p error estimates as well. Thus, in one space dimension, we have at least two alternative families of commuting interpolation operators that yield optimal p-estimates.

Exercise 3.6. Prove the regularity result (113). *Hint:* Use the Fourier series representation of fractional spaces.

3.3 Localization Argument

In the next sections, we will need to estimate the 1D interpolation errors over the boundary of a 2D element[8]. We will need the following fundamental result.

Proposition 3.2. *Let $I = (-1,1)$, $I_1 = (-1,0)$, $I_2 = (0,1)$. Let $0 \leq s \leq 1$, $s \neq \frac{1}{2}$. There exists a constant $C > 0$ such that,*

$$\|u\|_{H^s(I)} \leq C\left(\|u\|_{H^s(I_1)} + \|u\|_{H^s(I_2)}\right), \ \forall u \in H^s(I). \tag{122}$$

Here, $u|_{I_i} \in H^s(I_i)$, $i = 1,2$, and by $\|u\|_{H^s(I_i)}$ we understand the norm of the restriction of function u in $H^s(I_i)$. Moreover, for $s = \frac{1}{2} + \epsilon$, or $s = \frac{1}{2} - \epsilon$, $C = O(\epsilon^{-1})$.

Proof. See [28, p. 29–30] or [17].

The result enables estimating the interpolation error on the boundary of a 2D element. Let $\partial\Omega$ denote the boundary of a 2D polygon Ω, composed of edges e. Let $0 \leq s \leq 1, s \neq \frac{1}{2}, \quad r > s$. We have

$$\begin{aligned} \|u - \Pi^{\partial}_s u\|_{H^s(\partial\Omega)} &\leq C\sum_e \|u - \Pi^{\partial}_s u\|_{H^s(e)} \\ &\leq Cp^{-(r-s)}\sum_e \|u\|_{H^r(e)} \\ &\leq Cp^{-(r-s)}\|u\|_{H^r(\partial\Omega)}. \end{aligned} \tag{123}$$

Here p denotes the order of approximation on the element boundary. For an element of variable order, p is the minimum order for all edges, $p = \min_e p_e$.

For $s = \frac{1}{2}$, we need to utilize the information about the blow up of constant $C = c(s)$ with $s \to \frac{1}{2}$. We have,

$$\|u - \Pi^{\partial}_{\frac{1}{2}} u\|_{H^{\frac{1}{2}}(\partial\Omega)} \leq \|u - \Pi^{\partial}_{\frac{1}{2}} u\|_{H^{\frac{1}{2}+\epsilon}(\partial\Omega)} \leq C\epsilon^{-1}\sum_e \|u - \Pi^{\partial}_{\frac{1}{2}} u\|_{H^{\frac{1}{2}+\epsilon}(e)}. \tag{124}$$

But, using an inverse inequality, we get

$$\begin{aligned} \|u - \Pi^{\partial}_{\frac{1}{2}} u\|_{H^{\frac{1}{2}+\epsilon}(e)} &\leq \|u - P^{\partial}_{\frac{1}{2}} u\|_{H^{\frac{1}{2}+\epsilon}(e)} + \|P^{\partial}_{\frac{1}{2}} u - \Pi^{\partial}_{\frac{1}{2}} u\|_{H^{\frac{1}{2}+\epsilon}(e)} \\ &\leq \|u - P^{\partial}_{\frac{1}{2}} u\|_{H^{\frac{1}{2}+\epsilon}(e)} + Cp^{2\epsilon}\|P^{\partial}_{\frac{1}{2}} u - \Pi^{\partial}_{\frac{1}{2}} u\|_{H^{\frac{1}{2}}(e)}. \end{aligned} \tag{125}$$

The second term is then estimated in the same way as in the Sect. 3.2 resulting in an extra ϵ^{-1} blow-up factor. The extra epsilon in the norm of the projection error present in the first term can be eliminated by using the inverse inequality argument,

[8] Analogously, for 1D elliptic problems, we estimate the interpolation error over the entire finite element mesh.

$$
\begin{aligned}
\|u - P^{\partial}_{\frac{1}{2}} u\|_{H^{\frac{1}{2}+\epsilon}(e)} &\le \|u - P^{\partial}_{\frac{1}{2}+\epsilon} u\|_{H^{\frac{1}{2}+\epsilon}(e)} + C p^{2\epsilon} \|P^{\partial}_{\frac{1}{2}+\epsilon} u - P^{\partial}_{\frac{1}{2}} u\|_{H^{\frac{1}{2}}(e)} \\
&\le \|u - P^{\partial}_{\frac{1}{2}+\epsilon} u\|_{H^{\frac{1}{2}+\epsilon}(e)} \\
&\quad + C p^{2\epsilon} \left(\|P^{\partial}_{\frac{1}{2}+\epsilon} u - u\|_{H^{\frac{1}{2}+\epsilon}(e)} + \|u - P^{\partial}_{\frac{1}{2}} u\| \right)_{H^{\frac{1}{2}}(e)} \\
&\le C p^{3\epsilon} p^{-(r-\frac{1}{2})} \|u\|_{H^r(e)}
\end{aligned}
\tag{126}
$$

for $r > \frac{1}{2}$. The final estimate reads as follows.

$$
\begin{aligned}
\|u - \Pi^{\partial}_{\frac{1}{2}} u\|_{H^{\frac{1}{2}}(\partial\Omega)} &\le C \epsilon^{-2} p^{3\epsilon} p^{-(r-\frac{1}{2})} \sum_e \|u\|_{H^r(e)} \\
&\le C (\ln p)^2 p^{-(r-\frac{1}{2})} \sum_e \|u\|_{H^r(e)} \\
&\le C (\ln p)^2 p^{-(r-\frac{1}{2})} \|u\|_{H^r(\partial\Omega)}.
\end{aligned}
\tag{127}
$$

We can use the result now to get an estimate in the negative norms. Let $0 \le s \le 1$, $s \ne \frac{1}{2}$, $\quad r > s, r > \frac{1}{2}$, and let $E \in H^{r-1}(\partial\Omega)$. Let E_0 denote the average value of E, i.e.,

$$
< E - E_0, 1 >= 0, \quad \|E_0\|_{H^{r-1}(\partial\Omega)} \le C \|E\|_{H^{r-1}(\partial\Omega)}. \tag{128}
$$

Exercise 3.7. Prove that, for the closed curve $\partial\Omega$, $1 \in H^r(\partial\Omega)$, and the estimate (128) holds for any $0 \le r \le 1$.

Notice also that, for a closed curve, the range of the (tangential) derivative coincides with distributions with zero average. Consequently, there exists a potential $u \in H^r(\partial\Omega)$ such that $u' = E - E_0$ and, by the commutativity property,

$$
(\Pi^{\partial}_s u|_e)' = \Pi_{s-1} u'|_e = \Pi_{s-1} E|_e - E_0. \tag{129}
$$

We have,

$$
E - \Pi_{s-1} E = E - \left[(\Pi^{\partial}_s u|_e)' + E_0 \right] = (u - \Pi^{\partial}_s u)' \tag{130}
$$

and,

$$
\|E - \Pi_{s-1} E\|_{H^{s-1}(\partial\Omega)} \le C \|u - \Pi^{\partial}_s u\|_{H^s(\partial\Omega)} \le C \sum_e \|u - \Pi^{\partial}_s u\|_{H^s(e)}. \tag{131}
$$

Let u_0 denote the average value of potential u on edge e. The interpolation operator reproduces constants and this implies that,

$$
\begin{aligned}
\|u - \Pi^{\partial}_s u\|_{H^s(e)} &= \|u - u_0 - \Pi^{\partial}_s (u - u_0)\|_{H^s(e)} \\
&\le C p^{-(r-s)} \|u - u_0\|_{H^r(e)} \\
&\le C p^{-(r-s)} \|u'\|_{H^{r-1}(e)} \\
&\le C p^{-(r-s)} \|E - E_0\|_{H^{r-1}(e)}.
\end{aligned}
\tag{132}
$$

Recalling (128), we get,

$$
\|E - \Pi_{s-1} E\|_{H^{s-1}(\partial\Omega)} \le C p^{-(r-s)} \|E - E_0\|_{H^{r-1}(\partial\Omega)} \le C p^{-(r-s)} \|E\|_{H^{r-1}(\partial\Omega)}. \tag{133}
$$

Exercise 3.8. Let $r > \frac{1}{2}$. Use the commutativity argument, localization result Proposition 3.2, and Theorem 3.2 to prove that,

$$\|E - \Pi_{-\frac{1}{2}}E\|_{H^{-\frac{1}{2}}(\partial\Omega)} \leq C(\ln p)^2 p^{-(r-s)} \|E\|_{H^{r-1}(\partial\Omega)}. \tag{134}$$

Notice that, at this point, we have not claimed any localization results in the negative norms. We do have, however,

Proposition 3.3. *Let $I = (-1,1)$, $I_1 = (-1,0)$, $I_2 = (0,1)$. Let $0 \leq t < \frac{1}{2}$. There exists a constant $C > 0$ such that,*

$$\|E\|_{H^{-t}(I)} \leq C \left(\|E\|_{H^{-t}(I_1)} + \|E\|_{H^{-t}(I_2)} \right) \quad \forall E \in H^{-t}(I). \tag{135}$$

Moreover, for $t = \frac{1}{2} - \epsilon$, $C = O(\epsilon^{-\frac{1}{2}})$.

Proof. For the range $0 \leq t < \frac{1}{2}$,

$$\|\phi\|_{\widetilde{H}^t(I)} \leq C \|\phi\|_{H^t(I)} \tag{136}$$

with the equivalence constant $C = O(\epsilon^{-\frac{1}{2}})$ for $t = \frac{1}{2} - \epsilon$, see [32, p. 105]. By the duality argument,

$$\|E\|_{\widetilde{H}^{-t}(I)} \leq C \|E\|_{H^{-t}(I)}. \tag{137}$$

Let $\phi \in \mathcal{D}(I)$ be an arbitrary test function. Choose $\phi^i_n \in \mathcal{D}(I_i)$ converging to restriction $\phi|_{I_i}$ in $H^t(I_i)$-norm. Then,

$$| < E, \phi > | \leq \sum_{i=1}^{2} \|E\|_{\widetilde{H}^{-t}(I_i)} \|\phi^i_n\|_{H^t(I_i)} + \|E\|_{\widetilde{H}^{-t}(I)} \sum_{i=1}^{2} \|\phi|_{I_i} - \phi^i_n\|_{H^t(I_i)}. \tag{138}$$

Passing to the limit with $n \to \infty$, we get,

$$|< E, \phi >| \leq \left(\sum_{i=1}^{2} \|E\|_{\widetilde{H}^{-t}(I_i)} \right) \|\phi\|_{H^t(I)} \tag{139}$$

which, combined with (137), finishes the argument.

The localization result allows for an alternative proof of estimate of the interpolation error in the negative norm (133) for $s > \frac{1}{2}$. Localization in the dual norm $H^{-\frac{1}{2}}$ is impossible, but the information about the blow up rate can again be translated into an alternative proof of estimate (134). Combining Proposition 3.3 with Theorem 3.2, we get a slightly sharper result,

$$\|E - \Pi_{s-1}E\|_{H^{-\frac{1}{2}}(\partial\Omega)} \leq C \ln p \, p^{-(r-\frac{1}{2})} \|E\|_{H^{r-1}(\partial\Omega)}. \tag{140}$$

Localization in the dual norm H^{-t} for $\frac{1}{2} < t \leq 1$ requires extra compatibility conditions for the functional to be localized. This can be immediately seen by considering the delta functional,

$$< \delta, \phi > := \phi(0). \tag{141}$$

Obviously, the delta functional cannot be localized. But we have, for instance,

Exercise 3.9. Let $I = (-1,1)$, $I_1 = (-1,0)$, $I_2 = (0,1)$. Let $\frac{1}{2} < t \leq 1$. Let $\phi_0 \in \widetilde{H}^t(I)$ be a specific function such that $\phi_0(0) = 1$. Let $E \in H^{-t}(I)$ be an arbitrary functional such that $< E, \phi_0 >= 0$. Prove that there exists a constant $C > 0$ such that,

$$\|E\|_{H^{-t}(I)} \leq C \left(\|E\|_{H^{-t}(I_1)} + \|E\|_{H^{-t}(I_2)} \right), \tag{142}$$

where the constant C depends upon test function ϕ_0 but it is independent of E. Conversely, prove that if, for a particular E, the estimate above holds, then there must exist a function $\phi_0 \in \widetilde{H}^t(I), \phi_0(0) = 1$ such that $< E, \phi_0 >= 0$.

Another sufficient condition can be extracted from the reasoning leading to estimate (133).

Exercise 3.10. Let $I = (-1,1)$, $I_1 = (-1,0)$, $I_2 = (0,1)$. Let $\frac{1}{2} < t \leq 1$. Let $E \in H^{-t}(I)$ be such that there exists a potential $u \in H^{1-t}(I)$, $u' = E$, such that,

$$\int_{I_i} u = 0, \quad i = 1,2. \tag{143}$$

Then (142) holds.

4 Commuting Projections and Projection-Based Interpolation Operators in Two Space Dimensions

4.1 Definitions and Commutativity

We shall consider the following diagram.

$$\begin{array}{ccccccccc}
\mathbb{R} & \longrightarrow & H^r(\Omega) & \xrightarrow{\nabla} & \boldsymbol{H}^{r-1}(\mathrm{curl},\Omega) & \xrightarrow{\mathrm{curl}} & H^{r-1}(\Omega) & \longrightarrow & \{0\} \\
\big\downarrow & & P_s^{grad} \big\downarrow \Pi_s^{grad} & & P_{s-1}^{curl} \big\downarrow \Pi_{s-1}^{curl} & & P_{s-1} \big\downarrow \Pi_{s-1} & & \big\downarrow \\
\mathbb{R} & \longrightarrow & W_p & \xrightarrow{\nabla} & \boldsymbol{Q}_p & \xrightarrow{\mathrm{curl}} & Y_p & \longrightarrow & \{0\}.
\end{array} \tag{144}$$

Here $\frac{1}{2} \leq s \leq 1$ with $s \leq r$, $r > 1$, and curl denotes the scalar-valued curl operator in 2D. By $\boldsymbol{H}^{r-1}(\mathrm{curl}, \Omega)$ we understand the space of all vector-valued functions in $\boldsymbol{H}^{r-1}(\Omega)$ whose curl is in $H^{r-1}(\Omega)$. Ω stands for a 2D element, either a quad or a triangle, and $Q_p, \boldsymbol{W}_p, Y_p$ denote any of the exact polynomial sequences defined on element Ω, discussed in Sect. 2. The common property of those sequences is that the corresponding trace spaces for $Q_p, \boldsymbol{W}_p$ corresponding to any edge e, define the 1D exact polynomial sequence discussed in the previous section.

The projection operators $P_s^{grad}, P_{s-1}^{curl}$ are defined as follows.

$$\begin{cases} P_s^{grad} u =: u_p \in W_p \\ \|\nabla u_p - \nabla u\|_{H^{s-1}(\Omega)} \to \min \\ (u_p - u, 1)_{H^s(\Omega)} = 0 \end{cases} \tag{145}$$

$$\begin{cases} P^{curl}_{s-1}\boldsymbol{E} =: \boldsymbol{E}_p \in \boldsymbol{Q}_p \\ \|\mathrm{curl}\boldsymbol{E}_p - \mathrm{curl}\boldsymbol{E}\|_{H^{s-1}(\Omega)} \to \min \\ (\boldsymbol{E}_p - \boldsymbol{E}, \nabla\phi)_{H^{s-1}(\Omega)} = 0, \ \forall\phi \in W_p \end{cases} \tag{146}$$

and P_{s-1} denotes the orthogonal projection onto Y_p in the H^{s-1}-norm.

Exercise 4.1. Show that the projections defined above make the diagram (144) commute.

The projection-based interpolation operators are defined as follows.

$$\begin{cases} \Pi^{grad}_s u =: u_p \in W_p \\ u_p = \Pi^{\partial}_{s-\frac{1}{2}} u \text{ on } \partial\Omega \\ \|\nabla u_p - \nabla u\|_{H^{s-1}(\Omega)} \to \min \end{cases} \tag{147}$$

$$\begin{cases} \Pi^{curl}_{s-1}\boldsymbol{E} =: \boldsymbol{E}_p \in \boldsymbol{Q}_p \\ E_{t,p} = \Pi_{s-\frac{3}{2}} E_t \text{ on } \partial\Omega \\ \|\mathrm{curl}\boldsymbol{E}_p - \mathrm{curl}\boldsymbol{E}\|_{H^{s-1}(\Omega)} \to \min \\ (\boldsymbol{E}_p - \boldsymbol{E}, \nabla\phi)_{H^{s-1}(\Omega)} = 0, \ \forall\phi \in W_p \ : \ \phi = 0 \text{ on } \partial\Omega \end{cases} \tag{148}$$

and

$$\begin{cases} \Pi_{s-1} v =: v_p \in Y_p \\ < v_p - v, 1 >= 0 \\ \|v_p - v\|_{H^{s-1}(\Omega)} \to \min\,. \end{cases} \tag{149}$$

Here Π^{∂}_s, Π_s are the 1D interpolation operators discussed in the previous sections, and $E_t, E_{t,p}$ denote the tangential component of $\boldsymbol{E}, \boldsymbol{E}_p$, respectively. Notice that all minimization problems are *constrained minimization problems*—the boundary values of the interpolants in (147) and (148), and the average value of the interpolant in (149), are fixed. Similarly to 1D, the projection operators can be interpreted as local minimization problems with Neumann boundary conditions, while the interpolation operators employ local Dirichlet boundary conditions. Finally, remember that by the boundary values of fields $\boldsymbol{E} \in \boldsymbol{H}^{r-1}(\mathrm{curl}, T)$, we always understand the trace of the tangential component E_t. Definition of the tangential component E_t is nontrivial. For $\boldsymbol{E} \in \boldsymbol{H}^r$, $\frac{1}{2} < r < \frac{3}{2}$, the tangential component is understood in the sense of the trace theorem and we have, cf. [32, p. 102],

$$\|E_t\|_{H^{r-\frac{1}{2}}(\partial\Omega)} \leq C\|\boldsymbol{E}\|_{H^r(\Omega)}. \tag{150}$$

The definition for the range $-\frac{1}{2} < r < \frac{1}{2}$ is more complicated. We consider first a sufficiently regular field $\boldsymbol{E} \in \boldsymbol{H}^r(\Omega)$ and a test function $\phi \in H^{\frac{1}{2}-r}(\partial\Omega)$, to invoke the integration by parts formula:

$$\int_{\partial\Omega} E_t\phi = \int_{\Omega} (\text{curl}\boldsymbol{E})\Phi - \int_{\Omega} \boldsymbol{E}(\nabla \times \Phi). \tag{151}$$

Here $\Phi \in H^{1-r}(\Omega)$ is an extension of ϕ such that,

$$\|\Phi\|_{H^{1-r}(\Omega)} \leq C\|\phi\|_{H^{\frac{1}{2}-r}(\partial\Omega)} \tag{152}$$

see [32, p. 101], and $\nabla\times$ denotes the vector-valued curl operator in 2D. We have,

$$\begin{aligned}\left|\int_{\partial\Omega} E_t\phi\right| &\leq \|\text{curl}\boldsymbol{E}\|_{H^s(\Omega)}\|\Phi\|_{\widetilde{H}^{-s}(\Omega)} + \|\boldsymbol{E}\|_{H^r(\Omega)}\|\nabla \times \Phi\|_{\widetilde{H}^{-r}(\Omega)} \\ &\leq C\left(\|\text{curl}\boldsymbol{E}\|_{H^s(\Omega)}\|\Phi\|_{H^{1-r}(\Omega)} + \|\boldsymbol{E}\|_{H^r(\Omega)}\|\nabla \times \Phi\|_{H^{-r}(\Omega)}\right) \\ &\leq C\left(\|\text{curl}\boldsymbol{E}\|_{H^s(\Omega)} + \|\boldsymbol{E}\|_{H^r(\Omega)}\right)\|\phi\|_{H^{\frac{1}{2}-r}(\partial\Omega)}\end{aligned} \tag{153}$$

where the range of r secures the equivalence of $\widetilde{H}^r$- and H^r-norms, and $-\frac{1}{2} < s < \frac{1}{2}$ is an arbitrary, not necessarily related to r parameter (we may, of course, choose $s = r$). The range of s implies that $-s \leq 1 - r$. The density argument allows now to extend the notion of the tangential component for every field $\boldsymbol{E} \in \boldsymbol{H}^r$ such that $\text{curl}\boldsymbol{E} \in H^s(\Omega)$, with both r and s from interval $(-\frac{1}{2}, \frac{1}{2})$. We get the estimate,

$$\|E_t\|_{H^{r-\frac{1}{2}}(\partial\Omega)} \leq C\left(\|\text{curl}\boldsymbol{E}\|_{H^s(\Omega)} + \|\boldsymbol{E}\|_{\boldsymbol{H}^r(\Omega)}\right) \tag{154}$$

which can be seen as an equivalent of the Trace Theorem for the space,

$$\{\boldsymbol{E} \in \boldsymbol{H}^r(\Omega) \;:\; \text{curl}\,\boldsymbol{E} \in H^s(\Omega)\}. \tag{155}$$

Note that the blow up of the equivalence constants prohibits extending the definition to values $s, r = -\frac{1}{2}, \frac{1}{2}$, and that in both cases the constants are of order $O(\epsilon^{-\frac{1}{2}})$ for $s, r = -\frac{1}{2} + \epsilon$ or $s, r = \frac{1}{2} - \epsilon$.

Exercise 4.2. Show that the definition of E_t discussed above is independent of the choice of extension Φ.

Theorem 4.1. *The interpolation operators make the diagram (144) commute.*

Proof. The commutativity of the first block follows from the fact that operator Π_s^{grad} preserves constants. In order to show the commutativity of the second block, we need to demonstrate that,

$$\Pi_{s-1}^{curl}\nabla u = \nabla \Pi_s^{grad} u. \tag{156}$$

Let $\boldsymbol{E} = \nabla u$. By the commutativity of the 1D diagram, we have,

$$E_{t,p} = \Pi_{s-\frac{3}{2}}\frac{\partial u}{\partial t} = \frac{\partial u_p}{\partial t} \tag{157}$$

where $u_p = \Pi_{s-\frac{1}{2}}^{\partial} u$ and $\frac{\partial}{\partial t}$ denotes the tangential derivative on the boundary of the element. Consequently,

$$\int_{\Omega} \mathrm{curl} \boldsymbol{E}_p = \int_{\partial\Omega} E_{t,p} = \int_{\partial\Omega} \frac{\partial u_p}{\partial t} = 0. \tag{158}$$

At the same time,

$$(\mathrm{curl} \boldsymbol{E}_p, \mathrm{curl} \boldsymbol{F})_{H^{s-1}(\Omega)} = 0 \tag{159}$$

for every $\boldsymbol{F} \in \boldsymbol{Q}_p$, $F_t = 0$ on $\partial\Omega$. However, the image of such polynomials F under the curl operator, coincides exactly with polynomials in Y_p with zero average, where the curl of $\boldsymbol{E}_p$ lives. Consequently, $\mathrm{curl}\boldsymbol{E}_p = 0$ and $E_p = \nabla u_p$ for some $u_p \in W_p$. Substituting ∇u_p into $(148)_4$ we learn that $u_p = \Pi_s^{grad} u$.

To prove the last commutativity property, we need to show that,

$$\Pi_{s-1}(\mathrm{curl}\boldsymbol{E}) = \mathrm{curl}\,(\Pi_{s-1}^{curl} \boldsymbol{E}). \tag{160}$$

Let $\boldsymbol{E}_p = \Pi_{s-1}^{curl} \boldsymbol{E}$. It follows form the definition of the 1D interpolation operator that,

$$< \mathrm{curl}\boldsymbol{E}_p - \mathrm{curl}\boldsymbol{E}, 1 >=< E_{t,p} - E_t, 1 >= 0. \tag{161}$$

Finally, condition $(149)_3$ follows directly from condition $(148)_3$.

4.2 Polynomial Preserving Extension Operators

Let u_p be the trace of a polynomial from space W_p defined on the boundary of the element. Of fundamental importance for the presented theory is the existence of a polynomial extension $U_p \in W_p$, $U_p|_{\partial\Omega} = u_p$ such that,

$$\|U_p\|_{H^s(\Omega)} \leq C \|u_p\|_{H^{s-\frac{1}{2}}(\partial\Omega)} \tag{162}$$

with constant C *independent of p*. Here, we are interested in the range $\frac{1}{2} \leq s \leq 1$. A more demanding request is to look for a general extension operator,

$$\mathrm{Ext}\,:\, H^{s-\frac{1}{2}}(\partial\Omega) \ni u \to U \in H^s(\Omega) \tag{163}$$

that is continuous and polynomial preserving. For $s = 1$ and both triangular and rectangular elements, the operator of this type was first constructed by Babuška and Suri [4], see also Babuška et al. [3]. For a triangular element, different, explicit constructions were shown by Munoz-Sola [33], Ainsworth and Demkowicz [1], see also [22]. In particular, the explicit construction in [1] allows for an immediate proof of continuity in fractional norms. For a square element, see also the recent result of Costabel, Dauge and Demkowicz [15] motivated by discrete harmonic extensions studied by Pavarino and Widlund [38] and the results of Maday [31]. All these results are quite technical.

In two space dimensions, the existence of a polynomial extension for the H^s-space, implies immediately an analogous result for the $\boldsymbol{H}^{s-1}(\mathrm{curl})$ space. Indeed, let $E_{t,p}$ be the tangential trace of a polynomial in $\boldsymbol{Q}_p$. Let E_0 be the average value of $E_{t,p}$ on the boundary of the element. By the result of Exercise 3.7, the average E_0 depends continuously upon the $H^{s-\frac{3}{2}}$-norm of $E_{t,p}$. Let $u \in H^{s-\frac{1}{2}}(\partial\Omega)$ be a polynomial of zero average such that $u' = E_{t,p} - E_0$. Let U be then the polynomial extension of u discussed above, and let $\boldsymbol{E}_0$ be the lowest order extension of the constant average E_0. Define $Ext^{curl} E_{t,p} = \boldsymbol{E}_p := \nabla U_p + \boldsymbol{E}_0$. We have,

$$\begin{aligned}
\|\boldsymbol{E}_p\|_{\boldsymbol{H}^{s-1}(\text{curl},\Omega)} &\leq \|\nabla U_p\|_{\boldsymbol{H}^{s-1}(\text{curl},\Omega)} + \|\boldsymbol{E}_0\|_{\boldsymbol{H}^{s-1}(\text{curl},\Omega)} \\
&\leq C\left(\|U_p\|_{H^s(\Omega)} + |E_0|\right) \\
&\leq C\left(\|u_p\|_{H^{s-\frac{1}{2}}(\partial\Omega)} + |E_0|\right) \\
&\leq C\left(\|E_{t,p} - E_0\|_{H^{s-\frac{3}{2}}(\partial\Omega)} + |E_0|\right) \\
&\leq C\,\|E_{t,p}\|_{H^{s-\frac{3}{2}}(\partial\Omega)} .
\end{aligned} \tag{164}$$

4.3 Right-Inverse of the Curl Operator: Discrete Friedrichs Inequality

Let Ω denote the master triangle or rectangle. Recall the operator K introduced in the discussion of Nédélec's triangle of the first type,

$$K\psi(\boldsymbol{x}) = -\boldsymbol{x} \times \left(\int_0^1 t\psi(t\boldsymbol{x})\, dt \right) \boldsymbol{e}_3 , \tag{165}$$

where $\boldsymbol{x} = (x_1, x_2, 0), \boldsymbol{e}_3 = (0, 0, 1)$.

Exercise 4.3. Prove that the operator K maps space Y_p into the space $\boldsymbol{Q}_p$ for all three Nédélec elements: the triangles of the first and second type, as well as the rectangle of the first type. Verify that,

$$\text{curl}(K\psi) = \psi. \tag{166}$$

We will show now that operator K is a continuous operator from $H^{-s}(\Omega)$ into $\boldsymbol{H}^{-s}(\text{curl}, \Omega)$. We are interested in the range $0 \leq s \leq \frac{1}{2}$. In order to demonstrate the continuity in the negative exponent norm, we compute first the adjoint operator. Switching to polar coordinates (r, θ), we obtain,

$$K\psi(r, \theta) = r \int_0^1 t\psi(tr, \theta)\, dt\, \boldsymbol{e}_\theta = \frac{1}{r} \int_0^r s\psi(s, \theta)\, ds\, \boldsymbol{e}_\theta ,$$

where $\boldsymbol{e}_r, \boldsymbol{e}_\theta$ denote the unit vectors of the polar coordinate system. Representing the argument of the dual operator in the polar coordinates as $\boldsymbol{\phi} = \phi_r \boldsymbol{e}_r + \phi_\theta \boldsymbol{e}_\theta$, we get,

$$\begin{aligned}
\int_f K\psi\boldsymbol{\phi}\, d\boldsymbol{x} &= \int_0^{\frac{\pi}{2}} \int_0^{\hat{r}(\theta)} \frac{1}{r} \int_0^r s\psi(s, \theta)\, ds \phi_\theta\, r dr d\theta \\
&= \int_0^{\frac{\pi}{2}} \int_0^{\hat{r}(\theta)} s\psi(s, \theta) \int_s^{\hat{r}(\theta)} \phi_\theta(r, \theta)\, dr ds d\theta \\
&= \int_f \psi \underbrace{\int_s^{\hat{r}(\theta)} \phi_\theta(r, \theta)\, dr}_{K^*\boldsymbol{\phi}(s,\theta)}\, d\boldsymbol{x} .
\end{aligned}$$

Symbol $\hat{r}(\theta)$ is explained in Fig. 1.

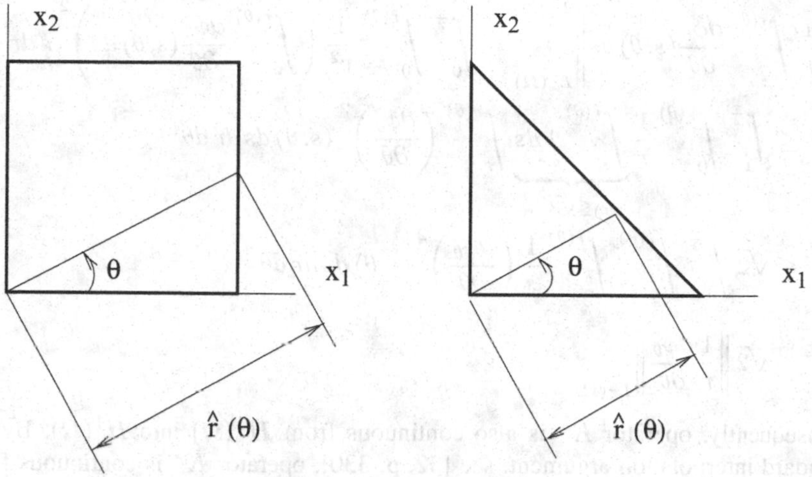

Fig. 1. Polar coordinates and integration over the master element

A "brute force" estimate follows.

$$\begin{aligned}\|K^*\phi\|^2_{0,f} &= \int_0^{\frac{\pi}{2}} \int_0^{\hat{r}(\theta)} \left(\int_r^{\hat{r}(\theta)} \phi_\theta(s,\theta)\, ds \right)^2 r dr d\theta \\ &\leq \int_0^{\frac{\pi}{2}} \int_0^{\hat{r}(\theta)} \underbrace{\int_r^{\hat{r}(\theta)} s^{-\frac{1}{2}}\, ds}_{2(\sqrt{\hat{r}(\theta)}-\sqrt{r})\leq 2} \int_r^{\hat{r}(\theta)} s\phi_\theta^2(s,\theta)\, ds\, r dr d\theta \\ &\leq 2 \int_0^{\frac{\pi}{2}} \underbrace{\int_0^{\hat{r}(\theta)} r\, dr}_{\leq \frac{1}{2}} \int_0^{\hat{r}(\theta)} s\phi_\theta^2(s,\theta)\, ds d\theta \\ &\leq \|\phi_\theta\|^2_{L^2(\Omega)} \,.\end{aligned}$$

Operator K^* is thus a continuous operator from $\boldsymbol{L}^2(\Omega)$ into $L^2(\Omega)$. We compute now the gradient,

$$\nabla K^*\phi(r,\theta) = -\phi_\theta(r,\theta)\, e_r + \frac{1}{r} \left(\underbrace{\phi_\theta(\hat{r}(\theta),\theta)}_{=0} \frac{d\hat{r}}{d\theta}(\theta) + \int_r^{\hat{r}(\theta)} \frac{\partial \phi_\theta}{\partial \theta}(s,\theta)\, ds \right) e_\theta \,,$$

where we have assumed that ϕ vanishes on the boundary. The first term estimates trivially, and for the second one we have,

$$\left\| \frac{1}{r} \int_r^{\hat{r}(\theta)} \frac{\partial \phi_\theta}{\partial \theta}(s,\theta)\, ds \right\|^2_{L^2(\Omega)} = \int_0^{\frac{\pi}{2}} \int_0^{\hat{r}(\theta)} \frac{1}{r^2} \left(\int_r^{\hat{r}(\theta)} \frac{\partial \phi_\theta}{\partial \theta}(s,\theta)\, ds \right)^2 r\, dr\, d\theta$$

$$\leq \int_0^{\frac{\pi}{2}} \int_0^{\hat{r}(\theta)} \frac{1}{r} \underbrace{\int_r^{\hat{r}(\theta)} 1\, ds}_{\leq \sqrt{2}} \int_r^{\hat{r}(\theta)} \left(\frac{\partial \phi_\theta}{\partial \theta} \right)^2 (s,\theta)\, ds\, dr\, d\theta$$

$$\leq \sqrt{2} \int_0^{\frac{\pi}{2}} \int_0^{\hat{r}(\theta)} \int_r^{\hat{r}(\theta)} \frac{1}{s} \left(\frac{\partial \phi_\theta}{\partial \theta} \right)^2 (s,\theta)\, ds\, dr\, d\theta$$

$$\leq \sqrt{2} \left\| \frac{1}{r} \frac{\partial \phi_\theta}{\partial \theta} \right\|^2_{L^2(\Omega)} .$$

Consequently, operator K^* is also continuous from $\boldsymbol{H}^1_0(\Omega)$ into $H^1_0(\Omega)$. By the standard interpolation argument, see [32, p. 330], operator K^* is continuous from $\widetilde{\boldsymbol{H}}^s(\Omega)$ into $\widetilde{H}^s(\Omega)$ and, consequently, operator K is continuous from $H^{-s}(\Omega)$ into $\boldsymbol{H}^{-s}(\Omega)$. Note that the continuity constant is independent of s.

For elements of variable order, the operator (165) still has to be modified to have a range in the right polynomial space. The issue is with the polynomial degree on the boundary. For $\psi \in Y_p$, the tangential trace of $K\psi$ vanishes on edges $x_1 = 0, x_2 = 0$ but it has, in general, a nonzero tangential trace on the rest of the boundary. We utilize the extension operator discussed above and define the ultimate operator as follows,

$$K^{mod}\psi = (I - Ext^{curl} Tr) K\psi_0 + \boldsymbol{E}_0 \, . \tag{167}$$

Here $\psi = \psi_0 + c$ is the decomposition of function ψ into a constant c, and a function ψ_0 with zero average. If all edge orders $p_e = -1$, then $c = 0$. Otherwise $\boldsymbol{E}_0$ denotes any linear combination of lowest order shape functions in $\boldsymbol{Q}_p$ whose curl reproduces the constant c. Notice that, due to the commuting diagram property, K^{mod} is still a right inverse of the curl operator (with operator norm independent of s). For $\psi \in Y_p$ with zero average, $K^{mod}\psi$ has a zero tangential trace.

Lemma 4.1 (Discrete Friedrichs Inequality for fractional spaces in 2D). *Let $0 \leq s \leq \frac{1}{2}$. There exist $C > 0$, independent of s and p, such that,*

$$\|\boldsymbol{E}\|_{\boldsymbol{H}^{-s}(\Omega)} \leq C \|\operatorname{curl} \boldsymbol{E}\|_{H^{-s}(\Omega)} \, , \tag{168}$$

for every discrete divergence free polynomial $\boldsymbol{E} \in \boldsymbol{Q}_p$, i.e.,

$$(\boldsymbol{E}, \nabla\phi)_{H^{-s}(\Omega)} = 0, \quad \forall \phi \in W_p \, . \tag{169}$$

Note that the result covers the case of polynomials with zero tangential trace.

Proof. We utilize the properties of the right-inverse K of the curl operator.

$$\begin{aligned} \|\boldsymbol{E}\|_{\boldsymbol{H}^{-s}(\Omega)} &= \inf_{\phi \in W_p} \|\boldsymbol{E} - \nabla\phi\|_{\boldsymbol{H}^{-s}(\Omega)} \\ &\leq \|\boldsymbol{E} - (\boldsymbol{E} - K^{mod}(\nabla \times \boldsymbol{E}))\|_{\boldsymbol{H}^{-s}(\Omega)} \\ &\leq \|K^{mod}\| \, \|\nabla \times \boldsymbol{E}\|_{\boldsymbol{H}^{-s}(\Omega)} \, . \end{aligned} \tag{170}$$

We shall also need a generalization of the classical Poincaré inequalities to the case of fractional Sobolev spaces.

Lemma 4.2 (Poincaré inequalities for fractional spaces in 2D).
Let $\frac{1}{2} < s \leq 1$. There exist $C > 0$ such that

$$\|u\|_{L^2(\Omega)} \leq C|u|_{H^s(\Omega)} \approx C\|\nabla u\|_{\boldsymbol{H}^{s-1}(\Omega)}, \tag{171}$$

for every function $u \in H^s(\Omega)$ belonging to either of the two families:
Case 1: $< u, 1 >= 0$, *C is independent of s,*
Case 2: $u = 0$ *on* $\partial\Omega$, $C = O(\epsilon^{-\frac{1}{2}})$, *for* $s = \frac{1}{2} + \epsilon$.

Proof. The equivalence of $|u|_{H^s(\Omega)}$ and $\|\nabla u\|_{\boldsymbol{H}^{s-1}(\Omega)}$ for $s = 0, 1$, follows from definitions of Sobolev norms, and for real s from the interpolation argument. Same arguments apply to Case 1. To prove Case 2, extend element Ω to a strip,

$$\{(x, y) \, : x \in \mathbb{R}, y \in (0, 1)\}$$

in such a way that one of the edges coincides with $y = 0$. Let u be a test function with support in Ω and U its extension by zero to the whole plane. We have,

$$\begin{aligned}
\int_\Omega |u(x, y)|^2 \, dydx &= \int_{\mathbb{R}} \int_0^1 |U(x, y)|^2 \, dydx \\
&\leq \int_{\mathbb{R}} \int_0^1 y^{-2s} |U(x, y)|^2 \, dydx \\
&\leq C_1 \int_{\mathbb{R}} \int_0^1 \int_0^1 \frac{|U(x, y) - U(x, z)|^2}{|y - z|^{1+2s}} \, dzdydx \quad \text{(see [32, p. 95}^1\text{])} \\
&\leq C_1 C_2 \int_{\mathbb{R}^2} |\boldsymbol{\xi}|^{2s} |\hat{U}(\boldsymbol{\xi})|^2 \, d\boldsymbol{\xi} \quad \text{(see [32, p. 95}^{4,5}\text{])} \\
&\leq C_1 C_2 C_3 \int_{\mathbb{R}^2} \int_{\mathbb{R}^2} \frac{|U(\boldsymbol{x}) - U(\boldsymbol{z})|^2}{|\boldsymbol{x} - \boldsymbol{z}|^{2+2s}} \quad \text{(see [32, Lemma 3.15])} \\
&\leq C_1 C_2 C_3 |u|^2_{H^s(\Omega)}.
\end{aligned} \tag{172}$$

Out of the three constants C_1, C_2, C_3 only the first one experiences the described blow up at $s = \frac{1}{2}$. The result follows then from the density argument.

4.4 Projection Error Estimates

As in the 1D case, it is illuminating to see that the definitions of commuting projection operators P_s^{grad}, P_{s-1}^{curl} are equivalent to the solution of mixed problems. The mixed formulation for determining $P_s^{grad} u$ looks as follows.

$$\begin{cases}
P_s^{grad} u =: u_p \in W_p, \ \lambda \in \mathbb{R} \\
(\nabla u_p - \nabla u, \nabla v)_{H^{s-1}(\Omega)} + (\lambda, v)_{H^s(\Omega)} = 0 \quad \forall v \in W_p \\
(u_p - u, \mu)_{H^s(\Omega)} \qquad\qquad\qquad\qquad = 0 \quad \forall \mu \in \mathbb{R}.
\end{cases} \tag{173}$$

Substituting $v = \text{const} = \lambda$ in the first equation, we learn that the Lagrange multiplier $\lambda = 0$. The fact that constants are included in space W_p implies the satisfaction of the first Brezzi's inf–sup condition. The inf–sup in kernel condition is implied by the Poincaré inequality, case 1.

The situation is similar with the mixed formulation for determining $P^{curl}_{s-1}\boldsymbol{E}$.

$$\begin{cases} P^{curl}_{s-1}\boldsymbol{E} =: \boldsymbol{E}_p \in \boldsymbol{Q}_p,\ \psi \in W_p \\ (\text{curl}\,\boldsymbol{E}_p - \text{curl}\,\boldsymbol{F}, \text{curl}\,\boldsymbol{F})_{H^{s-1}(\Omega)} + (\nabla\psi, \boldsymbol{F})_{H^{s-1}(\Omega)} = 0 \quad \forall \boldsymbol{F} \in \boldsymbol{Q}_p \\ (\boldsymbol{E}_p - \boldsymbol{E}, \nabla\phi)_{H^{s-1}(\Omega)} = 0 \quad \forall \phi \in W_p. \end{cases} \tag{174}$$

Substituting $\boldsymbol{F} = \nabla\psi$ into the first equation, we learn again that $\nabla\psi = 0$. The exact sequence property, i.e. the inclusion $\nabla W_p \subset \boldsymbol{Q}_p$ implies the automatic satisfaction of the first inf–sup condition with constant $\beta = 1$, and the discrete Friedrichs inequality proved in Lemma 4.1 implies the inf–sup in kernel condition. We can conclude the projection error estimates.

Theorem 4.2 (Commuting projection error estimates in 2D).
Let $\frac{1}{2} \le s \le 1$, $r > s, r > 1$. There exist constants $C > 0$, independent of p, such that,

$$\begin{aligned} \|u - P^{grad}_s u\|_{H^s(\Omega)} &\le C \inf_{u_p \in W_p} \|u - u_p\|_{H^s(\Omega)} \\ &\le C p^{-(r-s)} \|u\|_{H^r(\Omega)} \\ \|\boldsymbol{E} - P^{curl}_{s-1}\boldsymbol{E}\|_{\boldsymbol{H}^{s-1}(\text{curl},\Omega)} &\le C \inf_{\boldsymbol{E}_p \in \boldsymbol{Q}_p} \|\boldsymbol{E} - \boldsymbol{E}_p\|_{\boldsymbol{H}^{s-1}(\text{curl},\Omega)} \\ &\le C p^{-(r-s)} \|\boldsymbol{E}\|_{\boldsymbol{H}^{r-1}(\text{curl},\Omega)} \\ \|v - P_{s-1}v\|_{H^{s-1}(\Omega)} &= \inf_{v_p \in Y_p} \|v - v_p\|_{H^{s-1}(\Omega)} \\ &\le C p^{-(r-s)} \|v\|_{H^{r-1}(\Omega)} \end{aligned} \tag{175}$$

for every $u \in H^r(\Omega)$, $\boldsymbol{E} \in \boldsymbol{H}^{r-1}(\text{curl}, \Omega)$, $v \in H^{r-1}(\Omega)$.

Proof. For the best approximation results in the H^s-norm, see [40], and in the $\boldsymbol{H}^{r-1}(\text{curl}, \Omega)$-norm, see [19].

As in 1D, estimating the projection error with the best approximation error can be done directly by using the right-inverse of the curl operator and an analogous, polynomial preserving, right-inverse of the grad operator,

$$GE(\boldsymbol{x}) = \boldsymbol{x} \cdot \int_0^1 \boldsymbol{E}(t\boldsymbol{x})\, dt \tag{176}$$

where $\boldsymbol{x} = (x_1, x_2)$ and $\cdot$ denotes the dot product.

Exercise 4.4. Let Ω be the square or triangular master element, and $\frac{1}{2} \le s \le 1$. Let $\boldsymbol{E} \in \mathcal{R}(\nabla)$. Prove that:

- $\nabla(GE) = E$,
- $E_t = 0$ on $\partial\Omega$ implies $GE = 0$ on $\partial\Omega$,
- operator (176) is a continuous, polynomial preserving operator from $\boldsymbol{H}^{s-1}(\mathrm{curl}, \Omega)$ into $H^s(\Omega)$.

The two commuting projection operators can then be represented in the form [25],

$$\begin{aligned} P^{curl}_{s-1}\boldsymbol{E} &= P^{curl0}_{s-1}(\boldsymbol{E} - KP_{s-1}(\mathrm{curl}\,\boldsymbol{E})) + KP_{s-1}(\mathrm{curl}\,\boldsymbol{E}) \\ P^{grad}_{s}u &= P^{grad0}_{s}(u - GP^{curl0}_{s-1}(\nabla u)) + GP^{curl0}_{s-1}(\nabla u) \end{aligned} \tag{177}$$

where P^{curl0}_{s-1} and P^{grad0}_{s} denote the orthogonal projections in $\boldsymbol{H}^{s-1}$- and H^s-norms onto the subspaces of polynomials in $\boldsymbol{Q}_p$ and W_p with zero curl and gradient, respectively (i.e. onto the range of the gradient operator and constants). The continuity of the right-inverses, and the polynomial-preserving property imply then that the commuting projection errors are bounded by the best approximation errors.

4.5 Interpolation Error Estimates

The interpolation error estimates are derived by comparing the interpolation errors with the commuting projection errors following the same procedure as in Sect. 3.2.

Theorem 4.3 (Commuting interpolation error estimates in 2D). *Let $\frac{1}{2} < s \leq 1, r > s, r > 1$. There exist constants $C_1, C_2 > 0$, independent of p, such that,*

$$\begin{aligned} \|u - \Pi^{grad}_s u\|_{H^s(\Omega)} &\leq C_1 \left(\|u - P^{grad}_s u\|_{H^s(\Omega)} + \|u - \Pi^{\partial}_{s-\frac{1}{2}} u\|_{H^{s-\frac{1}{2}}(\partial\Omega)} \right) \\ \|\boldsymbol{E} - \Pi^{curl}_{s-1}\boldsymbol{E}\|_{\boldsymbol{H}^{s-1}(\mathrm{curl},\Omega)} &\leq C_2 \left(\|\boldsymbol{E} - P^{curl}_{s-1}\boldsymbol{E}\|_{\boldsymbol{H}^{s-1}(\mathrm{curl},\Omega)} \right. \\ &\qquad \left. + \|E_t - \Pi^{curl}_{s-\frac{3}{2}} E_t\|_{H^{s-\frac{3}{2}}(\partial\Omega)} \right) \\ \|v - \Pi_{s-1} v\|_{H^{s-1}(\Omega)} &\leq C_2 \|v - P_{s-1} v\|_{H^{s-1}(\Omega)} \end{aligned} \tag{178}$$

for every $u \in H^r(\Omega), \boldsymbol{E} \in \boldsymbol{H}^{r-1}(\mathrm{curl}, \Omega), v \in H^{r-1}(\Omega)$. For $s = \frac{1}{2} + \epsilon$, constant $C_1 = O(\epsilon^{-1})$ and constant $C_2 = O(\epsilon^{-\frac{1}{2}})$. Combined with estimates (123), (127), and (133), (134), we obtain the following error estimates for the case $s = 1$,

$$\begin{aligned} \|u - \Pi^{grad}_1 u\|_{H^1(\Omega)} &\leq C(\ln p)^2\, p^{-(r-1)} \|u\|_{H^r(\Omega)} \\ \|\boldsymbol{E} - \Pi^{curl}_{-1}\boldsymbol{E}\|_{\boldsymbol{H}(\mathrm{curl},\Omega)} &\leq C \ln p\, p^{-(r-1)} \|\boldsymbol{E}\|_{\boldsymbol{H}^{r-1}(\mathrm{curl},\Omega)} \end{aligned} \tag{179}$$

and case $s = \frac{1}{2}$,

$$\begin{aligned} \|u - \Pi^{grad}_{\frac{1}{2}} u\|_{H^{\frac{1}{2}}(\Omega)} &\leq C \ln p\, p^{-(r-\frac{1}{2})} \|u\|_{H^r(\Omega)} \\ \|\boldsymbol{E} - \Pi^{curl}_{-\frac{1}{2}}\boldsymbol{E}\|_{\boldsymbol{H}^{-\frac{1}{2}}(\mathrm{curl},\Omega)} &\leq C(\ln p)^{\frac{1}{2}}\, p^{-(r-\frac{1}{2})} \|\boldsymbol{E}\|_{\boldsymbol{H}^{r-1}(\mathrm{curl},\Omega)} \\ \|v - \Pi_{-\frac{1}{2}} v\|_{H^{-\frac{1}{2}}(\Omega)} &\leq C(\ln p)^{\frac{1}{2}}\, p^{-(r-\frac{1}{2})} \|v\|_{H^{r-1}(\Omega)}. \end{aligned} \tag{180}$$

Proof. **Operator Π^{grad}_s**. We begin with the triangle inequality,

$$\|u - \Pi^{grad}_s u\|_{H^s(\Omega)} \leq \|u - P^{grad}_s u\|_{H^s(\Omega)} + \|P^{grad}_s u - \Pi^{grad}_s u\|_{H^s(\Omega)}. \quad (181)$$

Polynomial $\psi = P^{grad}_s u - \Pi^{grad}_s u \in W_p$ satisfies,

$$(\nabla\psi, \nabla\phi)_{H^{s-1}(\Omega)} = 0, \quad \forall\phi \in W_p : \phi = 0 \text{ on } \partial\Omega \quad (182)$$

and, therefore, it is the polynomial discrete minimum energy extension with the energy defined by the $H^{s-1}(\Omega)$-norm of gradient $\nabla\psi$. The Poincaré inequality, see Lemma 4.2, case 2, implies that the H^s-norm of ψ is bounded by the norm of the polynomial minimum-norm extension ψ_0 of the boundary trace of ψ. Indeed,

$$\begin{aligned}
\|\psi\|_{H^s(\Omega)} &\leq \|\psi - \psi_0\|_{H^s(\Omega)} + \|\psi_0\|_{H^s(\Omega)} \\
&\qquad \text{(triangle inequality)} \\
&\leq C(\|\nabla(\psi - \psi_0)\|_{H^{s-1}(\Omega)} + \|\psi_0\|_{H^s(\Omega)}) \\
&\qquad \text{(Poincaré inequality)} \\
&\leq C(\|\nabla\psi\|_{H^{s-1}(\Omega)} + \|\psi_0\|_{H^s(\Omega)}) \\
&\qquad (\psi_0 \text{ term is absorbed into the second term}) \\
&\leq C(\|\nabla\psi_0\|_{H^{s-1}(\Omega)} + \|\psi_0\|_{H^s(\Omega)}) \\
&\qquad (\psi \text{ is the polynomial minimum seminorm extension}) \\
&\leq C\|\psi_0\|_{H^s(\Omega)}. \\
&\qquad \text{(continuity of gradient)}
\end{aligned} \quad (183)$$

Note that constant $C = O(\epsilon^{-\frac{1}{2}})$ for $s = \frac{1}{2} + \epsilon$.

Denoting the trace of ψ by $tr\psi$, we therefore have,

$$\begin{aligned}
\|\psi\|_{H^s(\Omega)} &\leq C\|\psi_0\|_{H^s(\Omega)} \\
&\leq C\|Ext\, tr\psi\|_{H^s(\Omega)} \\
&\leq C\|Ext\|\, \|\Pi^{grad}_s u - P_s u\|_{H^{s-\frac{1}{2}}(\partial\Omega)} \\
&\leq C\|Ext\| \left(\|u - P^{grad}_s u\|_{H^{s-\frac{1}{2}}(\partial\Omega)} + \|u - \Pi^{grad}_s u\|_{H^{s-\frac{1}{2}}(\partial\Omega)} \right) \\
&\leq C\|Ext\| \left(C_{tr}\|u - P^{grad}_s u\|_{H^s(\Omega)} + \|u - \Pi^{\partial}_{s-\frac{1}{2}} u\|_{H^{s-\frac{1}{2}}(\partial\Omega)} \right).
\end{aligned} \quad (184)$$

Here C_{tr} is the norm of the trace operator of order $O(\epsilon^{-\frac{1}{2}})$ for $s = \frac{1}{2} + \epsilon$, cf. [32, p. 100]. Combining the triangle inequality (181) with the result above, we see that the interpolation error is bounded by the projection error, and the interpolation error on the boundary. The final estimates follow then from the estimates for the 1D interpolation operators discussed in Sect. 3.

Operator Π_{s-1}^{curl}. We follow exactly the same arguments as for the first case. If $\boldsymbol{\psi} = \Pi_{s-1}^{curl}\boldsymbol{E} - P_{s-1}^{curl}\boldsymbol{E}$, the discrete Friedrichs inequality corresponding to the subspace,

$$\{\boldsymbol{E} \in \boldsymbol{Q}_p \,:\, E_t = 0 \text{ on } \partial\Omega, \quad (\boldsymbol{E}, \nabla\phi)_{H^{s-1}(\Omega)} = 0, \forall\phi \in W_p \,:\, \phi = 0 \text{ on } \partial\Omega\} \tag{185}$$

is needed to establish the bound:

$$\|\boldsymbol{\psi}\|_{\boldsymbol{H}^{s-1}(\mathrm{curl},\Omega)} \leq C\|\,\mathrm{curl}\,\boldsymbol{\psi}\|_{H^{s-1}(\Omega)} \leq C\|\,\mathrm{curl}(Ext^{curl}tr\boldsymbol{\psi})\|_{H^{s-1}(\Omega)}\,. \tag{186}$$

Contrary to the previous case, constant C does not experience any blow up at $s = \frac{1}{2}$. The trace constant C_{tr} in (155) experiences the same blow up as in the case of operator Π_s^{grad}. The final estimates follow then from the estimates for the 1D interpolation operators discussed in Sect. 3.

Operator Π_{s-1}. Let $v \in H^{r-1}(\Omega)$ and let $\boldsymbol{E} \in \boldsymbol{H}^{r-1}(\mathrm{curl},\Omega)$ be such that $\mathrm{curl}\boldsymbol{E} = v$ and $E_t = \text{const}$ on $\partial\Omega$. The result follows then from the commutativity of the operators and the result for operator Π_{s-1}^{curl}.

4.6 Localization Argument

Proposition 3.2 remains true in multiple space dimensions, see [28, p. 29–30] or [17], and it makes it possible to generalize the interpolation error estimates to the boundary of a 3D polyhedral domain consisting of triangular or rectangular faces. Let $\partial\Omega$ denote the boundary of a 3D polyhedron Ω, composed of faces f. Let $\frac{1}{2} < s \leq 1, \quad r > s$. We have

$$\begin{aligned}\|u - \Pi_s^{grad}u\|_{H^s(\partial\Omega)} &\leq C\sum_f \|u - \Pi_s^{grad}u\|_{H^s(f)}\\ &\leq Cp^{-(r-s)}\sum_f \|u\|_{H^r(f)}\\ &\leq Cp^{-(r-s)}\|u\|_{H^r(\partial\Omega)}.\end{aligned} \tag{187}$$

Here p denotes the order of approximation on the element boundary. For the element of variable order, p is the minimum order for all faces and edges. For $s = \frac{1}{2}$, we need to utilize the information about the blow up of constant $C = C(s)$ with $s \to \frac{1}{2}$. We have for $r > 1$,

$$\begin{aligned}\|u - \Pi_{\frac{1}{2}}^{grad}u\|_{H^{\frac{1}{2}}(\partial\Omega)} &\leq \|u - \Pi_{\frac{1}{2}}^{grad}u\|_{H^{\frac{1}{2}+\epsilon}(\partial\Omega)}\\ &\leq C\epsilon^{-1}\sum_f \|u - \Pi_{\frac{1}{2}}^{grad}u\|_{H^{\frac{1}{2}+\epsilon}(f)}\\ &\leq C\epsilon^{-2}p^{-(r-(\frac{1}{2}+\epsilon))}\sum_f \|u\|_{H^r(f)}\\ &\leq C(\ln p)^2 p^{-(r-\frac{1}{2})}\sum_f \|u\|_{H^r(f)}\\ &\leq C(\ln p)^2 p^{-(r-\frac{1}{2})}\|u\|_{H^r(\partial\Omega)}.\end{aligned} \tag{188}$$

Notice that one of the $O(\epsilon^{-\frac{1}{2}})$ contributions comes from the use of the Poincaré inequality. This could have been avoided if the $H^{\frac{1}{2}}$-norm rather then $H^{\frac{1}{2}}$-seminorm were used in the projection over faces, cf. [13, p. 365].[9]

We discuss next the localization argument for spaces $\boldsymbol{H}^s(\text{curl}, \Omega)$. First of all, Proposition 3.3 generalizes to the multidimensional case. We have,

$$\|v\|_{H^{-t}(\partial\Omega)} \leq C \sum_f \|v\|_{H^{-t}(f)}, \quad \forall v \in H^{-t}(\partial\Omega) \tag{189}$$

where $0 \leq t < \frac{1}{2}$, and constant $C = O(\epsilon^{-\frac{1}{2}})$ for $t = \frac{1}{2} - \epsilon$. For $\boldsymbol{E}_t \in \boldsymbol{H}^{-t}(\text{curl}, \partial\Omega)$ (see [12] for precise definitions) this immediately implies that,

$$\|\boldsymbol{E}_t\|_{\boldsymbol{H}^{-t}(\partial\Omega)} \leq C \sum_f \|\boldsymbol{E}_t\|_{\boldsymbol{H}^{-t}(f)} \tag{190}$$

and,

$$\|\operatorname{curl} \boldsymbol{E}_t\|_{H^{-t}(\partial\Omega)} \leq C \sum_f \|\operatorname{curl} \boldsymbol{E}_t\|_{H^{-t}(f)}. \tag{191}$$

It remains only to argue that the restriction of the curl coincides with the curl of the restriction, i.e.,

$$\operatorname{curl}(\boldsymbol{E}_t|_f) = (\operatorname{curl} \boldsymbol{E}_t)|_f. \tag{192}$$

But this follows immediately from the definition of the distributional derivatives. The last identity remains true for the limiting case $t = \frac{1}{2}$, cf. [32, p. 104].

Using the same arguments as for the H^s-norms, we obtain the following estimates.

$$\begin{aligned}
\|\boldsymbol{E} - \Pi^{curl}_{s-1}\boldsymbol{E}\|_{\boldsymbol{H}^{s-1}(\text{curl},\partial\Omega)} &\leq C \sum_f \|\boldsymbol{E} - \Pi^{curl}_{s-1}\boldsymbol{E}\|_{\boldsymbol{H}^{s-1}(\text{curl},f)} \\
&\leq C p^{-(r-s)} \sum_f \|\boldsymbol{E}\|_{\boldsymbol{H}^{r-1}(\text{curl},f)} \\
&\leq C p^{-(r-s)} \|\boldsymbol{E}\|_{\boldsymbol{H}^{r-1}(\text{curl},\partial\Omega)} \\
\|\boldsymbol{E} - \Pi^{curl}_{-\frac{1}{2}}\boldsymbol{E}\|_{\boldsymbol{H}^{-\frac{1}{2}}(\text{curl},\partial\Omega)} &\leq C \ln p\, p^{-(r-\frac{1}{2})} \|\boldsymbol{E}\|_{\boldsymbol{H}^{r-1}(\text{curl},\partial\Omega)}.
\end{aligned} \tag{193}$$

Finally, we have the same results for the last operator.

$$\begin{aligned}
\|v - \Pi_{s-1}v\|_{H^{s-1}(\partial\Omega)} &\leq C \sum_f \|v - \Pi_{s-1}v\|_{H^{s-1}(f)} \\
&\leq C p^{-(r-s)} \sum_f \|v\|_{H^{r-1}(f)} \\
&\leq C p^{-(r-s)} \|v\|_{H^{r-1}(\partial\Omega)} \\
\|v - \Pi_{-\frac{1}{2}}v\|_{H^{-\frac{1}{2}}(v\partial\Omega)} &\leq C \ln p\, p^{-(r-\frac{1}{2})} \|v\|_{H^{r-1}(\partial\Omega)}.
\end{aligned} \tag{194}$$

[9] The use of the norm in place of the seminorm, however, destroys the commutativity property.

5 Commuting Projections and Projection-Based Interpolation Operators in Three Space Dimensions

5.1 Definitions and Commutativity

We shall consider the following diagram.

$$\begin{array}{ccccccccccc}
\mathbb{R} & \longrightarrow & H^r(\Omega) & \xrightarrow{\nabla} & \boldsymbol{H}^{r-1}(\mathbf{curl},\Omega) & \xrightarrow{\nabla\times} & \boldsymbol{H}^{r-1}(\mathrm{div},\Omega) & \xrightarrow{\nabla\cdot} & H^{r-1}(\Omega) & \longrightarrow & \{0\} \\
\downarrow & & P^{grad}\Big\downarrow \Pi^{grad} & & P^{curl}\Big\downarrow \Pi^{curl} & & P^{div}\Big\downarrow \Pi^{div} & & P\Big\downarrow & & \downarrow \\
\mathbb{R} & \longrightarrow & W_p & \xrightarrow{\nabla} & \boldsymbol{Q}_p & \xrightarrow{\nabla\times} & \boldsymbol{V}_p & \xrightarrow{\nabla\cdot} & Y_p & \longrightarrow & \{0\}
\end{array} \tag{195}$$

Here $r > \frac{3}{2}$, $\nabla\times$ denotes the vector-valued curl operator, and $\nabla\cdot$ is the scalar-valued divergence operator. By $\boldsymbol{H}^{r-1}(\mathbf{curl},\Omega)$ we understand the space of all vector-valued functions in $\boldsymbol{H}^{r-1}(\Omega)$ whose curl is in $\boldsymbol{H}^{r-1}(\Omega)$. Ω stands for a 3D element, a hexahedron, prism or tetrahedron, and $Q_p, \boldsymbol{W}_p, \boldsymbol{V}_p, Y_p$ denote any of the exact polynomial sequences defined on the element Ω, discussed in Sect. 2. The common property of those sequences is that the corresponding trace spaces for $Q_p, \boldsymbol{W}_p, \boldsymbol{V}_p$ corresponding to any face f, define 2D exact polynomial sequences discussed in the previous section.

The projection operators P^{grad}, P^{curl} and P^{div} are defined as follows.

$$\begin{cases} P^{grad}u =: u_p \in W_p \\ \|\nabla u_p - \nabla u\|_{L^2(\Omega)} \to \min \\ (u_p - u, 1)_{L^2(\Omega)} = 0 \end{cases} \tag{196}$$

$$\begin{cases} P^{curl}\boldsymbol{E} =: \boldsymbol{E}_p \in \boldsymbol{Q}_p \\ \|\nabla\times\boldsymbol{E}_p - \nabla\times\boldsymbol{E}\|_{L^2(\Omega)} \to \min \\ (\boldsymbol{E}_p - \boldsymbol{E}, \nabla\phi)_{L^2(\Omega)} = 0,\ \forall\phi\in W_p \end{cases} \tag{197}$$

$$\begin{cases} P^{div}\boldsymbol{v} =: \boldsymbol{v}_p \in \boldsymbol{V}_p \\ \|\nabla\cdot\boldsymbol{v}_p - \nabla\cdot\boldsymbol{v}\|_{L^2(\Omega)} \to \min \\ (\boldsymbol{v}_p - \boldsymbol{v}, \nabla\times\boldsymbol{\phi})_{L^2(\Omega)} = 0,\ \forall\boldsymbol{\phi}\in\boldsymbol{Q}_p \end{cases} \tag{198}$$

and P denotes the orthogonal projection onto Y_p in the L^2-norm.

Exercise 5.1. Show that the projections defined above make the diagram (195) commute.

The projection-based interpolation operators are defined as follows.

$$\begin{cases} \Pi^{grad} u =: u_p \in W_p \\ u_p = \Pi^{grad}_{\frac{1}{2}} u \text{ on } \partial\Omega \\ \|\nabla u_p - \nabla u\|_{L^2(\Omega)} \to \min \end{cases} \tag{199}$$

$$\begin{cases} \Pi^{curl} \boldsymbol{E} =: \boldsymbol{E}_p \in \boldsymbol{Q}_p \\ \boldsymbol{E}_{t,p} = \Pi^{curl}_{-\frac{1}{2}} \boldsymbol{E}_t \text{ on } \partial\Omega \\ \|\nabla \times \boldsymbol{E}_p - \nabla \times \boldsymbol{E}\|_{L^2(\Omega)} \to \min \\ (\boldsymbol{E}_p - \boldsymbol{E}, \nabla\phi)_{L^2(\Omega)} = 0, \ \forall \phi \in W_p \ : \ \phi = 0 \text{ on } \partial\Omega \end{cases} \tag{200}$$

and

$$\begin{cases} \Pi^{div} \boldsymbol{v} =: \boldsymbol{v}_p \in \boldsymbol{V}_p \\ v_{n,p} = \Pi_{-\frac{1}{2}} v_n \text{ on } \partial\Omega \\ \|\nabla \cdot v_p - \nabla \cdot v\|_{L^2(\Omega)} \to \min \\ (\boldsymbol{v}_p - \boldsymbol{v}, \nabla \times \boldsymbol{\phi})_{L^2(\Omega)} = 0, \ \forall \boldsymbol{\phi} \in \boldsymbol{Q}_p \ : \ \phi_t = 0 \text{ on } \partial\Omega. \end{cases} \tag{201}$$

Here $\Pi^{grad}_{\frac{1}{2}}, \Pi^{curl}_{-\frac{1}{2}}, \Pi_{-\frac{1}{2}}$ are the 2D interpolation operators discussed in the previous section, $\boldsymbol{E}_t$, $\boldsymbol{E}_{t,p}$ denote the tangential component of $\boldsymbol{E}$, $\boldsymbol{E}_p$, and $v_n, v_{n,p}$ denote the normal component of $\boldsymbol{v}, \boldsymbol{v}_p$ on the boundary $\partial\Omega$ respectively. Notice again that all minimization problems are constrained-minimization problems—the boundary values of the interpolants are fixed. Similarly to 1D and 2D, the projection operators can be interpreted as local minimization problems with Neumann boundary conditions, while the interpolation operators employ local Dirichlet boundary conditions implemented by means of the 2D interpolation operators.

Definition of Tangential and Normal Traces on the Boundary

For $\boldsymbol{E}, \boldsymbol{v} \in \boldsymbol{H}^r, \frac{1}{2} < r < \frac{3}{2}$, the tangential and normal components are understood in the sense of the Trace Theorem. The definition for the range $-\frac{1}{2} < r < \frac{1}{2}$ is again more complicated. The starting point for defining the normal component is the Gauss Theorem,

$$\int_\Omega \nabla \cdot \boldsymbol{v} \, \phi = - \int_\Omega \boldsymbol{v} \cdot \nabla\phi + \int_{\partial\Omega} v_n \, \phi. \tag{202}$$

Taking a test function $\phi \in H^{\frac{1}{2}-r}(\partial\Omega)$, we consider an extension Φ that is bounded in $H^{1-r}(\Omega)$-norm by the $H^{\frac{1}{2}-r}(\partial\Omega)$ norm of ϕ. An argument identical to the one used when defining the tangential component E_t in the previous section, leads to the estimate

$$\|v_n\|_{H^{r-\frac{1}{2}}(\partial\Omega)} \le C\left(\|\mathrm{div}\boldsymbol{v}\|_{H^s(\Omega)} + \|\boldsymbol{v}\|_{\boldsymbol{H}^r(\Omega)}\right) \tag{203}$$

where $s > -\frac{1}{2}$ and can be taken equal to r. The estimate can again be considered as an equivalent of the Trace Theorem. Constant C is of order $O(\epsilon^{-\frac{1}{2}})$ for $r = -\frac{1}{2} + \epsilon$ or $r = \frac{1}{2} - \epsilon$. For $r = -\frac{1}{2}, \frac{1}{2}$, the normal trace cannot be defined.

The definition of the tangential component is more technical. Referring to [14, 12] for details, we sketch the main idea only. Again, we consider first a sufficiently regular field $\boldsymbol{E} \in \boldsymbol{H}^r(\Omega)$, and a test function $\boldsymbol{\phi} \in \boldsymbol{H}^{\frac{1}{2}-r}(\partial\Omega)$, to invoke the integration by parts formula:

$$\int_{\partial\Omega} \boldsymbol{E}_t(\boldsymbol{n} \times \boldsymbol{\phi}) = \int_\Omega (\nabla \times \boldsymbol{E})\boldsymbol{\Phi} - \int_\Omega \boldsymbol{E}(\nabla \times \boldsymbol{\Phi}). \tag{204}$$

Here $\boldsymbol{\Phi} \in \boldsymbol{H}^{1-r}(\Omega)$ is an extension of $\boldsymbol{\phi}$ (i.e. $\boldsymbol{n} \times \boldsymbol{\Phi}|_{\partial\Omega} = \boldsymbol{\phi}$) such that,

$$\|\boldsymbol{\Phi}\|_{\boldsymbol{H}^{1-r}(\Omega)} \le C\|\boldsymbol{\phi}\|_{\boldsymbol{H}^{\frac{1}{2}-r}(\partial\Omega)} \tag{205}$$

The following estimate follows,

$$\|\boldsymbol{E}_t\|_{\boldsymbol{H}^{r-\frac{1}{2}}(\partial\Omega)} \le C\left(\|\nabla \times \boldsymbol{E}\|_{\boldsymbol{H}^s(\Omega)} + \|\boldsymbol{E}\|_{\boldsymbol{H}^r(\Omega)}\right) \tag{206}$$

where $-\frac{1}{2} < s < \frac{1}{2}$. Next we employ a special test function $\boldsymbol{\phi} = \nabla_{\partial K}\phi, \phi \in H^{\frac{1}{2}-s}(\partial\Omega)$, and consider an extension $\Phi \in H^{1-s}(\Omega)$ of potential ϕ. Integrating the boundary term by parts, we get,

$$\int_{\partial\Omega} (\mathrm{curl}_{\partial\Omega}\, \boldsymbol{E}_t)\phi = \int_{\partial\Omega} \boldsymbol{E}_t \nabla_{\partial\Omega} \times \phi = \int_\Omega (\nabla \times \boldsymbol{E})\nabla\Phi. \tag{207}$$

This yields an additional estimate,

$$\|\,\mathrm{curl}_{\partial\Omega}\, \boldsymbol{E}_t\|_{H^{s-\frac{1}{2}}(\partial\Omega)} \le C\|\nabla \times \boldsymbol{E}\|_{\boldsymbol{H}^s(\Omega)}. \tag{208}$$

Setting $s = r$ and combining the two estimates, we get a "Trace Theorem" for the $\boldsymbol{H}(\mathbf{curl})$ space.

$$\|\boldsymbol{E}_t\|_{\boldsymbol{H}^{r-\frac{1}{2}}(\mathrm{curl},\partial\Omega)} \le C\left(\|\nabla \times \boldsymbol{E}\|_{\boldsymbol{H}^r(\Omega)} + \|\boldsymbol{E}\|_{\boldsymbol{H}^r(\Omega)}\right). \tag{209}$$

The blow up of the constants prohibits extending the definition to values $s, r = -\frac{1}{2}, \frac{1}{2}$ and the constants are of order $O(\epsilon^{-\frac{1}{2}})$ for $s, r = -\frac{1}{2} + \epsilon$ or $s, r = \frac{1}{2} - \epsilon$. In what follows, we shall use the inequalities (209) and (203) for the case of $r = 0$ only.

Theorem 5.1. *The interpolation operators make the diagram (195) commute.*

Proof. The commutativity of the first block follows from the fact that operator Π_s^{grad} preserves constants. In order to show the commutativity of the second block, we need to demonstrate that,

$$\Pi^{curl}(\nabla u) = \nabla(\Pi^{grad} u). \tag{210}$$

Let $\boldsymbol{E} = \nabla u$. By the commutativity of the 2D diagram, we have,

$$\boldsymbol{E}_{t,p} = \Pi^{curl}_{-\frac{1}{2}} \nabla_{\partial\Omega} u = \nabla_{\partial\Omega} u_p \tag{211}$$

where $u_p = \Pi^{grad}_{\frac{1}{2}} u$ and $\nabla_{\partial\Omega}$ denotes the tangential gradient on the boundary of the element. Consequently,

$$\int_{\Omega} \nabla \times \boldsymbol{E}_p = \int_{\partial\Omega} \boldsymbol{n} \times \nabla_{\partial\Omega} u_p = 0. \tag{212}$$

At the same time,

$$(\text{curl}\boldsymbol{E}_p, \text{curl}\boldsymbol{F})_{L^2(\Omega)} = 0 \tag{213}$$

for every $\boldsymbol{F} \in \boldsymbol{Q}_p, \boldsymbol{F}_t = 0$ on $\partial\Omega$. However, the image of such polynomials $\boldsymbol{F}$ under the curl operator, coincides exactly with polynomials in $\boldsymbol{V}_p$ with zero average and zero divergence, where the curl of $\boldsymbol{E}_p$ lives. Consequently, $\text{curl}\boldsymbol{E}_p = 0$ and $\boldsymbol{E}_p = \nabla u_p$ for some $u_p \in W_p$. Substituting ∇u_p into $(200)_4$ we learn that $u_p = \Pi^{grad} u$.

To prove the next commutativity property, we need to show that,

$$\Pi^{div}(\nabla \times \boldsymbol{E}) = \nabla \times (\Pi^{curl} \boldsymbol{E}) \tag{214}$$

Let $\boldsymbol{v} = \nabla \times \boldsymbol{E}$. By the commutativity of the 2D diagram, we have,

$$v_{n,p} = \Pi_{-\frac{1}{2}}(\text{curl}_{\partial\Omega} \boldsymbol{E}_t) = \text{curl}_{\partial\Omega} \boldsymbol{E}_{t,p} \tag{215}$$

where $\boldsymbol{E}_{t,p} = \Pi^{curl}_{-\frac{1}{2}} \boldsymbol{E}_t$ and $\text{curl}_{\partial\Omega}$ denotes the surface curl on the boundary of the element. Consequently,

$$\int_{\Omega} \nabla \cdot \boldsymbol{v}_p = -\int_{\partial\Omega} v_{n,p} = -\int_{\partial\Omega} \text{curl}_{\partial\Omega} \boldsymbol{E}_{t,p} = 0. \tag{216}$$

At the same time,

$$(\text{div}\, \boldsymbol{v}_p, \text{div}\, \boldsymbol{w})_{L^2(\Omega)} = 0 \tag{217}$$

for every $\boldsymbol{w} \in \boldsymbol{V}_p, w_n = 0$ on $\partial\Omega$. However, the image of such polynomials $\boldsymbol{w}$ under the div operator, coincides exactly with polynomials in Y_p with zero average, where the div of $\boldsymbol{v}_p$ lives. Consequently, $\text{div}\, \boldsymbol{v}_p = 0$ and $\boldsymbol{v}_p = \nabla \times \boldsymbol{E}_p$ for some $\boldsymbol{E}_p \in \boldsymbol{Q}_p$. Substituting $\nabla \times \boldsymbol{E}_p$ into $(201)_4$, we learn that $\boldsymbol{E}_p = \Pi^{curl} \boldsymbol{E}$.

To prove the last commutativity property, we need to show that,

$$P(\nabla \cdot \boldsymbol{v}) = \nabla \cdot (\Pi^{div} \boldsymbol{v}). \tag{218}$$

By the commutativity of the 2D diagram, we have,

$$\int_{\Omega} \text{div}(\boldsymbol{v}_p - \boldsymbol{v}) = -\int_{\partial\Omega} \boldsymbol{n} \cdot (\Pi^{div} \boldsymbol{v} - \boldsymbol{v}) = -\int_{\partial\Omega} (\Pi_{-\frac{1}{2}} v_n - v_n) = 0. \tag{219}$$

At the same time, the image of functions from $\boldsymbol{V}_p$ with zero normal traces coincides with the subspaces of functions from Y_p with zero average. Combining (219) with $(201)_4$ we obtain that,

$$\int_\Omega \operatorname{div}(\boldsymbol{v}_p - \boldsymbol{v})\, w_p = 0, \quad \forall w_p \in Y_p. \tag{220}$$

Thus (218) holds.

5.2 Polynomial Preserving Extension Operators

We shall utilize the existence of commuting and polynomial preserving extension operators,

$$\begin{array}{ccccccccc}
\mathbb{R} \longrightarrow & H^1(\Omega) & \xrightarrow{\nabla} & \boldsymbol{H}(\mathbf{curl}, \Omega) & \xrightarrow{\nabla\times} & \boldsymbol{H}(\operatorname{div}, \Omega) & \xrightarrow{\nabla\cdot} & L^2(\Omega) & \\
& \uparrow Ext & & \uparrow Ext^{curl} & & \uparrow Ext^{div} & & & \\
\mathbb{R} \longrightarrow & H^{\frac{1}{2}}(\partial\Omega) & \xrightarrow{\nabla_{\partial\Omega}} & \boldsymbol{H}^{-\frac{1}{2}}(\operatorname{curl}, \partial\Omega) & \xrightarrow{\operatorname{curl}_{\partial\Omega}} & H^{-\frac{1}{2}}(\partial\Omega) & \xrightarrow{\int_{\partial\Omega}} & \mathbb{R} &
\end{array} \tag{221}$$

that are right inverses of the trace operators for the energy spaces $H^1(\Omega)$, $\boldsymbol{H}(\mathbf{curl}, \Omega), \boldsymbol{H}(\operatorname{div}, \Omega)$.

I am aware of five existing contributions addressing the existence of 3D polynomial preserving, extension operators for the H^1 space. Belgacem was the first one to construct such an operator for the cube, see [5], and Munoz-Sola [33] constructed such an operator for a tetrahedral element. In [6] Bernardi, Dauge and Maday provided and alternative construction for a cube. An elementary construction for a tetrahedron has recently been shown in [22]. Finally, a family of polynomial extension operators with a bound *independent of* polynomial degree *p*, was constructed in [15].

To my best knowledge, construction of analogous extension operators for the $\boldsymbol{H}(\mathbf{curl}, \Omega)$ and $\boldsymbol{H}(\operatorname{div}, \Omega)$-spaces has been tackled only in two contributions. Costabel, Dauge and Demkowicz [15] constructed a family of such polynomial extension operators (with a bound independent of polynomial degree *p*) on a cube, and Schoeberl, Gopalakrishnan and Demkowicz [22] constructed a polynomial preserving extension operator for the $\boldsymbol{H}(\mathbf{curl})$-space on a tetrahedron. We hope to present an analogous, final result for the $\boldsymbol{H}(\operatorname{div})$-space in a forthcoming report.

In view of results mentioned above, construction of polynomial preserving extension operators for a prism seems to be feasible but has not been published yet.

5.3 Polynomial Preserving, Right-Inverses of Grad, Curl, and Div Operators: Discrete Friedrichs Inequalities

The following three operators, sometimes known as Poincare's maps, have been studied in [25].

$$G : \boldsymbol{H}(\mathbf{curl}, \Omega) \to H^1(\Omega) \qquad (GE)(\boldsymbol{x}) = \boldsymbol{x} \cdot \int_0^1 \boldsymbol{E}(t\boldsymbol{x})\, dt$$
$$K : \boldsymbol{H}(\mathrm{div}, \Omega) \to \boldsymbol{H}(\mathbf{curl}, \Omega) \; (K\boldsymbol{v})(\boldsymbol{x}) = -\boldsymbol{x} \times \int_0^1 t\boldsymbol{v}(t\boldsymbol{x})\, dt \tag{222}$$
$$D : L^2(\Omega) \to \boldsymbol{H}(\mathrm{div}, \Omega) \qquad (Dw)(\boldsymbol{x}) = \boldsymbol{x} \int_0^1 t^2 w(t\boldsymbol{x})\, dt$$

Exercise 5.2. Prove the following statements.

- The operators G, K, D satisfy the following identities,

$$w = \nabla \cdot (Dw), \qquad \forall w \in L^2(\Omega)$$
$$\boldsymbol{v} = \nabla \times (K\boldsymbol{v}) + D(\nabla \cdot \boldsymbol{v}), \quad \forall \boldsymbol{v} \in \boldsymbol{H}(\mathrm{div}, \Omega) \tag{223}$$
$$\boldsymbol{E} = \nabla(G\boldsymbol{E}) + K(\nabla \times \boldsymbol{E}), \quad \forall \boldsymbol{E} \in \boldsymbol{H}(\mathbf{curl}, \Omega)$$

- This implies that operators G, K, D, restricted to the range of operators grad, curl and div, respectively, are their right-inverses, i.e.,

$$\nabla \cdot (Dw) = w, \qquad \forall w \in L^2(\Omega)$$
$$\nabla \times (K\boldsymbol{v}) = \boldsymbol{v}, \quad \forall \boldsymbol{v} \in \boldsymbol{H}(\mathrm{div}, \Omega) : \nabla \cdot \boldsymbol{v} = 0 \tag{224}$$
$$\nabla(G\boldsymbol{E}) = \boldsymbol{E}, \qquad \forall \boldsymbol{E} \in \boldsymbol{H}(\mathbf{curl}, \Omega) : \nabla \times \boldsymbol{E} = \boldsymbol{0}$$

- Prove that the operators are continuous (you may assume that Ω is the master tetrahedron, hexahedron, or prism).
- Prove that, for all discussed[10] elements: the Nédélec tetrahedra of the first and second types, the Nédélec hexahedron of constant polynomial order, operators G, K, D map the corresponding "face element" spaces into the "edge element" spaces.

Lemma 5.1 (Discrete Friedrichs Inequalities for $\boldsymbol{H}(\mathbf{curl}, \Omega)$ space). *Let Ω be the master tetrahedron or hexahedron of uniform order p. There exists a constant $C > 0$ such that,*

$$\|\boldsymbol{E}\|_{L^2(\Omega)} \le C \|\nabla \times \boldsymbol{E}\|_{L^2(\Omega)}, \tag{225}$$

for every discrete divergence free polynomial $\boldsymbol{E} \in \boldsymbol{Q}_p$ belonging to one of the two families,

Case 1:

$$(\boldsymbol{E}, \nabla\phi)_{L^2(\Omega)} = 0, \quad \forall \phi \in W_p \tag{226}$$

Case 2: $\boldsymbol{E}_t = \boldsymbol{0}$ *on $\partial\Omega$, and*

$$(\boldsymbol{E}, \nabla\phi)_{L^2(\Omega)} = 0, \quad \forall \phi \in W_p : \phi = 0 \text{ on } \partial\Omega \tag{227}$$

[10] The result is true for the prismatic elements as well.

Proof. **Case 1** follows immediately from the continuity of the right-inverse K,

$$\begin{aligned}\|\boldsymbol{E}\|_{L^2(\Omega)} &= \inf_{\phi\in W_p} \|\boldsymbol{E} - \nabla\phi\|_{L^2(\Omega)} \\ &\leq \|\boldsymbol{E} - (\boldsymbol{E} - K(\nabla\times\boldsymbol{E}))\|_{L^2(\Omega)} \\ &\leq C\|K(\nabla\times\boldsymbol{E})\|_{L^2(\Omega)} \\ &\leq C\|\nabla\times\boldsymbol{E}\|_{L^2(\Omega)}.\end{aligned} \tag{228}$$

Case 2: In order to account for the homogeneous boundary conditions, operator K has to be modified. Let $E \in \boldsymbol{Q}_p$ be a divergence free polynomial with zero trace. Consider $\boldsymbol{E} - K(\nabla\times\boldsymbol{E})$ where K is the right-inverse of the curl operator defined above. There exists then a polynomial $\psi \in W_p$ such that, $\boldsymbol{E} - K(\nabla\times\boldsymbol{E}) = \nabla\psi$. Then,

$$\boldsymbol{n}\times(\nabla\psi) = -\boldsymbol{n}\times(K(\nabla\times\boldsymbol{E})) \tag{229}$$

as $\boldsymbol{E}$ has a zero trace. Let $\Psi = Ext(tr\psi)$ where Ext is the polynomial preserving extension operator for the H^1-space. We have,

$$\begin{aligned}\|\nabla\Psi\|_{L^2(\Omega)} &\leq C\|\psi\|_{H^{\frac{1}{2}}(\partial\Omega)} \leq C\|\boldsymbol{n}\times(\nabla\psi)\|_{H^{-\frac{1}{2}}(\partial\Omega)} \\ &\leq C\|K(\nabla\times\boldsymbol{E})\|_{\boldsymbol{H}(\mathrm{curl},\Omega)} \leq C\|\nabla\times\boldsymbol{E}\|_{L^2(\Omega)}.\end{aligned} \tag{230}$$

Finally,

$$\begin{aligned}\|\boldsymbol{E}\|_{L^2(\Omega)} &= \inf_{\phi\in W_p, \phi=0 \text{ on } \partial\Omega} \|\boldsymbol{E} - \nabla\phi\|_{L^2(\Omega)} \\ &\leq \|\boldsymbol{E} - (\boldsymbol{E} - K(\nabla\times\boldsymbol{E}) - \nabla\Psi)\|_{L^2(\Omega)} \\ &\leq C\|K(\nabla\times\boldsymbol{E}) + \nabla\Psi\|_{L^2(\Omega)} \\ &\leq C\|\nabla\times\boldsymbol{E}\|_{L^2(\Omega)}.\end{aligned} \tag{231}$$

The crucial fact is that the correction $\nabla\Psi$ is controlled only by the trace of the $K(\nabla\times\boldsymbol{E})$ and, consequently, by the L^2-norm of $\nabla\times\boldsymbol{E}$ only.

A generalization to elements of variable order is nontrivial. The following reasoning has been put forth in [20] for the case of the tetrahedral element of variable order with an additional technical assumption that the polynomial order for the element edges is set to the minimum of the order for neighboring faces. The idea is based on the observation that the operator K preserves some of the local properties of the polynomial order of approximation. Let $\boldsymbol{v} \in \boldsymbol{V}_p$, $\nabla\cdot\boldsymbol{v} = 0$. It is easy to verify that the tangential components of $K\boldsymbol{v}$ along the coordinate axes are zero and that the order of $K\boldsymbol{v}$ for three faces neighboring the origin matches the order of approximation in space $\boldsymbol{V}_p$ (increased by one). The order of the sloped face is implied by the order of the (interior of the) tetrahedron. By using the Piola transformation (60), we extend the construction of the operator to other vertices of the tetrahedron arriving at four maps $K_i, i = 0, \ldots, 3$, each corresponding to one of the vertices.

The second observation is the possibility of a stable decomposition

$$v = v_0 + \sum_{i=1}^{4} v_i, \text{ where } \quad v_i \cdot n = 0 \text{ on face } f_j, j = 1, \ldots, i-1, \quad i = 1, \ldots, 4 \tag{232}$$

and v_0 has a zero normal trace. Additionally,

$$\|v_i\|_{H(\mathrm{div},\Omega)} \leq C\|v\|_{H(\mathrm{div},\Omega)}, \quad i = 0, \ldots, 4. \tag{233}$$

Such a decomposition follows from the construction of polynomial-preserving extension operators for the tetrahedron, and the possibility of extending boundary values from one, two, three, and four faces (one-face, two-face, etc. extension operators). The logic of the decomposition is as follows. We pick a face f_1. Then,

$$\|v_n\|_{H^{-\frac{1}{2}}(f_1)} \leq \|v_n\|_{H^{-\frac{1}{2}}(\partial\Omega)} \leq C\|v\|_{H(\mathrm{div},\Omega)}. \tag{234}$$

Let v_1 be then a stable extension of the restriction of v_n to face f_1. We subtract v_1 from v and apply the same procedure to the union of two faces $f_1 \cup f_2$. Continuing in this manner we end up with the decomposition above.

Enumerating the faces in the order of increasing polynomial order for the normal component on the face, we construct the final right-inverse in the form,

$$Kv = \sum_{i=1}^{4} K_i v_i + K_0 v_0 \tag{235}$$

where K_0 is the operator constructed in the proof of Lemma 5.1 for the case with homogeneous boundary conditions. Notice that each component v_i shares the order of the face and polynomial $K_i v_i$ has zero trace on faces of lower order.

The idea does not extend to the hedrahedral element.

Lemma 5.2 (Discrete Friedrichs Inequality for (div) space).

Let Ω be a tetrahedral or hexahedral element of an arbitrary variable order. There exists a constant $C > 0$ such that,

$$\|v\|_{L^2(\Omega)} \leq C\|\nabla \cdot v\|_{L^2(\Omega)}, \tag{236}$$

for every discrete curl free polynomial v belonging to either of the two families:
Case 1: $v \in V_p$ *and,*

$$(v, \nabla \times \phi)_{L^2(\Omega)} = 0, \quad \forall \phi \in Q_p. \tag{237}$$

Case 2: $v \in V_p$, $v \cdot n = 0$ *on* $\partial\Omega$, *and,*

$$(v, \nabla \times \phi)_{L^2(\Omega)} = 0, \quad \forall \phi \in Q_p : \phi_t = 0 \text{ on } \partial\Omega. \tag{238}$$

Constant C is independent of polynomial order p.

Proof. Let $w \in L^2(\Omega)$. We begin by decomposing w into a constant and a function with zero average,

$$w = c + w_0, \qquad \int_\Omega w_0 = 0. \tag{239}$$

The normal component of function Dw_0 vanishes at the three faces adjacent to the origin but it may be nonzero on the remaining faces. Notice that the normal trace of Dw_0 has a zero average over the boundary and, therefore, it may be identified with a (surface) curl of an element from the trace space of $\boldsymbol{Q}_p$. Due to the commutativity of extension operators (221), extension $Ext^{div} Tr Dw_0$ has zero divergence. Consequently, map,

$$w_0 \to Dw_0 - Ext^{div} Tr Dw_0 \tag{240}$$

maps functions of zero average into functions with zero trace, and it is bounded. Let $\boldsymbol{v}_c \in \boldsymbol{V}_p$ be now any first order polynomial with divergence equal one. Map,

$$D^{mod} w = c\boldsymbol{v}_c + Dw_0 - Ext^{div} Tr Dw_0 \tag{241}$$

is a bounded, polynomial preserving right-inverse of the div operator. We conclude,

$$\begin{aligned} \|\boldsymbol{v}\|_{L^2(\Omega)} &= \inf_{\boldsymbol{\phi}} \|\boldsymbol{E} - \nabla \times \boldsymbol{\phi}\|_{L^2(\Omega)} \\ &\leq \|\boldsymbol{v} - (\boldsymbol{v} - D^{mod}(\nabla \cdot \boldsymbol{v}))\|_{L^2(\Omega)} \\ &\leq C\|D^{mod}(\nabla \cdot \boldsymbol{v})\|_{L^2(\Omega)} \\ &\leq C\|\nabla \cdot \boldsymbol{v}\|_{L^2(\Omega)} . \end{aligned} \tag{242}$$

In Case 1, the infimum is taken over the whole space $\boldsymbol{W}_p$, in Case 2, the infimum is taken over the subspace of $\boldsymbol{W}_p$ of functions with zero trace.

5.4 Projection and Interpolation Error Estimates

We record yet the classical result.

Lemma 5.3 (Poincaré's inequalities).
There exist $C > 0$ such that,

$$\|u\|_{L^2(\Omega)} \leq C\|\nabla u\|_{L^2(\Omega)} ,$$

for every function $u \in H^1(\Omega)$ belonging to either of the two families:
Case 1: $(u, 1)_{L^2(\Omega)} = 0$,
Case 2: $u = 0$ *on* $\partial\Omega$.

In particular, both inequalities hold on the discrete level, for polynomials $u \in W_p$.

Equipped with the Poincaré and discrete Friedrichs inequalities, we can reproduce the arguments used in the previous sections, to conclude the error estimates for both the commuting projection and the projection-based interpolation operators.

Determining each of the projections $P^{grad} u, P^{curl} \boldsymbol{E}, P^{div} \boldsymbol{v}$ can be interpreted as the solution of a constrained minimization problem that leads to a mixed formulation with the Lagrange multiplier equal zero.

Exercise 5.3. Write out the mixed formulations corresponding to the definition of the commuting projections.

In each of the three cases, the first Brezzi's inf–sup condition is automatically implied by the discrete exact sequence property, with constant $\beta = 1$. The inf–sup in kernel condition is implied by the Poincaré and discrete Friedrichs inequalities, Case 1 (with no boundary conditions).

Equivalently, one can prove the projection error estimates by using directly the constructed right-inverses of operators grad, curl, div,

$$\begin{array}{ccccccc}
H^1(\Omega) & \xrightarrow{\nabla} & \boldsymbol{H}(\mathrm{curl},\Omega) & \xrightarrow{\nabla\times} & \boldsymbol{H}(\mathrm{div},\Omega) & \xrightarrow{\nabla\cdot} & L^2(\Omega) \\
P^{grad}\big\downarrow P^{grad0} & & P^{curl}\big\downarrow P^{curl0} & & P^{div}\big\downarrow P^{div0} & & \big\downarrow P \\
W_p & \xleftarrow{G} & \boldsymbol{Q}_p & \xleftarrow{K} & \boldsymbol{V}_p & \xleftarrow{D} & Y_p .
\end{array} \tag{243}$$

Here $P^{grad0}, P^{curl0}, P^{div0}$ denote projections onto the subspaces of polynomials with zero grad, curl or div, i.e. on constants, gradients and curls of polynomials. With those inverses in hand, one can represent the projection operators in somehow less intuitive but more compact form [25],

$$\begin{aligned}
P^{div}\boldsymbol{F} &= P_0^{div}(\boldsymbol{F} - DP(\nabla\cdot\boldsymbol{F})) + DP(\nabla\cdot\boldsymbol{F}) \\
P^{curl}\boldsymbol{E} &= P_0^{curl}(\boldsymbol{E} - KP_0^{div}(\nabla\times\boldsymbol{E})) + KP_0^{div}(\nabla\times\boldsymbol{E}) \\
P^{grad}\boldsymbol{F} &= P_0^{grad}(u - GP_0^{curl}(\nabla u)) + GP_0^{curl}(\nabla u) .
\end{aligned} \tag{244}$$

The representations imply the continuity of the commuting projections. As all of them are also preserving the polynomial spaces, this implies in turn their optimality—the projection errors can be bounded by the *best approximation errors* in norms $H^1, \boldsymbol{H}(\mathrm{curl}), \boldsymbol{H}(\mathrm{div})$, respectively.

Theorem 5.2. *There exist constants $C > 0$, independent of p such that,*

$$\begin{aligned}
\|u - P^{grad}u\|_{H^1(\Omega)} &\le C \inf_{u_p\in W_p} \|u - u_p\|_{H^1(\Omega)} \\
&\le Cp^{-(r-1)}\|u\|_{H^r(\Omega)}, \\
&\qquad \forall u \in H^r(\Omega),\ r > 1 \\
\|\boldsymbol{E} - P^{curl}\boldsymbol{E}\|_{\boldsymbol{H}(\mathbf{curl},\Omega)} &\le C \inf_{\boldsymbol{E}_p\in\boldsymbol{Q}_p} \|\boldsymbol{E} - \boldsymbol{E}_p\|_{\boldsymbol{H}(\mathbf{curl},\Omega)} \\
&\le Cp^{-(r-1)}\|\boldsymbol{E}\|_{\boldsymbol{H}^{r-1}(\mathbf{curl},\Omega)}, \\
&\qquad \forall \boldsymbol{E} \in \boldsymbol{H}^{r-1}(\mathbf{curl},\Omega),\ r > 1 \\
\|\boldsymbol{v} - P^{div}\boldsymbol{v}\|_{\boldsymbol{H}(\mathrm{div},\Omega)} &\le C \inf_{\boldsymbol{v}_p\in\boldsymbol{V}_p} \|\boldsymbol{v} - \boldsymbol{v}_p\|_{\boldsymbol{H}(\mathrm{div},\Omega)} \\
&\le Cp^{-(r-1)}\|\boldsymbol{v}\|_{\boldsymbol{H}^{r-1}(\mathrm{div},\Omega)}, \\
&\qquad \forall \boldsymbol{v} \in \boldsymbol{H}^{r-1}(\mathrm{div},\Omega),\ r > 1.
\end{aligned} \tag{245}$$

Proof. For the best approximation results, see [40].

The reasoning leading to the interpolation error estimates is identical for all three cases. We shall discuss the $\boldsymbol{H}(\mathbf{curl})$ case, with the remaining two being fully analogous. We begin by comparing the projection and interpolation errors.

$$\begin{aligned}
&\|\boldsymbol{E} - \Pi^{curl}\boldsymbol{E}\|_{\boldsymbol{H}(\mathbf{curl},\Omega)} \\
&\quad \leq \|\boldsymbol{E} - P^{curl}\boldsymbol{E}\|_{\boldsymbol{H}(\mathbf{curl},\Omega)} + \|P^{curl}\boldsymbol{E} - \Pi^{curl}\boldsymbol{E}\|_{\boldsymbol{H}(\mathbf{curl},\Omega)}.
\end{aligned} \tag{246}$$

It follows from the definitions of the projection and interpolation operators that function $\boldsymbol{\psi} = P^{curl}\boldsymbol{E} - \Pi^{curl}\boldsymbol{E}$ is a discrete divergence-free, minimum energy extension of its boundary values, with the energy measured using the L^2-norm of the curl. The discrete Friedrichs inequality, case 2, implies that the $\boldsymbol{H}(\mathbf{curl})$-norm of $\boldsymbol{\psi}$ is bounded by the norm of an analogous discrete divergence-free, minimum energy extension with the energy measured using the full $\boldsymbol{H}(\mathbf{curl})$-norm. Indeed, let $\boldsymbol{\phi}$ be such an extension. We have,

$$\begin{aligned}
\|\boldsymbol{\psi}\|_{\boldsymbol{H}(\mathbf{curl},\Omega)} &\leq C(\|\boldsymbol{\psi}\|_{L^2(\Omega)} + \|\nabla \times \boldsymbol{\psi}\|_{L^2(\Omega)}) \\
&\leq C(\|\boldsymbol{\psi} - \boldsymbol{\phi}\|_{L^2(\Omega)} + \|\boldsymbol{\phi}\|_{L^2(\Omega)} + \|\nabla \times \boldsymbol{\psi}\|_{L^2(\Omega)}) \\
&\leq C(\|\nabla \times (\boldsymbol{\psi} - \boldsymbol{\phi})\|_{L^2(\Omega)} + \|\boldsymbol{\phi}\|_{L^2(\Omega)} + \|\nabla \times \boldsymbol{\phi}\|_{L^2(\Omega)}) \\
&\leq C(\|\nabla \times \boldsymbol{\psi}\|_{L^2(\Omega)} + \|\nabla \times \boldsymbol{\phi}\|_{L^2(\Omega)} \\
&\quad + \|\boldsymbol{\phi}\|_{L^2(\Omega)} + \|\nabla \times \boldsymbol{\phi}\|_{L^2(\Omega)}) \\
&\leq C(\|\nabla \times \boldsymbol{\phi}\|_{L^2(\Omega)} + \|\boldsymbol{\phi}\|_{L^2(\Omega)} + \|\nabla \times \boldsymbol{\phi}\|_{L^2(\Omega)}) \\
&\leq C\|\boldsymbol{\phi}\|_{\boldsymbol{H}(\mathbf{curl},\Omega)}.
\end{aligned} \tag{247}$$

We can invoke now the argument with polynomial preserving extension operators (which use the full norms...) to arrive at the final conclusion.

$$\begin{aligned}
\|\boldsymbol{\psi}\|_{\boldsymbol{H}(\mathbf{curl},\Omega)} &\leq C\|Ext^{curl}\|\|Tr(P^{curl}\boldsymbol{E} - \Pi^{curl}\boldsymbol{E})\|_{\boldsymbol{H}^{-\frac{1}{2}}(\mathrm{curl},\partial\Omega)} \\
&\leq C\|Ext^{curl}\|\Big(\|Tr(P^{curl}\boldsymbol{E} - \boldsymbol{E})\|_{\boldsymbol{H}^{-\frac{1}{2}}(\mathrm{curl},\partial\Omega)} \\
&\quad + \|Tr(\boldsymbol{E} - \Pi^{curl}\boldsymbol{E})\|_{\boldsymbol{H}^{-\frac{1}{2}}(\mathrm{curl},\partial\Omega)}\Big) \\
&\leq C\|Ext^{curl}\|\Big(C_{tr}\|(P^{curl}\boldsymbol{E} - \boldsymbol{E})\|_{\boldsymbol{H}(\mathrm{curl},\Omega)} \\
&\quad + \|\boldsymbol{E}_t - \Pi^{curl}_{-\frac{1}{2}}\boldsymbol{E}_t)\|_{\boldsymbol{H}^{-\frac{1}{2}}(\mathrm{curl},\partial\Omega)}\Big)
\end{aligned} \tag{248}$$

where C_{tr} is the trace constant corresponding to estimate (209).

Exercise 5.4. Reproduce the reasoning above for the H^1 and $\boldsymbol{H}(\mathrm{div})$-spaces.

Combining the result above with Theorem 5.2 and estimates (188), (193), (194), we get our final result.

Theorem 5.3. *There exist constants $C > 0$, independent of p such that,*

$$\|u - \Pi^{grad} u\|_{H^1(\Omega)}$$
$$\leq C \left(\inf_{u_p \in W_p} \|u - u_p\|_{H^1(\Omega)} + \|u - \Pi^{grad}_{\frac{1}{2}} u)\|_{\boldsymbol{H}^{\frac{1}{2}}(\partial\Omega)} \right)$$
$$\leq C(\ln p)^2 p^{-(r-1)} \|u\|_{H^r(\Omega)}, \quad \forall u \in H^r(\Omega),\ r > \frac{3}{2}$$

$$\|\boldsymbol{E} - \Pi^{curl} \boldsymbol{E}\|_{\boldsymbol{H}(\mathbf{curl},\Omega)}$$
$$\leq C \left(\inf_{\boldsymbol{E}_p \in \boldsymbol{Q}_p} \|\boldsymbol{E} - \boldsymbol{E}_p\|_{\boldsymbol{H}(\mathbf{curl},\Omega)} + \|\boldsymbol{E}_t - \Pi^{curl}_{-\frac{1}{2}} \boldsymbol{E}_t)\|_{\boldsymbol{H}^{-\frac{1}{2}}(\mathrm{curl},\partial\Omega)} \right)$$
$$\leq C \ln p\, p^{-r} \|\boldsymbol{E}\|_{\boldsymbol{H}^r(\mathbf{curl},\Omega)}, \quad \forall \boldsymbol{E} \in \boldsymbol{H}^r(\mathbf{curl}, \Omega),\ r > \frac{1}{2}$$

$$\|\boldsymbol{v} - \Pi^{div} \boldsymbol{v}\|_{\boldsymbol{H}(\mathrm{div},\Omega)}$$
$$\leq C \left(\inf_{\boldsymbol{v}_p \in \boldsymbol{V}_p} \|\boldsymbol{v} - \boldsymbol{v}_p\|_{\boldsymbol{H}(\mathrm{div},\Omega)} + \|v_n - \Pi_{-\frac{1}{2}} v_n)\|_{H^{-\frac{1}{2}}(\partial\Omega)} \right)$$
$$\leq C \ln p\, p^{-r} \|\boldsymbol{v}\|_{\boldsymbol{H}^r(\mathrm{div},\Omega)}, \quad \forall \boldsymbol{v} \in \boldsymbol{H}^r(\mathrm{div}, \Omega),\ r > 0.$$

Remark 5.1. Since all interpolation operators are polynomial-preserving, the classical Bramble-Hilbert argument allows to generalize the p estimates in Theorem 5.3 to corresponding *hp*-estimates.

6 Application to Maxwell Equations: Open Problems

6.1 Time-Harmonic Maxwell Equations

We shall consider the time-harmonic Maxwell equations in a bounded domain $\Omega \subset \mathbb{R}^n$, $n = 2, 3$.

- Faraday's law,

$$\frac{1}{\mu} \nabla \times \boldsymbol{E} = -i\omega \boldsymbol{H}, \tag{249}$$

- Ampere's law,

$$\nabla \times \boldsymbol{H} = \boldsymbol{J}^{imp} + \sigma \boldsymbol{E} + \epsilon i\omega \boldsymbol{E}. \tag{250}$$

Here μ, σ, ϵ denote the material data: permeability, conductivity and permittivity, assumed to be piecewise constant, ω is the angular frequency, and $\boldsymbol{J}^{imp}$ stands for the impressed current. We can derive two alternative variational formulations by choosing one of the equations to be satisfied in a weak, distributional sense, and the other one pointwise. The choice is dictated usually by the nature of source terms and/or boundary conditions. Choosing e.g. the Ampere's law to be satisfied in the weak sense, we multiply (250) with a test function $\boldsymbol{F}$, integrate over the domain and integrate by parts to obtain,

$$\int_\Omega \boldsymbol{H}\,\nabla\times\boldsymbol{F}+\int_{\partial\Omega}\boldsymbol{n}\times\boldsymbol{H}\,\boldsymbol{F}-\int_\Omega(\sigma\boldsymbol{E}+\epsilon i\omega\boldsymbol{E})\boldsymbol{F}=\int_\Omega \boldsymbol{J}^{imp}\boldsymbol{F}. \tag{251}$$

Notice that equations (249) and (251) imply implicitly the satisfaction of the Gauss law for magnetism (in the strong sense) and the continuity equation (in the weak sense). Eliminating $\boldsymbol{H}$ using (249) and employing appropriate boundary conditions, we get the classical variational formulation.

$$\left\{\begin{array}{l}\boldsymbol{E}\in H(\mathrm{curl},\Omega),\ \boldsymbol{n}\times\boldsymbol{E}=\boldsymbol{n}\times\boldsymbol{E}_D \text{ on } \Gamma_D\\[2mm] \displaystyle\int_\Omega\left\{\frac{1}{\mu}(\nabla\times\boldsymbol{E})(\nabla\times\boldsymbol{F})-(\omega^2\epsilon-i\omega\sigma)\boldsymbol{E}\boldsymbol{F}\right\}d\boldsymbol{x}+i\omega\int_{\Gamma_C}\gamma\boldsymbol{E}_t\boldsymbol{F}\,dS\\[2mm] \displaystyle=-i\omega\int_\Omega \boldsymbol{J}^{imp}\boldsymbol{F}\,d\boldsymbol{x}+i\omega\int_{\Gamma_N\cup\Gamma_C}\boldsymbol{J}_S^{imp}\boldsymbol{F}\,dS\\ \qquad\qquad\text{for every } \boldsymbol{F}\in H(\mathrm{curl},\Omega),\ \boldsymbol{n}\times\boldsymbol{F}=0 \text{ on } \Gamma_D\,.\end{array}\right. \tag{252}$$

Here Γ_D, Γ_N and Γ_C stand for the parts of the boundary where the Dirichlet (perfect conductor), Neumann (prescribed magnetic current) and Cauchy (prescribed impedance) boundary conditions have been set up, $\boldsymbol{E}_D$ stands for the Dirichlet data, γ is the impedance constant, and $\boldsymbol{J}_S^{imp}$ is a surface current prescribed on both Neumann and Cauchy parts of the boundary.

Weak Form of the Continuity Equation

Employing a special test function, $\boldsymbol{F}=\nabla q, q\in H^1(\Omega), q=0$ on Γ_D, we learn that the solution to the variational problem satisfies automatically the weak form of the continuity equation,

$$\begin{array}{l}\displaystyle\int_\Omega -(\omega^2\epsilon-i\omega\sigma)\boldsymbol{E}\nabla q\,d\boldsymbol{x}+i\omega\int_{\Gamma_C}\gamma\boldsymbol{E}_t\nabla q\,dS\\ \displaystyle\qquad=-i\omega\int_\Omega \boldsymbol{J}^{imp}\nabla q\,d\boldsymbol{x}+i\omega\int_{\Gamma_N\cup\Gamma_C}\boldsymbol{J}_S^{imp}\nabla q\,dS\\ \qquad\qquad\text{for every } q\in H^1(\Omega),\ q=0 \text{ on } \Gamma_D\,.\end{array} \tag{253}$$

Upon integrating by parts, we learn that solution $\boldsymbol{E}$ satisfies the continuity equation,

$$\mathrm{div}\left((\omega^2\epsilon-i\omega\sigma)\boldsymbol{E}\right)=i\omega\,\mathrm{div}\boldsymbol{J}^{imp}\qquad(=\omega^2\rho)\,,$$

plus additional boundary conditions on Γ_N, Γ_C, and interface conditions across material interfaces.

Maxwell Eigenvalue Problem

Related to the time-harmonic problem (252) is the eigenvalue problem,

$$\left\{\begin{array}{l}\boldsymbol{E}\in H(\mathrm{curl},\Omega),\ \boldsymbol{n}\times\boldsymbol{E}=0 \text{ on } \Gamma_D,\quad \lambda\in\mathbb{R}\\[2mm] \displaystyle\int_\Omega\frac{1}{\mu}(\nabla\times\boldsymbol{E})(\nabla\times\boldsymbol{F})\,d\boldsymbol{x}=\lambda\int_\Omega\epsilon\boldsymbol{E}\boldsymbol{F}\,d\boldsymbol{x}\\ \qquad\qquad\text{for every } \boldsymbol{F}\in H(\mathrm{curl},\Omega),\ \boldsymbol{n}\times\boldsymbol{F}=0 \text{ on } \Gamma_D\,.\end{array}\right. \tag{254}$$

The curl–curl operator is self-adjoint, its spectrum consists of $\lambda = 0$ with an infinite-dimensional eigenspace consisting of all gradients $\nabla p, p \in H^1(\Omega), p = 0$ on Γ_D, and a sequence of positive eigenvalues $\lambda_1 < \lambda_2 < \dots \lambda_n \to \infty$ with corresponding eigenspaces of finite dimension. Only the eigenvectors corresponding to positive eigenvalues are physical. Repeating the reasoning with the substitution $\boldsymbol{F} = \nabla q$, we learn that they satisfy automatically the continuity equation.

Stabilized Variational Formulation

The standard variational formulation (252) is *not* uniformly stable with respect to frequency ω. As $\omega \to 0$, we loose the control over gradients. This corresponds to the fact that, in the limiting case $\omega = 0$, the problem is ill-posed as the gradient component remains undetermined. A remedy to this problem is to enforce the continuity equation explicitly at the expense of introducing a Lagrange multiplier p. The so called *stabilized variational formulation* looks as follows.

$$
\left\{
\begin{aligned}
&\boldsymbol{E} \in H(\text{curl}, \Omega), p \in H^1(\Omega), \quad \boldsymbol{n} \times \boldsymbol{E} = \boldsymbol{n} \times \boldsymbol{E}_0, p = 0 \text{ on } \Gamma_D, \\
&\int_\Omega \frac{1}{\mu}(\nabla \times \boldsymbol{E})(\nabla \times \boldsymbol{F}) d\boldsymbol{x} - \int_\Omega (\omega^2\epsilon - i\omega\sigma)\boldsymbol{E} \cdot \boldsymbol{F} d\boldsymbol{x} + i\omega \int_{\Gamma_C} \gamma \boldsymbol{E}_t \boldsymbol{F}\, dS \\
&- \int_\Omega (\omega^2\epsilon - i\omega\sigma)\nabla p \cdot \boldsymbol{F} d\boldsymbol{x} = -i\omega \int_\Omega \boldsymbol{J}^{imp} \cdot \boldsymbol{F} d\boldsymbol{x} + i\omega \int_{\Gamma_N \cup \Gamma_C} \boldsymbol{J}_S^{imp} \cdot \boldsymbol{F} dS \\
&\qquad\qquad\qquad\qquad \forall \boldsymbol{F} \in H(\text{curl}, \Omega), \ \boldsymbol{n} \times \boldsymbol{F} = 0 \text{ on } \Gamma_D, \\
&- \int_\Omega (\omega^2\epsilon - i\omega\sigma)\boldsymbol{E} \cdot \nabla q \, d\boldsymbol{x} + i\omega \int_{\Gamma_C} \gamma \boldsymbol{E}_t \nabla q \, dS = \\
&\qquad\qquad -i\omega \int_\Omega \boldsymbol{J}^{imp} \cdot \nabla q \, d\boldsymbol{x} + i\omega \int_{\Gamma_N} \boldsymbol{J}_S^{imp} \cdot \nabla q dS \\
&\qquad\qquad\qquad\qquad \forall q \in H^1(\Omega), \ q = 0 \text{ on } \Gamma_D .
\end{aligned}
\right.
\tag{255}
$$

By repeating the reasoning with the substitution $\boldsymbol{F} = \nabla q$ in the first equation, we learn that the Lagrange multiplier p satisfies the weak form of a Laplace-like equation with homogeneous boundary conditions and, therefore, it *identically vanishes*. For that reason, it is frequently called the *hidden variable*. The stabilized formulation has improved stability properties for small ω. In the case of $\sigma = 0$ and right hand side of (253) vanishing, we can rescale the Lagrange multiplier, $p = \omega^2 p, q = \omega^2 q$, to obtain a symmetric mixed variational formulation with stability constant converging to one as $\omega \to 0$. In the general case we cannot avoid a degeneration as $\omega \to 0$ but we can still rescale the Lagrange multiplier with ω ($p = \omega p, q = \omega q$), to improve the stability of the formulation for small ω. The stabilized formulation is possible because gradients of the scalar-valued potentials from $H^1(\Omega)$ form precisely the null space of the curl–curl operator.

The point about the stabilized (mixed) formulation is that, whether we use it or not in the actual computations (the improved stability is one good reason to do it. . .), the original variational problem is *equivalent* to the mixed problem. This suggests that we cannot escape from the theory of mixed formulations when analyzing the problem.

6.2 So Why Does the Projection-Based Interpolation Matter?

The classical result of the numerical analysis for linear problems states that *discrete stability and approximability imply convergence*. For Finite Element (FE) approximations of mixed problems this translates into the control of the two inf–sup constants and best approximation error estimates. As for the mixed formulations of commuting projections, the exact sequence property implies the automatic satisfaction of the first Brezzi's inf–sup condition, with constant $\beta = 1$. The exact sequence is now understood at the level of the whole FE mesh. The satisfaction of the inf–sup in kernel condition is implied by the convergence of Maxwell eigenvalues, see [24] for the analysis of the lossless case $\sigma = 0$, and [10, 7, 8] for the related work. In this context, the projection-based interpolation enters the picture in two places. The best approximation error (over the whole mesh) is estimated with the interpolation error for the exact solution. The minimal regularity assumptions allow for estimating the error for solutions of "real" problems exhibiting multiple singularities due to the presence of reentrant corners and edges, and material interfaces.

The second use of the projection-based interpolation error has been recorded in the only existing proof on the *hp*-convergence of Maxwell eigenvalues for 2D Nédélec quads of the second type and 1-irregular meshes with hanging nodes, see [9]. Contrary to the 3D case, the 2D interpolant $\Pi_0^{curl}\boldsymbol{E}$ requires only an increased regularity in the field itself, $\boldsymbol{E} \in \boldsymbol{H}^r$, $r > 0$, but with $\mathrm{curl}\boldsymbol{E} \in L^2(\Omega)$ only. This leads to the possibility of estimating the error in L^2-norm,

$$\|\boldsymbol{E} - \Pi_0^{curl}\boldsymbol{E}\| \le C\left(\frac{h}{p}\right)^r (\|\boldsymbol{E}\|_{\boldsymbol{H}^r(\Omega)} + \|\mathrm{curl}\boldsymbol{E}\|_{L^2(\Omega)}). \qquad (256)$$

The estimate does not follow from the classical duality argument and exceeds the scope of these notes. Its use has been essential in proving the discrete compactness result in [9] which leads to the convergence result for the Maxwell eigenvalue problem and, in the end, the stability result for the 2D time-harmonic Maxwell equations.

Finally, the projection-based interpolation has been the driving idea behind the fully automatic *hp*-adaptivity producing a sequence of *hp* meshes that deliver exponential convergence, see [18, 23] and the literature therein. The concept of the projection-based interpolation extends naturally to element patches and spaces of piecewise polynomials.

6.3 Open Problems

We finish by summarizing the major open problems related to the theory of the projection-based interpolation and the grad-curl-div sequence for elements of higher order.

Extension Operators

With Part 3 of the contribution [22] in place, the task of constructing the commuting, polynomial preserving extension operators for the hexahedral and terahedral elements will be completed. Based on the ideas from [15] and [22], construction of extension operators for a prism seems to be a straightforward exercise but it should be completed.

Construction of the right-inverse of the curl operator for a hexahedron of an arbitrary (variable) order remains open. The construction for the tetrahedral element presented in this work, should be freed from the technical assumption on the polynomial degree for edges.

Pyramids

A successful three-dimensional FE code for Maxwell equations must include all four kinds of geometrical shapes: teds, hexas, prisms, and *pyramids*. The theory of the exact sequence and higher order elements for the pyramid element remains to be one of the most urgent research issues. Nigam and Phillips [36] have recently tackled the problem, extending the original construction of Gradinaru and Hiptmair [27] for the lowest order pyramid.

Discrete Compactness, L^2-Estimates, Nonlocal Interpolation

It is not clear whether the techniques used in [9], can be generalized to 3D. The fact that, in the discrete compactness argument, $\nabla \times \boldsymbol{E}$ lives only on $L^2(\Omega)$, eliminates the use of the projection-based interpolant in the argument analogous to the one used in [9]. The minimum regularity assumptions for the projection-based interpolation are identical with those for the classical Lagrange, Nédélec and Raviart–Thomas interpolants. The use of nonlocal interpolation techniques like the one proposed by Schoeberl [39] seems to be essential.

Acknowledgment

The work was done in collaboration and with help of many friends: Annalisa Buffa, Weiming Cao, Mark Ainsworth, Ivo Babuška, Jay Gopalakrishnan, Peter Monk, Waldek Rachowicz, and Joachim Schoeberl. The author would also like to thank Patrick Rabier for discussions on the subject.

References

1. M. Ainsworth and L. Demkowicz. Explicit polynomial preserving trace liftings on a triangle. *Math. Nachr.*, in press. ICES Report 03-47.
2. D.N. Arnold, B. Boffi, and R.S. Falk. Quadrilateral $H(div)$ finite element. *SIAM J. Numer. Anal.*, 42:2429–2451, 2005.

3. I. Babuška, A. Craig, J. Mandel, and J. Pitkaränta. Efficient preconditioning for the p-version finite element method in two dimensions. *SIAM J. Numer. Anal*, 28(3): 624–661, 1991.
4. I. Babuška and M. Suri. The optimal convergence rate of the p-version of the finite element method. *SIAM J. Numer. Anal.*, 24:750–776, 1987.
5. F. Ben Belgacem, Polynomial extensions of compatible polynomial traces in three dimensions. *Comput. Methods Appl. Mech. Engrg.*, 116:235–241, 1994. ICOSAHOM'92, Montpellier, 1992.
6. C. Bernardi, M. Dauge, and Y. Maday. Polynomials in the Sobolev world. Technical Report R 03038, Laboratoire Jacques-Louis Lions, Université Pierre at Marie Curie, 2003. http://www.ann.jussieu.fr/.
7. D. Boffi. Fortin operator and discrete compactness for edge elements. *Numer. Math.*, 87(2):229–246, 2000.
8. D. Boffi. A note on the de Rham complex and a discrete compactness property. *Appl. Math. Lett.*, 14(1):33–38, 2001.
9. D. Boffi, M. Dauge, M. Costabel, and L. Demkowicz. Discrete compactness for the *hp* version of rectangular edge finite elements. *SIAM J. Numer. Anal.*, 44(3):979–1004, 2006.
10. D. Boffi, P. Fernandes, L. Gastaldi, and I. Perugia. Computational models of electromagnetic resonators: analysis of edge element approximation. *SIAM J. Numer. Anal.*, 36(4):1264–1290, 1999.
11. F. Brezzi. On the existence, uniqueness and approximation of saddle-point problems arising from Lagrange multipliers. *R.A.I.R.O.*, 8(R2):129–151, 1974.
12. A. Buffa and P. Ciarlet. On traces for functional spaces related to Maxwell's equations. Part i: an integration by parts formula in Lipschitz polyhedra. *Math. Methods Appl. Sci.*, 24:9–30, 2001.
13. W. Cao and L. Demkowicz. Optimal error estimate for the projection based interpolation in three dimensions. *Comput. Math. Appl.*, 50:359–366, 2005.
14. M. Cessenat. *Mathematical Methods in Electromagnetism.* World Scientific, Singapore, 1996.
15. M. Costabel, M. Dauge, and L. Demkowicz. Polynomial extension operators for H^1, $\boldsymbol{H}(\mathbf{curl})$ and $\boldsymbol{H}(\mathrm{div})$ spaces on a cube. *Math. Comput.*, accepted. IRMAR Rennes Report 07-15.
16. L. Demkowicz. Edge finite elements of variable order for Maxwell's equations. In D. Hecht, U. van Rienen, M. Gunther, editors, *Scientific Computing in Electrical Engineering, Lecture Notes in Computational Science and Engineering*, vol. 18, pp. 15–34. Springer, Berlin Heidelberg New York, 2000. Proceedings of the 3rd International Workshop, August 20-23, Warnemuende, Germany.
17. L. Demkowicz. Projection based interpolation. In *Transactions on Structural Mechanics and Materials*. Cracow University of Technology Publications, Cracow, 2004. Monograph 302, A special issue in honor of 70th Birthday of Prof. Gwidon Szefer. ICES Report 04-03.
18. L. Demkowicz. *Computing with hp Finite Elements. I. One- and Two-Dimensional Elliptic and Maxwell Problems.* Chapman & Hall/CRC, 2006.
19. L. Demkowicz and I. Babuška. p interpolation error estimates for edge finite elements of variable order in two dimensions. *SIAM J. Numer. Anal.*, 41(4):1195–1208 (electronic), 2003.
20. L. Demkowicz and A. Buffa. H^1, $H(\mathrm{curl})$ and $H(\mathrm{div})$-conforming projection-based interpolation in three dimensions. Quasi-optimal p-interpolation estimates. *Comput. Methods Appl. Mech. Eng.*, 194:267–296, 2005.

21. L. Demkowicz, P. Monk, L. Vardapetyan, and W. Rachowicz. De Rham diagram for *hp* finite element spaces. *Comput. Math. Appl.*, 39(7-8):29–38, 2000.
22. L. Demkowicz, J. Gopalakrishnan and J. Schoeberl. Polynomial Extension Operators. Part I and II. *SIAM J. Numer. Anal*, submitted. RICAM Reports 07-15, 07-16.
23. L. Demkowicz, J. Kurtz, D. Pardo, M. Paszyński, W. Rachowicz and A. Zdunek. *Computing with hp Finite Elements. II. Frontiers: Three-Dimensional Elliptic and Maxwell Problems with Applications*. Chapman & Hall/CRC, 2007.
24. L. Demkowicz and L. Vardapetyan. Modeling of electromagnetic absorption/scattering problems using *hp*-adaptive finite elements. *Comput. Methods Appl. Mech. Eng.*, 152 (1–2):103–124, 1998.
25. J. Gopalakrishnan and L. Demkowicz. Quasioptimality of some spectral mixed methods. *J. Comput. Appl. Math.*, 167(1):163–182, 2004.
26. J. Gopalakrishnan, L.E. García-Castillo, and L. Demkowicz. Nédélec spaces in affine ccordinates. *Comput. Math. Appl.*, 49:1285–1294, 2005.
27. V. Gradinaru and R. Hiptmair. Whitney elements on pyramids. *ETNA*, 8:154–168, 1999. Report 113, SFB 382, Universitt Tbingen, March 1999.
28. P. Grisvard. *Singularities in Boundary Value Problems, Recherches en Mathématiques Appliquées [Research in Applied Mathematics]*, vol. 22. Masson, Paris, 1992.
29. R. Hiptmair. Canonical construction of finite elements. *Math. Comput.*, 68:1325–1346, 1999.
30. R. Hiptmair. Higher order Whitney forms. Technical Report 156, Universität Tübingen, 2000.
31. Y. Maday. Relévement de traces polynomiales et interpolations hilbertiennes entre espaces de polynomes. *C. R. Acad. Sci. Paris*, 309:463–468, 1989.
32. W. McLean. *Strongly Elliptic Systems and Boundary Integral Equations*. Cambridge University Press, 2000.
33. R. Munoz-Sola. Polynomial lifting on a tetrahedron and application to the *hp*-version of the finite element method. *SIAM J. Numer. Anal.*, 34:282–314, 1997.
34. J.C. Nédélec. Mixed finite elements in $\mathbb{R}^3$. *Numer. Math.*, 35:315–341, 1980.
35. J.C. Nédélec. A new family of mixed finite elements in $\mathbb{R}^3$. *Numer. Math.*, 50:57–81, 1986.
36. N. Nigam and J. Phillips. Higher-order finite flements on pyramids. LanL arXiV preprint: arXiv:math/0610206v3.
37. J.T. Oden, L. Demkowicz, R. Rachowicz, and T.A. Westermann. Toward a universal *hp* adaptive finite element strategy. Part 2: a posteriori error estimation. *Comput. Methods Appl. Mech. Eng.*, 77:113–180, 1989.
38. L.F. Pavarino and O. B. Widlund. A polylogarithmic bound for iterative subtructuring method for spectral elements in three dimensions. *SIAM J. Numer. Anal.*, 33(4):1303–1335, 1996.
39. J. Schoeberl. Commuting quasi-interpolation operators for mixed finite elements. Technical report, Texas A&M University, 2001. Preprint ISC-01-10-MATH.
40. Ch. Schwab. *p and hp-Finite Element Methods*. Clarendon Press, Oxford, 1998.
41. J.P. Webb. Hierarchical vector based funtions of arbitrary order for triangular and tetrahedral finite elements. *IEEE Antennas Propagat. Mag.*, 47(8):1244–1253, 1999.
42. H. Whitney. *Geometric Integration Theory*. Princeton University Press, 1957.

Finite Element Methods for Linear Elasticity

Richard S. Falk*

Department of Mathematics – Hill Center, Rutgers, The State University of New Jersey, 110 Frelinghuysen Rd., Piscataway, NJ 08854-8019, USA
falk@math.rutgers.edu

1 Introduction

The equations of linear elasticity can be written as a system of equations of the form

$$A\sigma = \varepsilon(u), \qquad \operatorname{div}\sigma = f \quad \text{in } \Omega. \tag{1}$$

Here the unknowns σ and u denote the stress and displacement fields caused by a body force f acting on a linearly elastic body which occupies a region $\Omega \subset \mathbb{R}^n$, with boundary $\partial\Omega$. Then σ takes values in the space $\mathbb{S} = \mathbb{R}^{n\times n}_{\text{sym}}$ of symmetric $n \times n$ matrices and u takes values in $\mathbb{V} = \mathbb{R}^n$. The differential operator ε is the symmetric part of the gradient, (i.e., $(\varepsilon(u))_{ij} = (\partial u_i/\partial x_j + \partial u_j/\partial x_i)/2$), div denotes the divergence operator, applied row-wise, and the compliance tensor $A = A(\tau) : \mathbb{S} \to \mathbb{S}$ is a bounded and symmetric, uniformly positive definite operator reflecting the properties of the material at each point. In the isotropic case, the mapping $\sigma \mapsto A\sigma$ has the form

$$A\sigma = \frac{1}{2\mu}\left(\sigma - \frac{\lambda}{2\mu + n\lambda}\operatorname{tr}(\sigma)I\right),$$

where $\lambda(x), \mu(x)$ are positive scalar coefficients, the Lamé coefficients, and tr denotes the trace. If the body is clamped on the boundary $\partial\Omega$, then the proper boundary condition for the system (1) is $u = 0$ on $\partial\Omega$. For simplicity, this boundary condition will be assumed throughout the discussion here. However, there are issues that arise when other boundary conditions are assumed (e.g., traction boundary conditions $\sigma n = 0$). The modifications needed to deal with such boundary conditions are discussed in detail in [9].

In the case when A is invertible, i.e., $\sigma = A^{-1}\varepsilon(u) = C\varepsilon(u)$, then for isotropic elasticity, $C\tau = 2\mu(\tau + \lambda \operatorname{tr}\tau I)$. We may then formulate the elasticity system weakly in the form: Find $\sigma \in L^2(\Omega, \mathbb{S})$, $u \in \mathring{H}^1(\Omega; \mathbb{V})$ such that

* This work supported by NSF grants DMS03-08347 and DMS06-09755. 9/8/06.

$$\int_\Omega \sigma : \tau \, dx - \int_\Omega C\varepsilon(u) : \tau \, dx = 0, \ \tau \in L^2(\Omega, \mathbb{S}),$$
$$\int_\Omega \sigma : \varepsilon(v) \, dx = \int_\Omega f \cdot v \, dx, \ v \in \mathring{H}^1(\Omega; \mathbb{V}),$$

where $\sigma : \tau = \sum_{i,j=1}^n \sigma_{ij}\tau_{ij}$. Note that in this case, we may eliminate σ completely to obtain the pure displacement formulation: Find $u \in \mathring{H}^1(\Omega; \mathbb{V})$ such that

$$\int_\Omega C\varepsilon(u) : \varepsilon(v) \, dx = \int_\Omega f \cdot v \, dx, \ v \in \mathring{H}^1(\Omega; \mathbb{V}).$$

As the material becomes incompressible, i.e., $\lambda \to \infty$, this will not be a good formulation, since the operator norm of C is also approaching infinity. Instead, we can consider a formulation involving u and a new variable $p = (\lambda/[2\mu+n\lambda]) \operatorname{tr} \sigma$. Taking the trace of the equation $A\sigma = \varepsilon(u)$, we find that $\operatorname{div} u = \lambda^{-1}p$. Then we may write $\sigma = 2\mu\varepsilon(u) + pI$, and thus obtain the variational formulation: Find $u \in \mathring{H}^1(\Omega; \mathbb{V})$, $p \in L^2_0(\Omega) = \{p \in L^2(\Omega) : \int_\Omega p \, dx = 0\}$, such that

$$\int_\Omega 2\mu \, \varepsilon(u) : \varepsilon(v) \, dx + \int_\Omega p \operatorname{div} v \, dx = \int_\Omega f \cdot v \, dx, \ v \in \mathring{H}^1(\Omega; \mathbb{V}),$$
$$\int_\Omega \operatorname{div} u \, q \, dx = \int_\Omega \lambda^{-1} p \, q \, dx, \ q \in L^2_0(\Omega).$$

This formulation makes sense even for the limit $\lambda \to \infty$ and in that case gives the stationary Stokes equations. Even in the case of nearly incompressible elasticity, one should apply methods that are stable for the Stokes equations. Since such methods will be considered in other lectures, we will not consider them here. Instead, we now turn to other types of weak formulations involving both σ and u. One of these is to seek $\sigma \in H(\operatorname{div}, \Omega; \mathbb{S})$, the space of square-integrable symmetric matrix fields with square-integrable divergence, and $u \in L^2(\Omega; \mathbb{V})$, satisfying

$$\int_\Omega (A\sigma : \tau + \operatorname{div} \tau \cdot u) \, dx = 0, \ \tau \in H(\operatorname{div}, \Omega; \mathbb{S}), \tag{2}$$
$$\int_\Omega \operatorname{div} \sigma \cdot v \, dx = \int_\Omega f \cdot v \, dx, \ v \in L^2(\Omega; \mathbb{V}).$$

A second weak formulation, that enforces the symmetry weakly, seeks $\sigma \in H(\operatorname{div}, \Omega; \mathbb{M})$, $u \in L^2(\Omega; \mathbb{V})$, and $p \in L^2(\Omega; \mathbb{K})$ satisfying

$$\int_\Omega (A\sigma : \tau + \operatorname{div} \tau \cdot u + \tau : p) \, dx = 0, \quad \tau \in H(\operatorname{div}, \Omega; \mathbb{M}),$$
$$\int_\Omega \operatorname{div} \sigma \cdot v \, dx = \int_\Omega f \cdot v \, dx, \quad v \in L^2(\Omega; \mathbb{V}), \tag{3}$$
$$\int_\Omega \sigma : q \, dx = 0, \quad q \in L^2(\Omega; \mathbb{K}),$$

where $\mathbb{M}$ is the space of $n \times n$ matrices, $\mathbb{K}$ the subspace of skew-symmetric matrices, and the compliance tensor $A(x)$ is now considered as a symmetric and positive definite operator mapping $\mathbb{M}$ into $\mathbb{M}$.

Stable finite element discretizations with reasonable computational complexity based on the variational formulation (2) have proved very difficult to construct. In particular, it is not possible to simply take multiple copies of standard finite elements for scalar elliptic problems, since the resulting stress matrix will not be symmetric. One successful approach has been to use composite elements, in which the approximate displacement space consists of piecewise polynomials with respect to one triangulation of the domain, while the approximate stress space consists of piecewise polynomials with respect to a different, more refined, triangulation [22, 30, 24, 4]. In two space dimensions, the first stable finite elements with polynomial shape functions were presented in [10]. The simplest and lowest order spaces in the family of spaces constructed consist of discontinuous piecewise linear vector fields for displacements and a stress space which is locally the span of piecewise quadratic matrix fields and the cubic matrix fields that are divergence-free. Hence, it takes 24 stress and 6 displacement degrees of freedom to determine an element on a given triangle. A simpler first-order element pair with 21 stress and 3 displacement degrees of freedom per triangle is also constructed in [10]. All of these elements require vertex degrees of freedom. To obtain simpler elements, the same authors also considered nonconforming elements in [12]. One element constructed there approximates the stress by a nonconforming piecewise quadratic with 15 degrees of freedom and approximates the displacement field by discontinuous linear vectors (6 local degrees of freedom). A second element reduces the number of degrees of freedom to 12 and 3, respectively. See also [11] for an overview. In three dimensions, a piecewise quartic stress space is constructed with 162 degrees of freedom on each tetrahedron in [1].

Because of the lack of suitable mixed elasticity elements that strongly impose the symmetry of the stresses, a number of authors have developed approximation schemes based on the weak symmetry formulation (3): see [22], [2], [3], [27], [28], [29], [5], [25], [26], [21]. Although (2) and (3) are equivalent on the continuous level, an approximation scheme based on (3) may not produce a symmetric approximation to the stress tensor, depending on the choices of finite element spaces.

These notes will mainly concentrate on the development and analysis of finite element approximations of the equations of linear elasticity based on the mixed formulation (3) with weak symmetry. Using a generalization of an approach first developed in [8] in two dimensions and [6] in three dimensions, and then expanded further in [9], we establish a systematic way to obtain stable finite element approximation schemes. The families of methods developed in [8] and [6] are the prototype examples and we show that they satisfy the conditions we develop for stability. However, the somewhat more general approach we present here allows us to analyze some of the previously proposed schemes discussed above in the same systematic manner and also leads to a new scheme. Before considering weakly symmetric schemes, we first discuss some methods based on the strong symmetry formulation (2).

2 Finite Element Methods with Strong Symmetry

In this section, we consider finite element methods based on the variational formulation (2). Thus, we let $\boldsymbol{\Sigma}_h \subset H(\mathrm{div}, \Omega; \mathbb{S})$ and $\boldsymbol{V}_h \subset L^2(\Omega; \mathbb{V})$ and seek $\sigma_h \in \boldsymbol{\Sigma}_h$ and $u_h \in \boldsymbol{V}_h$ satisfying

$$\int_\Omega (A\sigma_h : \tau + \mathrm{div}\, \tau \cdot u_h)\, \mathrm{d}x = 0, \tau \in \boldsymbol{\Sigma}_h, \quad \int_\Omega \mathrm{div}\, \sigma_h \cdot v\, \mathrm{d}x = \int_\Omega f \cdot v\, \mathrm{d}x, \ v \in \boldsymbol{V}_h.$$

This is in a form to which one may apply the standard analysis of mixed finite element theory (e.g., [14, 15, 20, 18]. We note that in the case of isotropic elasticity, if we write $\sigma = \sigma_D + (1/n)\,\mathrm{tr}\,\sigma I$, where $\mathrm{tr}\,\sigma_D = 0$, then $\|\sigma\|_0^2 = \|\sigma_D\|_0^2 + (1/n)\|\,\mathrm{tr}\,\sigma\|_0^2$ and so

$$\int_\Omega A\sigma : \sigma\, \mathrm{d}x = \int_\Omega \left[\frac{1}{2\mu}\sigma_D : \sigma_D + \frac{1}{2\mu + n\lambda}(\mathrm{tr}\,\sigma)^2\right] \mathrm{d}x.$$

Thus, this form is not uniformly coercive as $\lambda \to \infty$. However, for all σ satisfying

$$\int_\Omega \mathrm{tr}\,\sigma\, \mathrm{d}x = 0, \qquad \mathrm{div}\,\sigma = 0, \tag{4}$$

one can show (cf. [15]) that $\|\,\mathrm{tr}\,\sigma\|_0 \le C\|\sigma_D\|_0$, and hence $(A\sigma, \sigma) \ge \alpha\|\sigma\|^2_{H(\mathrm{div})}$ for all σ satisfying (4), with α independent of λ. This is what is needed to satisfy the first Brezzi condition with a constant independent of λ. A simple result of mixed finite element theory, giving conditions under which the second Brezzi condition is satisfied, and that fits the methods that we will consider here, is the following.

Theorem 2.1 *Suppose that for every $\tau \in \boldsymbol{H}^1(\Omega)$, there exists $\Pi_h \tau \in \boldsymbol{\Sigma}_h$ satisfying*

$$\int_\Omega \mathrm{div}(\tau - \Pi_h\tau) \cdot v\, \mathrm{d}x = 0, \quad v \in \boldsymbol{V}_h, \qquad \|\Pi_h \tau\|_{H(\mathrm{div})} \le C\|\tau\|_{H(\mathrm{div})}.$$

Further suppose that for all $\tau \in \boldsymbol{\Sigma}_h$ satisfying $\int_\Omega \mathrm{div}\,\tau \cdot v\, \mathrm{d}x = 0$, $v \in \boldsymbol{V}_h$, that $\mathrm{div}\,\tau = 0$. Then for all $v_h \in \boldsymbol{V}_h$,

$$\|\sigma - \sigma_h\|_0 \le C\|\sigma - \Pi_h\sigma\|_0, \qquad \|u - u_h\|_0 \le C(\|u - v_h\|_0 + \|\sigma - \sigma_h\|_0).$$

To describe some finite element methods based on the strong symmetry formulation, we let $\mathcal{P}_k(X,Y)$ denote the space of polynomial functions on X of degree at most k and taking values in Y.

2.1 Composite Elements

One of the first methods based on the symmetric formulation was the method of [30] analyzed in [24]. We describe below only the triangular element (there was also a similar quadrilateral element). The basic idea is to approximate the stresses by a composite finite element. Starting from a mesh $\mathcal{T}_h$ of triangles, one connects the barycenter of each triangle K to the three vertices to form a composite element made up of three triangles, i.e., $K = T_1 \cup T_2 \cup T_3$. We then define

$$\begin{aligned}\boldsymbol{\Sigma}_h &= \{\tau \in H(\operatorname{div}, \Omega; \mathbb{S}) : \tau|_{T_i} \in \mathcal{P}_1(T_i, \mathbb{S})\},\\ \boldsymbol{V}_h &= \{v \in \boldsymbol{L}^2(\Omega) : v|_K \in \mathcal{P}_1(K, \mathbb{R}^2\}.\end{aligned}$$

Composite Element

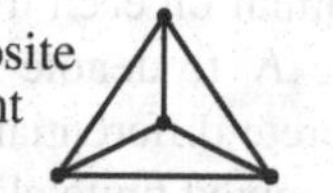

Thus the displacements are defined on the coarse mesh $\mathcal{T}_h$. By the definition of $\boldsymbol{\Sigma}_h|_K$, we start from a space of 27 degrees of freedom, on which we impose at most 12 constraints that require that τn be continuous across each of the three internal edges of K. In fact, these constraints are all independent. Then, a key point is to show that on each K, τ is uniquely determined by the following 15 degrees of freedom (i) the values of $\tau \cdot n$ at two points on each edge of K and (ii) $\int_K \tau_{ij}\,\mathrm{d}x$, $i, j = 1, 2$. It is then easy to check that if $\int_K \operatorname{div} \tau \cdot v\,\mathrm{d}x = 0$ for $v \in \mathcal{P}_1(K, \mathbb{R}^2)$, then $\operatorname{div} \tau = 0$. If we define Π_h to correspond to the degrees of freedom, then it is also easy to check that $\int_K \operatorname{div}(\tau - \Pi_h\tau) \cdot v\,\mathrm{d}x = 0$ for $v \in \mathcal{P}_1(K, \mathbb{R}^2)$. After establishing the $H(\operatorname{div}, \Omega)$ norm bound on $\Pi_h\sigma$, one easily obtains the error estimates:

$$\|\sigma - \sigma_h\|_0 \le Ch^2\|\sigma\|_2, \qquad \|u - u_h\|_0 \le Ch^2(\|\sigma\|_2 + \|u\|_2).$$

The use of composite finite elements to approximate the stress tensor was extended to a family of elements in [4]. For $k \ge 2$,

$$\begin{aligned}\boldsymbol{\Sigma}_h &= \{\tau \in H(\operatorname{div}, \Omega; \mathbb{S}) : \tau|_{T_i} \in \mathcal{P}_k(T_i, \mathbb{S})\},\\ \boldsymbol{V}_h &= \{v \in \boldsymbol{L}^2(\Omega) : v|_K \in \mathcal{P}_{k-1}(K, \mathbb{R}^2)\}.\end{aligned}$$

The space $\boldsymbol{\Sigma}_h$ is constructed so that if $\tau \in \boldsymbol{\Sigma}_h|_K$, then τn will be continuous across internal edges, and in addition $\operatorname{div} \tau \in \mathcal{P}_{k-1}(K, \mathbb{R}^2)$, i.e., it is a vector polynomial on K, not just on each of the T_i.

The degrees of freedom for an element $\tau \in \boldsymbol{\Sigma}_h$ on the triangle K are chosen to be

$$\int_e (\tau n) \cdot p\,\mathrm{d}s, \quad p \in \mathcal{P}_k(e, \mathbb{R}^2), \qquad \text{for each edge } e,$$

$$\int_K \tau : \varrho\,\mathrm{d}x, \quad \varrho \in \varepsilon(\mathcal{P}_{k-1}(K, \mathbb{R}^2)) + \operatorname{airy}(\lambda_1^2\lambda_1^2\lambda_3^2 P_{k-4}(K, \mathbb{R})),$$

where the λ_i are the barycentric coordinates of K and

$$J\phi \equiv \operatorname{airy}\phi = \begin{pmatrix} \partial^2\phi/\partial y^2 & -\partial^2\phi/\partial x\partial y \\ -\partial^2\phi/\partial x\partial y & \partial^2\phi/\partial x^2 \end{pmatrix}.$$

One can show that $\dim \boldsymbol{\Sigma}_h|_K = (3/2)k^2 + (9/2)k + 6$. In the lowest order case $k = 2$, there are 18 edge degrees of freedom and 3 interior degrees of freedom on each macro-triangle K. For the general case $k \ge 2$, it is shown that

$$\begin{aligned}\|u - u_h\|_0 &\le Ch^r\|u\|_r, \quad 2 \le r \le k,\\ \|\sigma - \sigma_h\|_0 &\le Ch^r\|u\|_{r+1}, \quad 1 \le r \le k+1,\\ \|\operatorname{div}(\sigma - \sigma_h)\|_0 &\le Ch^r\|\operatorname{div}\sigma\|_r, \quad 0 \le r \le k.\end{aligned}$$

2.2 Noncomposite Elements of Arnold and Winther

We now turn to the more recent methods that produce approximations to both stresses and displacements that are polynomial on each triangle $T \in \mathcal{T}_h$ (since there are no macro triangles, we no longer use K to denote a generic triangle). The approach of [10] is based on the use of discrete differential complexes and the close relation between the construction of stable mixed finite element methods for the approximation of the Laplacian and discrete versions of the de Rham complex

$$\mathbb{R} \xrightarrow{\subset} C^\infty(\Omega) \xrightarrow{\mathbf{curl}} C^\infty(\Omega;\mathbb{R}^2) \xrightarrow{\mathrm{div}} C^\infty(\Omega) \to 0.$$

If we assume that Ω is simply-connected, this sequence is exact (i.e., the range of each map is the kernel of the following one). As discussed later in this paper, many of the standard spaces leading to stable mixed finite element methods for Laplace's equation have the property that the following diagram commutes

$$\begin{array}{ccccccccc} \mathbb{R} & \xrightarrow{\subset} & C^\infty(\Omega) & \xrightarrow{\mathbf{curl}} & C^\infty(\Omega,\mathbb{R}^2) & \xrightarrow{\mathrm{div}} & C^\infty(\Omega) & \longrightarrow & 0 \\ \mathrm{id}\big\downarrow & & I_h\big\downarrow & & \Pi_h\big\downarrow & & P_h\big\downarrow & & \\ \mathbb{R} & \xrightarrow{\subset} & Q_h & \xrightarrow{\mathbf{curl}} & \Sigma_h & \xrightarrow{\mathrm{div}} & V_h & \longrightarrow & 0 \end{array}, \tag{5}$$

where I_h, Π_h, P_h are the natural interpolation operators into the corresponding finite element spaces Q_h, Σ_h, and V_h. For example, the simplest case is when Q_h is the space of continuous piecewise linear functions, Σ_h the space of lowest order Raviart–Thomas elements, and V_h the space of piecewise constants. The right half of the commuting diagram, involving Π_h and P_h is a key result in establishing the second Brezzi stability condition. See [7, 9] for further discussion of this idea.

The starting point of [10] is that there is also an elasticity differential complex, which summarizes important aspects of the structure of the plane elasticity system, i.e.,

$$P_1(\Omega) \xrightarrow{\subset} C^\infty(\Omega) \xrightarrow{J} C^\infty(\Omega,\mathbb{S}) \xrightarrow{\mathrm{div}} C^\infty(\Omega,\mathbb{R}^2) \to 0. \tag{6}$$

Again assuming that Ω is simply-connected, this sequence is also exact. Thus this sequence encodes the fact that every smooth vector-field is the divergence of a smooth symmetric matrix-field, that the divergence-free symmetric matrix-fields are precisely those that can be written as the Airy stress-field associated to some scalar potential, and that the only potentials for which the corresponding Airy stress vanishes are the linear polynomials. The result stated above is in terms of smooth functions, but analogous results hold with less smoothness. For example, the sequence

$$P_1(\Omega) \xrightarrow{\subset} H^2(\Omega) \xrightarrow{J} H(\mathrm{div},\Omega;\mathbb{S}) \xrightarrow{\mathrm{div}} L^2(\Omega,\mathbb{R}^2) \to 0 \tag{7}$$

is also exact. The well-posedness of the continuous problem, i.e., that for every $f \in L^2(\Omega,\mathbb{R}^2)$, there exists a unique $(\sigma,u) \in H(\mathrm{div},\Omega;\mathbb{S}) \times L^2(\Omega,\mathbb{R}^2)$ which is a critical point of (1.1), follows from this.

Just as there is a close relation between the construction of stable mixed finite element methods for the approximation of the Laplacian and discrete versions of the de Rham complex, there is also a close relation between mixed finite elements for linear elasticity and discretization of the elasticity complex, given above. The stable pairs of finite element spaces $(\boldsymbol{\Sigma}_h, \boldsymbol{V}_h)$ introduced in [10] have the property that $\operatorname{div} \boldsymbol{\Sigma}_h = \boldsymbol{V}_h$, i.e., the short sequence

$$\boldsymbol{\Sigma}_h \xrightarrow{\operatorname{div}} \boldsymbol{V}_h \to 0 \tag{8}$$

is exact. Moreover, if there are projections $P_h : C^\infty(\Omega, \mathbb{R}^2) \mapsto \boldsymbol{V}_h$ and $\Pi_h : C^\infty(\Omega, \mathbb{S}) \mapsto \boldsymbol{\Sigma}_h$ defined by the degrees of freedom that determine the finite element spaces, it can be shown that the following diagram commutes:

$$\begin{array}{ccc} C^\infty(\Omega, \mathbb{S}) & \xrightarrow{\operatorname{div}} & C^\infty(\Omega, \mathbb{R}^2) \\ \Pi_h \big\downarrow & & P_h \big\downarrow \\ \boldsymbol{\Sigma}_h & \xrightarrow{\operatorname{div}} & \boldsymbol{V}_h \end{array} \tag{9}$$

The stability of the mixed method follows from the exactness of (8), the commutativity of (9), and the well-posedness of the continuous problem.

Information about the construction of such finite element spaces can be gained by completing the sequence (8) to a sequence analogous to (6). For this purpose, we set $Q_h = \{q \in H^2(\Omega) : Jq \in \boldsymbol{\Sigma}_h\}$. Note Q_h is a finite element approximation of $H^2(\Omega)$. Moreover, there is a natural interpolation operator $I_h : C^\infty(\Omega) \mapsto Q_h$ so that the following diagram with exact rows commutes:

$$\begin{array}{ccccccccc} P_1(\Omega) & \xrightarrow{\subset} & C^\infty(\Omega) & \xrightarrow{J} & C^\infty(\Omega, \mathbb{S}) & \xrightarrow{\operatorname{div}} & C^\infty(\Omega, \mathbb{R}^2) & \longrightarrow & 0 \\ \operatorname{id} \big\downarrow & & I_h \big\downarrow & & \Pi_h \big\downarrow & & P_h \big\downarrow & & \\ P_1(\Omega) & \xrightarrow{\subset} & Q_h & \xrightarrow{J} & \boldsymbol{\Sigma}_h & \xrightarrow{\operatorname{div}} & \boldsymbol{V}_h & \longrightarrow & 0 \end{array}$$

For a description of this construction, see [10]. As discussed there, under quite general conditions, the existence of a stable pair of spaces $(\boldsymbol{\Sigma}_h, \boldsymbol{V}_h)$ approximating $H(\operatorname{div}, \Omega; \mathbb{S}) \times L^2(\Omega, \mathbb{R}^2)$, implies the existence of a finite element approximation Q_h of $H^2(\Omega)$ related to $\boldsymbol{\Sigma}_h$ and $\boldsymbol{V}_h$ through the diagram above. The fact that the space Q_h requires $C^1(\Omega)$ finite elements represents a substantial obstruction to the construction of stable mixed elements, and in part accounts for their slow development. In fact, the lowest order element proposed in [10] corresponds to choosing Q_h to be the Argyris space of C^1 piecewise quintic polynomials (the simplest choice). Since $JQ_h \subset \boldsymbol{\Sigma}_h$, one then sees that $\boldsymbol{\Sigma}_h$ must be a piecewise cubic space, and since the Argyris space has second derivative degrees of freedom at the vertices, the degrees of freedom for $\boldsymbol{\Sigma}_h$ will include vertex degrees of freedom, not usually expected for subspaces of $H(\operatorname{div}; \Omega)$.

The family of elements developed in [10] chooses for $k \geq 1$, the local degrees of freedom for $\boldsymbol{\Sigma}_h$ to be

$$\begin{aligned} \Sigma_T &= \mathcal{P}_{k+1}(T, \mathbb{S}) + \{\tau \in \mathcal{P}_{k+2}(T, \mathbb{S}) : \operatorname{div} \tau = 0\} \\ &= \{\tau \in \mathcal{P}_{k+2}(T, \mathbb{S}) : \operatorname{div} \tau \in \mathcal{P}_k(T, \mathbb{R}^2)\}, \qquad V_T = \mathcal{P}_k(T, \mathbb{R}^2). \end{aligned}$$

Now $\dim V_T = (k+2)(k+1)$ and it is shown in [10] that $\dim \Sigma_T = (3k^2 + 17k + 28)/2$ and that a unisolvent set of local degrees of freedom is given by

- the values of three components of $\tau(x)$ at each vertex x of T (9 degrees of freedom)
- the values of the moments of degree at most k of the two normal components of τ on each edge e of T ($6k + 6$ degrees of freedom)
- the value of the moments $\int_T \tau : \phi \, dx$, $\phi \in \mathcal{P}_k(T, \mathbb{R}^2) + \operatorname{airy}(b_T^2 \mathcal{P}_{k-2}(T, \mathbb{R}))$.

For this family of elements, it is shown in [10] that

$$\begin{aligned} \|\sigma - \sigma_h\|_0 &\le C h^r \|\sigma\|_r, \quad 1 \le r \le k+2, \\ \|\operatorname{div}(\sigma - \sigma_h)\|_0 &\le C h^r \|\operatorname{div} \sigma\|_r, \quad 0 \le r \le k+1, \\ \|u - u_h\|_0 &\le C h^r \|u\|_{r+1}, \quad 1 \le r \le k+1. \end{aligned}$$

There is a variant of the lowest degree ($k = 1$) element involving fewer degrees of freedom. In this element, one chooses V_T to be the space of infinitesimal rigid motions on T , i.e., vector functions of the form $(a - by, c + bx)$. Then $\Sigma_T = \{\tau \in \mathcal{P}_3(T, \mathbb{S}) : \operatorname{div} \tau \in V_T\}$.

The element diagram for the choice $k = 1$ and a simplified element are depicted in Fig. 1.

In [12], the authors obtain simpler elements with fewer degrees of freedom, and also avoid the use of vertex degrees of freedom by developing nonconforming elements. Corresponding to the choice $V_T = \mathcal{P}_1(T, \mathbb{R}^2)$, one chooses for the stress shape functions

$$\Sigma_T = \{\tau \in \mathcal{P}_2(T, \mathbb{S}) : n \cdot \tau n \in \mathcal{P}_1(e, \mathbb{R}), \text{ for each edge } e \text{ of } T\}.$$

The space Σ_T has dimension 15, with degrees of freedom given by

- the values of the moments of degree 0 and 1 of the two normal components of τ on each edge e of T (12 degrees of freedom),
- the value of the three components of the moment of degree 0 of τ on T (3 degrees of freedom).

Note that this element is a nonconforming approximation of $H(\operatorname{div}, \Omega; \mathbb{S})$, since although $t \cdot \tau n$ may be quadratic on an edge, only its two lowest order moments are determined on each edge. Hence, τn may not be continuous across element boundaries. This space may be simplified in a manner similar to the lowest order conforming element, i.e., the displacement space may be chosen to be piecewise rigid

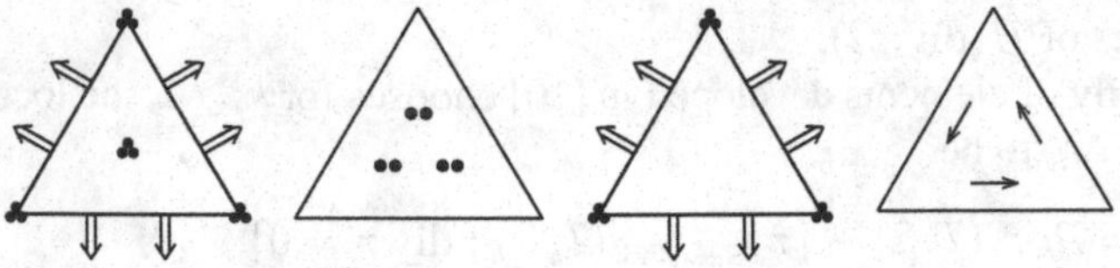

Fig. 1. $k = 1$ and simplified Arnold–Winther elements

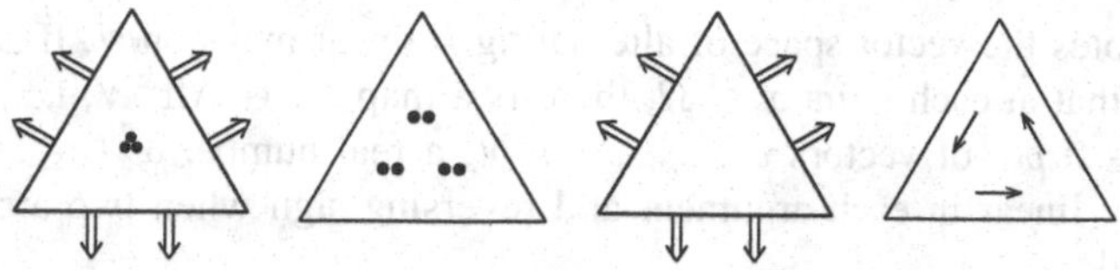

Fig. 2. Two nonconforming Arnold–Winther elements

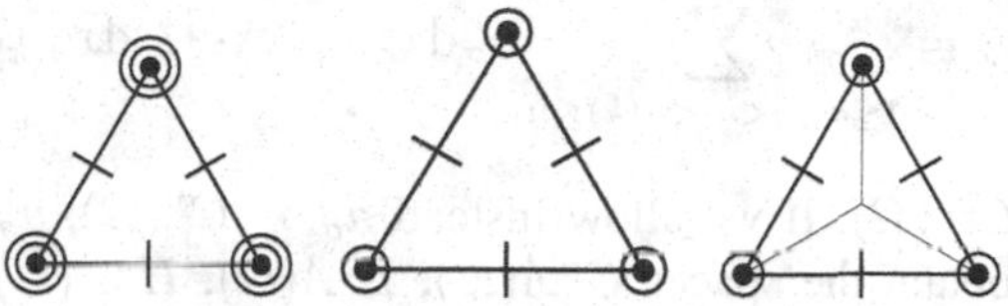

Fig. 3. Q_h spaces for $k = 1$ conforming element, nonconforming element, and composite element of [24]

motions and the stress space then reduced by requiring that the divergence be a rigid motion on each triangle. The local dimension of the resulting space is 12 and the first two moments of the normal traction on each edge form a unisolvent set of degrees of freedom (see Fig. 2).

As noted earlier, for $k = 1$, the corresponding space Q_h is the Argyris space consisting of C^1 piecewise quintic polynomials. There is also an analogous relationship for the composite elements discussed earlier. For the element of [24], the space Q_h is the Clough–Tocher composite H^2 element and for the element family of [4], the Q_h spaces are the higher order composite elements of [17] (see Fig. 3).

The remainder of these notes will be devoted to the development and analysis of mixed finite element methods based on the formulation (3) of the equations of elasticity with weak symmetry. An important advantage of such an approach is that it allows us to approximate the stress matrix by n copies of standard finite element approximations of $\boldsymbol{H}(\mathrm{div}, \Omega)$ used to discretize scalar second order elliptic problems. In fact, to develop our approximation schemes for (3), we will heavily exploit the many close connections between these two problems. Although there is some overhead to the development, much of the structure of these connections is most clearly seen in the language of differential forms. Thus, we devote the next section to a brief overview of the necessary background material.

3 Exterior Calculus on $\mathbb{R}^n$

To simplify matters, we will consider exterior calculus on $\mathbb{R}^n$, and summarize only the specific results we will need.

3.1 Differential Forms

Suppose that Ω is an open subset of $\mathbb{R}^n$. For $0 \le k \le n$, we let Λ^k denote the space of smooth differential k-forms of Ω, i.e., $\Lambda^k = \Lambda^k(\Omega) = C^\infty(\Omega; \mathrm{Alt}^k\,\mathbb{V})$, where

$\mathrm{Alt}^k\,\mathbb{V}$ denotes the vector space of alternating k-linear maps on $\mathbb{V}$. If $\omega \in \Lambda^k(\Omega)$, this means that at each point $x \in \Omega$, there is a map $\omega_x \in \mathrm{Alt}^k\,\mathbb{V}$, i.e., ω_x assigns to each $k - tuple$ of vectors $v_1, \dots, v_k$ of $\mathbb{V}$, a real number $\omega_x(v_1, \dots, v_k)$ with the mapping linear in each argument and reversing sign when two arguments are interchanged.

A general element of $\Lambda^k(\Omega)$ may be written

$$\omega_x = \sum_{1 \le \sigma(1) < \cdots < \sigma(k) \le n} a_\sigma \mathrm{d}x_{\sigma(1)} \wedge \cdots \wedge \mathrm{d}x_{\sigma(k)},$$

where the $a_\sigma \in C^\infty(\Omega)$. If we allow instead $a_\sigma \in C^p(\Omega)$, $a_\sigma \in L^2(\Omega)$, $a_\sigma \in H^s(\Omega)$, etc., we obtain the spaces $C^p\Lambda(\Omega)$, $L^2\Lambda(\Omega)$, $H^s\Lambda(\Omega)$, etc. Thus, when $n = 2$, for $k = 0, 1, 2$, $\omega \in \Lambda^k(\Omega)$ will have the respective forms

$$w, \qquad w_1 \mathrm{d}x_1 + w_2 \mathrm{d}x_2, \qquad w\,\mathrm{d}x_1 \wedge \mathrm{d}x_2.$$

To see the connection between differential forms and scalar- and vector-valued functions, we may identify $w \in \Lambda^0(\Omega)$ and $w\,\mathrm{d}x_1 \wedge \mathrm{d}x_2 \in \Lambda^2(\Omega)$ with the function $w \in C^\infty(\Omega)$ and $w_1\mathrm{d}x_1 + w_2\mathrm{d}x_2 \in \Lambda^1(\Omega)$ with the vector (w_1, w_2) or the vector $(-w_2, w_1) \in C^\infty(\Omega; \mathbb{R}^2)$. The associated fields are called *proxy fields* for the forms.

When $n = 3$, for $k = 0, 1, 2, 3$, $\omega \in \Lambda^k(\Omega)$ will have the respective forms

$$w, \quad w_1\mathrm{d}x_1 + w_2\mathrm{d}x_2 + w_3\mathrm{d}x_3, \quad w_1\mathrm{d}x_2 \wedge \mathrm{d}x_3 - w_2\mathrm{d}x_1 \wedge \mathrm{d}x_3 + w_3\mathrm{d}x_1 \wedge \mathrm{d}x_2,$$
$$w\,\mathrm{d}x_1 \wedge \mathrm{d}x_2 \wedge \mathrm{d}x_3.$$

In this case, we may identify $w \in \Lambda^0(\Omega)$ and $w\,\mathrm{d}x_1 \wedge \mathrm{d}x_2 \wedge \mathrm{d}x_3 \in \Lambda^3(\Omega)$ with the function $w \in C^\infty(\Omega)$ and $w_1\mathrm{d}x_1 + w_2\mathrm{d}x_2 + w_3\mathrm{d}x_3$ or $w_1\mathrm{d}x_2 \wedge \mathrm{d}x_3 - w_2\mathrm{d}x_1 \wedge \mathrm{d}x_3 + w_3\mathrm{d}x_1 \wedge \mathrm{d}x_2$ with the vector $(w_1, w_2, w_3) \in C^\infty(\Omega; \mathbb{R}^2)$. The correspondences are listed in Table 1.

To evaluate $\omega_x(v_1, \dots, v_k)$, we need a formula for evaluating the k-form $\mathrm{d}x_{\sigma(1)} \wedge \cdots \wedge \mathrm{d}x_{\sigma(k)}(v_1, \dots, v_k)$. Rather than presenting the general case, we note that for $v, w, z \in \mathbb{R}^n$,

$$\mathrm{d}x_i(v) = v_i, \qquad \mathrm{d}x_i \wedge \mathrm{d}x_j(v, w) = v_i w_j - v_j w_i,$$
$$\text{and, for } n = 3, \qquad \mathrm{d}x_1 \wedge \mathrm{d}x_2 \wedge \mathrm{d}x_3(v, w, z) = \det(v|w|z).$$

For $\omega \in \mathrm{Alt}^j\,V$ and $\eta \in \mathrm{Alt}^k\,V$, the *exterior product* or *wedge product* $\omega \wedge \eta \in \mathrm{Alt}^{j+k}\,V$ is bilinear and associative, and satisfies the anti-commutativity condition

Table 1. Correspondence between alternating algebraic forms on $\mathbb{R}^3$ and scalars/vectors

$\mathrm{Alt}^0\,\mathbb{R}^3 = \mathbb{R}$	$c \leftrightarrow c$
$\mathrm{Alt}^1\,\mathbb{R}^3 \xrightarrow{\cong} \mathbb{R}^3$	$u_1\,\mathrm{d}x_1 + u_2\,\mathrm{d}x_2 + u_3\,\mathrm{d}x_3 \leftrightarrow u$
$\mathrm{Alt}^2\,\mathbb{R}^3 \xrightarrow{\cong} \mathbb{R}^3$	$u_3\,\mathrm{d}x_1 \wedge \mathrm{d}x_2 - u_2\,\mathrm{d}x_1 \wedge \mathrm{d}x_3 + u_1\,\mathrm{d}x_2 \wedge \mathrm{d}x_3 \leftrightarrow u$
$\mathrm{Alt}^3\,\mathbb{R}^3 \xrightarrow{\cong} \mathbb{R}$	$c\,\mathrm{d}x_1 \wedge \mathrm{d}x_2 \wedge \mathrm{d}x_3 \leftrightarrow c$

$$\eta \wedge \omega = (-1)^{jk} \omega \wedge \eta, \quad \omega \in \mathrm{Alt}^j V, \quad \eta \in \mathrm{Alt}^k V.$$

Thus, $\mathrm{d}x_i \wedge \mathrm{d}x_j = -\mathrm{d}x_j \wedge \mathrm{d}x_i$ and so $\mathrm{d}x_i \wedge \mathrm{d}x_i = 0$.

If $\omega = \sum_{i=1}^n w_i dx_i \in \Lambda^1(\Omega)$ and $\eta \in \Lambda^0(\Omega)$, then $\omega \wedge \eta$ simply multiplies each of the coefficients w_i by η. If $\eta = \sum_{i=1}^n \eta_i dx_i \in \Lambda^1(\Omega)$, then from the bilinearity and antisymmetry, we have

$$\begin{aligned}
\omega \wedge \eta &= w_1\eta_1 \mathrm{d}x_1 \wedge \mathrm{d}x_1 + w_1\eta_2 \mathrm{d}x_1 \wedge \mathrm{d}x_2 + w_2\eta_1 \mathrm{d}x_2 \wedge \mathrm{d}x_1 + w_2\eta_1 \mathrm{d}x_2 \wedge \mathrm{d}x_2 \\
&= (w_1\eta_2 - w_2\eta_1)\mathrm{d}x_1 \wedge \mathrm{d}x_2, \qquad n = 2, \\
\omega \wedge \eta &= (w_1\eta_2 - w_2\eta_1)\mathrm{d}x_1 \wedge \mathrm{d}x_2 + (w_1\eta_3 - w_3\eta_1)\mathrm{d}x_1 \wedge \mathrm{d}x_3 \\
&\quad + (w_2\eta_3 - w_3\eta_2)\mathrm{d}x_2 \wedge \mathrm{d}x_3, \qquad n = 3.
\end{aligned}$$

Finally, if $\eta \in \Lambda^2(\Omega) = \eta_1 \mathrm{d}x_2 \wedge \mathrm{d}x_3 - \eta_2 \mathrm{d}x_1 \wedge \mathrm{d}x_3 + \eta_3 \mathrm{d}x_1 \wedge \mathrm{d}x_2$, then

$$\omega \wedge \eta = (w_1\eta_1 + w_2\eta_2 + w_3\eta_3)\mathrm{d}x_1 \wedge \mathrm{d}x_2 \wedge \mathrm{d}x_3.$$

One can give a general formula for the wedge product, which we omit here.

If ω_x and $\eta_x \in \Lambda^k(\Omega)$ are given by

$$\sum_{1 \le \sigma(1) < \cdots < \sigma(k) \le n} a_\sigma \mathrm{d}x_{\sigma(1)} \wedge \cdots \wedge \mathrm{d}x_{\sigma(k)}, \qquad \sum_{1 \le \sigma(1) < \cdots < \sigma(k) \le n} b_\sigma \mathrm{d}x_{\sigma(1)} \wedge \cdots \wedge \mathrm{d}x_{\sigma(k)},$$

respectively, we can define the inner products

$$\langle \omega_x, \eta_x \rangle = \sum_{1 \le \sigma(1) < \cdots < \sigma(k) \le n} a_\sigma b_\sigma, \qquad \langle \omega, \eta \rangle = \int_\Omega \langle \omega_x, \eta_x \rangle \mathrm{d}x_1 \wedge \cdots \wedge \mathrm{d}x_n,$$

where $\mathrm{d}x_1 \wedge \cdots \wedge \mathrm{d}x_n$ is the volume form.

A key object in our presentation is the exterior derivative $\mathrm{d} = \mathrm{d}_k : \Lambda^k(\Omega) \to \Lambda^{k+1}(\Omega)$, defined by

$$\mathrm{d} \sum a_\sigma \mathrm{d}x_{\sigma(1)} \wedge \cdots \wedge \mathrm{d}x_{\sigma(k)} = \sum_\sigma \sum_{i=1}^n \frac{\partial a_\sigma}{\partial x_i} \mathrm{d}x_i \wedge \mathrm{d}x_{\sigma(1)} \wedge \cdots \wedge \mathrm{d}x_{\sigma(k)}.$$

As we shall see below, the exterior derivative operator d corresponds to the standard differential operators **grad**, **curl**, div, and rot.

When $n = 2$, if $\omega \in \Lambda^0(\Omega)$, then $\mathrm{d}_0\omega = \partial w/\partial x_1 \mathrm{d}x_1 + \partial w/\partial x_2 \mathrm{d}x_2 \in \Lambda^1(\Omega)$. Identifying $\partial w/\partial x_1 \mathrm{d}x_1 + \partial w/\partial x_2 \mathrm{d}x_2$ with the vector $(\partial w/\partial x_1, \partial w/\partial x_2)$, d_0 corresponds to **grad**. If instead, we identify $\partial w/\partial x_1 \mathrm{d}x_1 + \partial w/\partial x_2 \mathrm{d}x_2$ with the vector $(-\partial w/\partial x_2, \partial w/\partial x_1)$, then d_0 corresponds to **curl**.

If $\mu = w_1 \mathrm{d}x_1 + w_2 \mathrm{d}x_2 \in \Lambda^1(\Omega)$, then $\mathrm{d}_1\mu = (\partial w_2/\partial x_1 - \partial w_1/\partial x_2)\mathrm{d}x_1 \wedge \mathrm{d}x_2 \in \Lambda^2(\Omega)$. If we identify $w_1 \mathrm{d}x_1 + w_2 \mathrm{d}x_2$ with the vector (w_1, w_2), then d_1 corresponds to rot. If instead, we identify $w_1 \mathrm{d}x_1 + w_2 \mathrm{d}x_2$ with the vector$(-w_2, w_1)$, then d_1 corresponds to – div.

When $n = 3$, if $\omega \in \Lambda^0(\Omega)$, then $\mathrm{d}_0\omega = \partial w/\partial x_1 \mathrm{d}x_1 + \partial w/\partial x_2 \mathrm{d}x_2 + \partial w/\partial x_3 \mathrm{d}x_3 \in \Lambda^1(\Omega)$. Identifying $\partial w/\partial x_1 \mathrm{d}x_1 + \partial w/\partial x_2 \mathrm{d}x_2 + \partial w/\partial x_3 \mathrm{d}x_3$ with

Table 2. Correspondences between differential forms ω on $\Omega \subset \mathbb{R}^3$ and scalar/vector fields w on Ω

k	$\Lambda^k(\Omega)$	$H\Lambda^k(\Omega)$	$\mathrm{d}\omega$
0	$C^\infty(\Omega)$	$H^1(\Omega)$	$\mathbf{grad}\, w$
1	$C^\infty(\Omega;\mathbb{R}^3)$	$H(\mathbf{curl},\Omega;\mathbb{R}^3)$	$\mathbf{curl}\, w$
2	$C^\infty(\Omega;\mathbb{R}^3)$	$H(\mathrm{div},\Omega;\mathbb{R}^3)$	$\mathrm{div}\, w$
3	$C^\infty(\Omega)$	$L^2(\Omega)$	0

$(\partial w/\partial x_1, \partial w/\partial x_2, \partial w/\partial x_3)$, d_0 corresponds to **grad**. If $\mu = w_1 \mathrm{d}x_1 + w_2 \mathrm{d}x_2 + w_3 \mathrm{d}x_3 \in \Lambda^1(\Omega)$, then $\mathrm{d}_1\mu = (\partial w_3/\partial x_2 - \partial w_2/\partial x_3)\mathrm{d}x_2 \wedge \mathrm{d}x_3 - (\partial w_1/\partial x_3 - \partial w_3/\partial x_1)\mathrm{d}x_1 \wedge \mathrm{d}x_3 + (\partial w_2/\partial x_1 - \partial w_1/\partial x_2)\mathrm{d}x_1 \wedge \mathrm{d}x_2 \in \Lambda^2(\Omega)$. Identifying $w_1 \mathrm{d}x_1 + w_2 \mathrm{d}x_2 + w_3 \mathrm{d}x_3$ with the vector (w_1, w_2, w_3), d_1 corresponds to **curl**. Finally, if $\mu = w_1 \mathrm{d}x_2 \wedge \mathrm{d}x_3 - w_2 \mathrm{d}x_1 \wedge \mathrm{d}x_3 + w_3 \mathrm{d}x_1 \wedge \mathrm{d}x_2 \in \Lambda^2(\Omega)$, then $\mathrm{d}_2\mu = (\partial w_1/\partial x_1 + \partial w_2/\partial x_2 + \partial w_3/\partial x_3)\mathrm{d}x_1 \wedge \mathrm{d}x_2 \wedge \mathrm{d}x_3 \in \Lambda^3(\Omega)$. Identifying μ with (w_1, w_2, w_3), d_2 corresponds to div. Table 2 summarizes correspondences between differential forms and their proxy fields in the case $\Omega \subset \mathbb{R}^3$.

An important role in our analysis is played by the de Rham sequence, the sequence of spaces and mappings given in the notation of differential forms by:

$$\mathbb{R} \xrightarrow{\subset} \Lambda^0(\Omega) \xrightarrow{\mathrm{d}_0} \Lambda^1(\Omega) \xrightarrow{\mathrm{d}_1} \cdots \xrightarrow{\mathrm{d}_{n-1}} \Lambda^n(\Omega) \to 0.$$

By introducing proxy fields and the usual differential operators, the de Rham complex (and its L^2 version) take the following forms. For $\Omega \subset \mathbb{R}^3$,

$$\mathbb{R} \xrightarrow{\subset} C^\infty(\Omega) \xrightarrow{\mathbf{grad}} C^\infty(\Omega;\mathbb{R}^3) \xrightarrow{\mathbf{curl}} C^\infty(\Omega;\mathbb{R}^3) \xrightarrow{\mathrm{div}} C^\infty(\Omega) \to 0,$$

$$\mathbb{R} \xrightarrow{\subset} H^1(\Omega) \xrightarrow{\mathbf{grad}} H(\mathbf{curl},\Omega;\mathbb{R}^3) \xrightarrow{\mathbf{curl}} H(\mathrm{div},\Omega;\mathbb{R}^3) \xrightarrow{\mathrm{div}} L^2(\Omega) \to 0.$$

For $\Omega \subset \mathbb{R}^2$, the de Rham complex becomes

$$\mathbb{R} \xrightarrow{\subset} C^\infty(\Omega) \xrightarrow{\mathbf{grad}} C^\infty(\Omega;\mathbb{R}^2) \xrightarrow{\mathbf{rot}} C^\infty(\Omega) \to 0,$$

or

$$\mathbb{R} \xrightarrow{\subset} C^\infty(\Omega) \xrightarrow{\mathbf{curl}} C^\infty(\Omega;\mathbb{R}^2) \xrightarrow{\mathrm{div}} C^\infty(\Omega) \to 0,$$

depending on whether we identify $w_1 \mathrm{d}x_1 + w_2 \mathrm{d}x_2 \in \Lambda^1(\Omega)$ with the vector (w_1, w_2) or the vector $(-w_2, w_1)$. There are also analogous L^2 complexes.

4 Basic Finite Element Spaces and their Properties

We now turn to the definition of the finite element spaces we shall use in our approximation schemes and their properties. For this we follow the approach developed in [9]. We begin by defining $\mathcal{P}_r$ as the space of polynomials in n variables of degree at most r and $\mathcal{P}_r\Lambda^k$ as the space of differential k-forms with coefficients belonging to $\mathcal{P}_r$. Let $\mathcal{T}_h$ be a triangulation of Ω by $n+1$ simplices T and set

$$\mathcal{P}_r\Lambda^k(\mathcal{T}_h) = \{\omega \in H\Lambda^k(\Omega) : \omega_T \in \mathcal{P}_r\Lambda^k(T)\ \forall T \in \mathcal{T}_h\}, \quad r \geq 0$$
$$\mathcal{P}_r^-\Lambda^k(\mathcal{T}_h) = \{\omega \in H\Lambda^k(\Omega) : \omega_T \in \mathcal{P}_r^-\Lambda^k(T)\ \forall T \in \mathcal{T}_h\}, \quad r \geq 1,$$

where $\mathcal{P}_r^-\Lambda^k(T) := \mathcal{P}_{r-1}\Lambda^k(T) + \kappa\mathcal{P}_{r-1}\Lambda^{k+1}(T)$ and $\kappa = \kappa_{k+1} : \Lambda^{k+1}(T) \to \Lambda^k(T)$ is the *Koszul differential* defined for $\omega = \sum_\sigma a_\sigma \mathrm{d}x_{\sigma(1)} \wedge \cdots \wedge \mathrm{d}x_{\sigma(k+1)} \in \Lambda^{k+1}$ by

$$\kappa\omega = \sum_\sigma \sum_{i=1}^{k+1} (-1)^{i+1} a_\sigma x_{\sigma(i)} \mathrm{d}x_{\sigma(1)} \wedge \cdots \wedge \hat{\mathrm{d}x_{\sigma(i)}} \wedge \cdots \mathrm{d}x_{\sigma(k+1)},$$

where the notation $\hat{\mathrm{d}x_{\sigma(i)}}$ means that the term is omitted in the sum. Note that $\kappa_k\kappa_{k+1} = 0$, and one can show that the Koszul complex

$$0 \to \mathcal{P}_{r-n}\Lambda^n(\Omega) \xrightarrow{\kappa_n} \mathcal{P}_{r-n+1}\Lambda^{n-1}(\Omega) \xrightarrow{\kappa_{n-1}} \cdots \xrightarrow{\kappa_1} \mathcal{P}_r\Lambda^0(\Omega) \to 0,$$

is exact. For $\Omega \subset \mathbb{R}^3$, this complex becomes

$$0 \to \mathcal{P}_{r-3}(\Omega) \xrightarrow{x} \mathcal{P}_{r-2}(\Omega;\mathbb{R}^3) \xrightarrow{\times x} \mathcal{P}_{r-1}(\Omega;\mathbb{R}^3) \xrightarrow{\cdot x} \mathcal{P}_r(\Omega) \to 0.$$

Comparing to the corresponding polynomial de Rham complex

$$0 \to \mathcal{P}_r(\Omega) \xrightarrow{\mathbf{grad}} \mathcal{P}_{r-1}(\Omega;\mathbb{R}^3) \xrightarrow{\mathbf{curl}} \mathcal{P}_{r-2}(\Omega;\mathbb{R}^3) \xrightarrow{\mathrm{div}} \mathcal{P}_{r-3}(\Omega) \to 0,$$

we see that the Koszul differential increases polynomial degree and decreases the order of the differential form, while exterior differentiation does exactly the opposite.

We note that $\mathcal{P}_r\Lambda^0(\mathcal{T}_h) = \mathcal{P}_r^-\Lambda^0(\mathcal{T}_h)$, $r \geq 1$ and $\mathcal{P}_r\Lambda^n(\mathcal{T}_h) = \mathcal{P}_{r+1}^-\Lambda^n(\mathcal{T}_h)$, $r \geq 0$. Using proxy fields, we can identify these spaces of finite element differential forms with finite element spaces of scalar and vector functions. In Tables 3 and 4, we summarize the correspondences between spaces of finite element differential forms and classical finite element spaces in two and three dimensions.

Degrees of freedom for these spaces are given as follows. For the space $\mathcal{P}_r\Lambda^k(\mathcal{T})$, we use

Table 3. Correspondences between finite element differential forms and the classical finite element spaces for $n = 2$

k	$\Lambda_h^k(\Omega)$	Classical finite element space
0	$\mathcal{P}_r\Lambda^0(\mathcal{T}_h)$	Lagrange elements of degree $\leq r$
1	$\mathcal{P}_r\Lambda^1(\mathcal{T}_h)$	Brezzi–Douglas–Marini $H(\mathrm{div})$ elements of degree $\leq r$
2	$\mathcal{P}_r\Lambda^2(\mathcal{T}_h)$	discontinuous elements of degree $\leq r$
0	$\mathcal{P}_r^-\Lambda^0(\mathcal{T}_h)$	Lagrange elements of degree $\leq r$
1	$\mathcal{P}_r^-\Lambda^1(\mathcal{T}_h)$	Raviart–Thomas $H(\mathrm{div})$ elements of order $r-1$
2	$\mathcal{P}_r^-\Lambda^2(\mathcal{T}_h)$	discontinuous elements of degree $\leq r-1$

Table 4. Correspondences between finite element differential forms and the classical finite element spaces for $n = 3$

k	$\Lambda_h^k(\Omega)$	Classical finite element space
0	$\mathcal{P}_r\Lambda^0(\mathcal{T}_h)$	Lagrange elements of degree $\le r$
1	$\mathcal{P}_r\Lambda^1(\mathcal{T}_h)$	Nédélec 2nd-kind $H(\mathbf{curl})$ elements of degree $\le r$
2	$\mathcal{P}_r\Lambda^2(\mathcal{T}_h)$	Nédélec 2nd-kind $H(\mathrm{div})$ elements of degree $\le r$
3	$\mathcal{P}_r\Lambda^3(\mathcal{T}_h)$	discontinuous elements of degree $\le r$
0	$\mathcal{P}_r^-\Lambda^0(\mathcal{T}_h)$	Lagrange elements of degree $\le r$
1	$\mathcal{P}_r^-\Lambda^1(\mathcal{T}_h)$	Nédélec 1st-kind $H(\mathbf{curl})$ elements of order $r-1$
2	$\mathcal{P}_r^-\Lambda^2(\mathcal{T}_h)$	Nédélec 1st-kind $H(\mathrm{div})$ elements of order $r-1$
3	$\mathcal{P}_r^-\Lambda^3(\mathcal{T}_h)$	discontinuous elements of degree $\le r-1$

$$\int_f \mathrm{Tr}_f\, \omega \wedge \nu, \quad \nu \in \mathcal{P}^-_{r-j+k}\Lambda^{j-k}(f), \quad f \in \Delta_j(\mathcal{T}),$$

for $k \le j \le \min(n, r+k-1)$, where $\mathrm{Tr}_f\,\omega$ denotes the trace of ω on the face f and $\Delta_j(\mathcal{T})$ is the set of all j-dimensional subsimplices generated by $\mathcal{T}_h$. For example, when $n = 3$, $\Delta_j(\mathcal{T})$ is the set of vertices, edges, faces, or tetrahedra in the mesh $\mathcal{T}_h$ for $j = 0, 1, 2, 3$. In this case, when $j = 0$, i.e., f is a vertex, $\int_f \mathrm{Tr}_f\,\omega$ means $w(f)$, where w is the function associated with $\omega \in \Lambda^0(\Omega)$. When $j = 1$, i.e., f is an edge of a tetrahedron, $\int_f \mathrm{Tr}_f\,\omega = \int_f w \cdot t\, d\mu$, where w is the vector associated to $\omega \in \Lambda^1(\Omega)$ and t is the unit tangent vector to f. When $j = 2$, i.e., f is a face of a tetrahedron, $\int_f \mathrm{Tr}_f\,\omega = \int_f w \cdot n\, d\mu$, where w is the vector associated to $\omega \in \Lambda^2(\Omega)$ and n is the unit outward normal to f. Finally, when $j = 3$, i.e., f is a tetrahedron, $\int_f \mathrm{Tr}_f\,\omega = \int_f w\, d\mu$, where w is the function associated to $\omega \in \Lambda^3(\Omega)$.

Analogously, the degrees of freedom for the space $\mathcal{P}_r^-\Lambda^k(\mathcal{T})$ are given by

$$\int_f \mathrm{Tr}_f\, \omega \wedge \nu, \quad \nu \in \mathcal{P}_{r-j+k-1}\Lambda^{j-k}(f), \quad f \in \Delta_j(\mathcal{T}),$$

for $k \le j \le \min(n, r+k-1)$. Note the key property that the degrees of freedom for each space are defined in terms of wedge products with elements of the other space.

An important property of these finite element spaces is that they form discrete de Rham sequences. In fact, as shown in [9], in n dimensions, there are exactly 2^{n-1} distinct sequences. When $n = 2$ and $r \ge 0$, these are

$$0 \to \mathcal{P}_{r+2}\Lambda^0(\mathcal{T}_h) \xrightarrow{\mathrm{d}_0} \mathcal{P}_{r+1}\Lambda^1(\mathcal{T}_h) \xrightarrow{\mathrm{d}_1} \mathcal{P}_r\Lambda^2(\mathcal{T}_h) \to 0,$$

$$0 \to \mathcal{P}_{r+1}\Lambda^0(\mathcal{T}_h) \xrightarrow{\mathrm{d}_0} \mathcal{P}^-_{r+1}\Lambda^1(\mathcal{T}_h) \xrightarrow{\mathrm{d}_1} \mathcal{P}_r\Lambda^2(\mathcal{T}_h) \to 0.$$

When $n = 3$ and $r \ge 0$, we have the four sequences

$$0 \to \mathcal{P}_{r+3}\Lambda^0(\mathcal{T}_h) \xrightarrow{d_0} \mathcal{P}_{r+2}\Lambda^1(\mathcal{T}_h) \xrightarrow{d_1} \mathcal{P}_{r+1}\Lambda^2(\mathcal{T}_h) \xrightarrow{d_2} \mathcal{P}_r\Lambda^3(\mathcal{T}_h) \to 0,$$
$$0 \to \mathcal{P}_{r+2}\Lambda^0(\mathcal{T}_h) \xrightarrow{d_0} \mathcal{P}_{r+1}\Lambda^1(\mathcal{T}_h) \xrightarrow{d_1} \mathcal{P}^-_{r+1}\Lambda^2(\mathcal{T}_h) \xrightarrow{d_2} \mathcal{P}_r\Lambda^3(\mathcal{T}_h) \to 0,$$
$$0 \to \mathcal{P}_{r+2}\Lambda^0(\mathcal{T}_h) \xrightarrow{d_0} \mathcal{P}^-_{r+2}\Lambda^1(\mathcal{T}_h) \xrightarrow{d_1} \mathcal{P}_{r+1}\Lambda^2(\mathcal{T}_h) \xrightarrow{d_2} \mathcal{P}_r\Lambda^3(\mathcal{T}_h) \to 0,$$
$$0 \to \mathcal{P}_{r+1}\Lambda^0(\mathcal{T}_h) \xrightarrow{d_0} \mathcal{P}^-_{r+1}\Lambda^1(\mathcal{T}_h) \xrightarrow{d_1} \mathcal{P}^-_{r+1}\Lambda^2(\mathcal{T}_h) \xrightarrow{d_2} \mathcal{P}_r\Lambda^3(\mathcal{T}_h) \to 0.$$

The first and last of these are exact sequences involving only the $\mathcal{P}_r\Lambda^k(\mathcal{T}_h)$ or $\mathcal{P}_r^-\Lambda^k(\mathcal{T}_h)$ spaces alone, while the middle two mix the two spaces. As we shall see, to obtain mixed finite element methods for elasticity when $n = 3$, it is one of these middle sequences that will play a key role.

To each of the spaces $\mathcal{P}_r\Lambda^k(\mathcal{T}_h)$, we may associate a canonical projection operator $\Pi(= \Pi_{\mathcal{T}_h}) : C^0\Lambda^k(\Omega) \to \mathcal{P}_r\Lambda^k(\mathcal{T}_h)$ defined by the equations:

$$\int_f \operatorname{Tr}_f \Pi\omega \wedge \nu = \int_f \operatorname{Tr}_f \omega \wedge \nu, \quad \nu \in \mathcal{P}^-_{r-j+k}\Lambda^{j-k}(f), \quad f \in \Delta_j(\mathcal{T}),$$

for $k \le j \le \min(n, r+k-1)$. Similarly, to each of the spaces $\mathcal{P}_r^-\Lambda^k(\mathcal{T}_h)$, we may associate a canonical projection operator $\Pi(= \Pi_{\mathcal{T}_h}) : C^0\Lambda^k(\Omega) \to \mathcal{P}_r^-\Lambda^k(\mathcal{T}_h)$ defined by the equations

$$\int_f \operatorname{Tr}_f \Pi\omega \wedge \nu = \int_f \operatorname{Tr}_f \omega \wedge \nu, \quad \nu \in \mathcal{P}_{r-j+k-1}\Lambda^{j-k}(f), \quad f \in \Delta_j(\mathcal{T}),$$

for $k \le j \le \min(n, r+k-1)$. A key property of these projection operators is that they commute with the exterior derivative, i.e., the following four diagrams commute.

$$\begin{array}{ccc} \Lambda^k(\Omega) & \xrightarrow{d_k} & \Lambda^{k+1}(\Omega) \\ \Pi\downarrow & & \Pi\downarrow \\ \mathcal{P}_r\Lambda^k(\mathcal{T}) & \xrightarrow{d_k} & \mathcal{P}_{r-1}\Lambda^{k+1}(\mathcal{T}) \end{array} \qquad \begin{array}{ccc} \Lambda^k(\Omega) & \xrightarrow{d_k} & \Lambda^{k+1}(\Omega) \\ \Pi\downarrow & & \Pi\downarrow \\ \mathcal{P}_r\Lambda^k(\mathcal{T}) & \xrightarrow{d_k} & \mathcal{P}_r^-\Lambda^{k+1}(\mathcal{T}) \end{array}$$

$$\begin{array}{ccc} \Lambda^k(\Omega) & \xrightarrow{d_k} & \Lambda^{k+1}(\Omega) \\ \Pi\downarrow & & \Pi\downarrow \\ \mathcal{P}_r^-\Lambda^k(\mathcal{T}) & \xrightarrow{d_k} & \mathcal{P}_r^-\Lambda^{k+1}(\mathcal{T}) \end{array} \qquad \begin{array}{ccc} \Lambda^k(\Omega) & \xrightarrow{d_k} & \Lambda^{k+1}(\Omega) \\ \Pi\downarrow & & \Pi\downarrow \\ \mathcal{P}_r^-\Lambda^k(\mathcal{T}) & \xrightarrow{d_k} & \mathcal{P}_{r-1}\Lambda^{k+1}(\mathcal{T}). \end{array}$$

These commuting diagrams will also play an essential role in the construction of stable mixed finite element approximation schemes for the equations of elasticity.

4.1 Differential Forms with Values in a Vector Space

To study the equations of linear elasticity in the language of differential forms, we will need to use differential forms with values in a vector space. Let V and W be finite-dimensional vector spaces. We then define the space $\Lambda^k(V; W)$ of differential forms on V with values in W. The two examples we have in mind are when $V = \mathbb{V} = \mathbb{R}^n$ and $W = \mathbb{V}$ or $W = \mathbb{K}$, the set of anti-symmetric matrices. When $n = 2$, $\omega \in \Lambda^k(\mathbb{V};\mathbb{V})$, $k = 0, 1, 2$ will have the respective forms

$$\begin{pmatrix} w_1 \\ w_2 \end{pmatrix}, \qquad \begin{pmatrix} w_{11} \\ w_{21} \end{pmatrix} dx_1 + \begin{pmatrix} w_{12} \\ w_{22} \end{pmatrix} dx_2, \qquad \begin{pmatrix} w_1 \\ w_2 \end{pmatrix} dx_1 \wedge dx_2,$$

while $\omega \in \Lambda^k(\mathbb{V};\mathbb{K})$ will have the respective forms

$$w\chi, \qquad w_1\chi dx_1 + w_2\chi dx_2, \qquad w\chi dx_1 \wedge dx_2, \qquad \text{where} \qquad \chi = \begin{pmatrix} 0 & -1 \\ 1 & 0 \end{pmatrix}.$$

Recalling that the 1-form $w_1 dx_1 + w_2 dx_2$ can be identified either with the vector (w_1, w_2) or the vector $(-w_2, w_1)$, we will have the analogous possibilities in the case of vector- or matrix-valued forms. Since we will be interested in de Rham sequences involving the operator div, we choose the second identification. Hence, $\begin{pmatrix} w_{11} \\ w_{21} \end{pmatrix} dx_1 + \begin{pmatrix} w_{12} \\ w_{22} \end{pmatrix} dx_2 \in \Lambda^1(\mathbb{V};\mathbb{V})$ will be identified with the matrix

$$\begin{pmatrix} W_{11} & W_{12} \\ W_{21} & W_{22} \end{pmatrix} = \begin{pmatrix} -w_{12} & w_{11} \\ -w_{22} & w_{21} \end{pmatrix}, \tag{10}$$

and $w_1\chi dx_1 + w_2\chi dx_2 \in \Lambda^1(\mathbb{V};\mathbb{K})$ with the vector $(-w_2, w_1)$. When $n = 3$, $\omega \in \Lambda^k(\mathbb{V};\mathbb{V})$ will have the respective forms

$$\begin{pmatrix} w_1 \\ w_2 \\ w_3 \end{pmatrix}, \qquad \begin{pmatrix} w_{11} \\ w_{21} \\ w_{31} \end{pmatrix} dx_1 + \begin{pmatrix} w_{12} \\ w_{22} \\ w_{32} \end{pmatrix} dx_2 + \begin{pmatrix} w_{13} \\ w_{23} \\ w_{33} \end{pmatrix} dx_3,$$

$$\begin{pmatrix} w_{11} \\ w_{21} \\ w_{31} \end{pmatrix} dx_2 \wedge dx_3 - \begin{pmatrix} w_{12} \\ w_{22} \\ w_{32} \end{pmatrix} dx_1 \wedge dx_3 + \begin{pmatrix} w_{13} \\ w_{23} \\ w_{33} \end{pmatrix} dx_1 \wedge dx_2,$$

$$\begin{pmatrix} w_1 \\ w_2 \\ w_3 \end{pmatrix} dx_1 \wedge dx_2 \wedge dx_3.$$

Hence, $\Lambda^0(\mathbb{V};\mathbb{V})$ and $\Lambda^3(\mathbb{V};\mathbb{V})$ have obvious identifications with the space of three-dimensional vectors and $\Lambda^1(\mathbb{V};\mathbb{V})$ and $\Lambda^2(\mathbb{V};\mathbb{V})$ have obvious identifications with the space of 3×3 matrices (i.e, $W_{ij} = w_{ij}$ in both cases). In fact, in treating the equations of elasticity on a domain $\Omega \subset \mathbb{R}^n$, we shall represent the stress as an element of $\Lambda^{n-1}(\Omega, \mathbb{V})$. To describe $\Lambda^k(\mathbb{V};\mathbb{K})$, it will be convenient to introduce the operator Skw taking a 3-vector to a skew-symmetric matrix. i.e.,

$$\operatorname{Skw}(w_1, w_2, w_3) = \begin{pmatrix} 0 & -w_3 & w_2 \\ w_3 & 0 & -w_1 \\ -w_2 & w_1 & 0 \end{pmatrix}.$$

Then $\omega \in \Lambda^k(\mathbb{V};\mathbb{K})$ will have the respective forms

$$\begin{gathered}\mathrm{Skw}(w_1, w_2, w_3),\\ \mathrm{Skw}(w_{11}, w_{21}, w_{31})\mathrm{d}x_1 + \mathrm{Skw}(w_{12}, w_{22}, w_{32})\mathrm{d}x_2 + \mathrm{Skw}(w_{13}, w_{23}, w_{33})\mathrm{d}x_3,\\ \mathrm{Skw}(w_{11}, w_{21}, w_{31})\mathrm{d}x_2 \wedge \mathrm{d}x_3 - \mathrm{Skw}(w_{12}, w_{22}, w_{32})\mathrm{d}x_1 \wedge \mathrm{d}x_3\\ + \mathrm{Skw}(w_{13}, w_{23}, w_{33})\mathrm{d}x_1 \wedge \mathrm{d}x_2,\\ \mathrm{Skw}(w_1, w_2, w_3)\mathrm{d}x_1 \wedge \mathrm{d}x_2 \wedge \mathrm{d}x_3.\end{gathered}$$

Note that from the above formulas, there is an obvious identification of $\Lambda^0(\mathbb{V};\mathbb{K})$ and $\Lambda^3(\mathbb{V};\mathbb{K})$ with the space of three-dimensional vectors and of $\Lambda^1(\mathbb{V};\mathbb{K})$ and $\Lambda^2(\mathbb{V};\mathbb{K})$ with 3×3 matrices (again with $W_{ij} = w_{ij}$ in both cases).

In the mixed formulation of elasticity, we shall need a special operator $S = S_k : \Lambda^k(\mathbb{V},\mathbb{V}) \to \Lambda^{k+1}(\mathbb{V},\mathbb{K})$ defined as follows: First define $K_k : \Lambda^k(\Omega;\mathbb{V}) \to \Lambda^k(\Omega;\mathbb{K})$ by

$$K_k\omega = X\omega^T - \omega X^T,$$

where $X = (x_1, \cdots, x_n)^T$. Then define

$$S_k = \mathrm{d}_k K_k - K_{k+1}\mathrm{d}_k : \Lambda^k(\Omega;\mathbb{V}) \to \Lambda^{k+1}(\Omega;\mathbb{K}).$$

Using the definition of the exterior derivative, the definition of K, and the Leibniz rule, one can show that for any vector $(v_1, \ldots, v_{k+1})$,

$$\begin{aligned}&(S_k\omega)_x(v_1, \ldots, v_{k+1})\\ &= \sum_{j=1}^{k+1}(-1)^{j+1}[v_j\omega^T(v_1, \cdots, \hat{v}_j, \cdots, v_{k+1}) - \omega(v_1, \cdots, \hat{v}_j, \cdots, v_{k+1})v_j^T],\end{aligned}$$

where the notation $\hat{v}_j$ means that this argument is omitted. Thus, S_k is a purely algebraic operator.

More specifically, we shall need this operator when $k = n - 2$ and $k = n - 1$. We examine these cases below for $n = 2$ and $n = 3$. When $n = 2$, we get for $\omega = (w_1, w_2)^T$, $K_0\omega = (w_1x_2 - w_2x_1)\chi$, and, after a simple computation,

$$S_0\omega = (\mathrm{d}_0K_0 - K_1\mathrm{d}_0)\omega = -w_2\chi\mathrm{d}x_1 + w_1\chi\mathrm{d}x_2.$$

Note that S_0 is invertible with

$$S_0^{-1}[\mu_1\chi\mathrm{d}x_1 + \mu_2\chi\mathrm{d}x_2] = (\mu_2, -\mu_1)^T.$$

If $\omega \in \Lambda^1(\mathbb{V};\mathbb{V})$ is given by:

$$\omega = w_1\mathrm{d}x_1 + w_2\mathrm{d}x_2, \quad w_1 = (w_{11}, w_{21})^T, \quad w_2 = (w_{12}, w_{22})^T,$$

then $S_1\omega = -(w_{11} + w_{22})\chi\mathrm{d}x_1 \wedge \mathrm{d}x_2$. If we identity ω with a matrix W by

$$\begin{pmatrix} W_{11} & W_{12}\\ W_{21} & W_{22}\end{pmatrix} = \begin{pmatrix} -w_{12} & w_{11}\\ -w_{22} & w_{21}\end{pmatrix},$$

then we can identify $S_1\omega$ with the matrix

$$\begin{pmatrix} 0 & W_{12} - W_{21} \\ W_{21} - W_{12} & 0 \end{pmatrix} = W - W^T \equiv 2\,\mathrm{skw}\, W.$$

When $n = 3$, we get for $\omega = w_1 \mathrm{d}x_1 + w_2 \mathrm{d}x_2 + w_3 \mathrm{d}x_3$, with $w_j = (w_{1j}, w_{2j}, w_{3j})^T$,

$$\begin{aligned} S_1\omega = \mathrm{Skw}(-w_{33} - w_{22}, w_{12}, w_{13}) \mathrm{d}x_2 \wedge \mathrm{d}x_3 \\ - \mathrm{Skw}(w_{21}, -w_{11} - w_{33}, w_{23}) \mathrm{d}x_1 \wedge \mathrm{d}x_3 \\ + \mathrm{Skw}(w_{31}, w_{32}, -w_{11} - w_{22}) \mathrm{d}x_1 \wedge \mathrm{d}x_2. \end{aligned}$$

If we identify $\omega \in \Lambda^1(\mathbb{V}; \mathbb{V})$ with a matrix W by $W_{ij} = w_{ij}$, and identify $S_1\omega \in \Lambda^2(\mathbb{V}; \mathbb{K})$ with the matrix U given by

$$U = \begin{pmatrix} -w_{33} - w_{22} & w_{21} & w_{31} \\ w_{12} & -w_{11} - w_{33} & w_{32} \\ w_{13} & w_{23} & -w_{11} - w_{22} \end{pmatrix},$$

then, W and U are related by the equations

$$U = \Xi W \equiv W^T - \mathrm{tr}(W) I, \qquad W = \Xi^{-1} U \equiv U^T - \frac{1}{2}\mathrm{tr}(U) I.$$

Hence, S_1 is invertible.

If $\omega = w_1 \mathrm{d}x_2 \wedge \mathrm{d}x_3 - w_2 \mathrm{d}x_1 \wedge \mathrm{d}x_3 + w_3 \mathrm{d}x_1 \wedge \mathrm{d}x_2$, then

$$S_2\omega = \begin{pmatrix} 0 & w_{21} - w_{12} & w_{31} - w_{13} \\ w_{12} - w_{21} & 0 & w_{32} - w_{23} \\ w_{13} - w_{31} & w_{23} - w_{32} & 0 \end{pmatrix} \mathrm{d}x_1 \wedge \mathrm{d}x_2 \wedge \mathrm{d}x_3.$$

If we identify ω with the matrix W given by $W_{ij} = w_{ij}$, then by the above, $S_2\omega$ may be identified with the matrix $-2\,\mathrm{skw}\, W$.

We easily obtain from the fact that $\mathrm{d}_{k+1}\mathrm{d}_k = 0$ and the definition $S_k = \mathrm{d}_k K_k - K_{k+1}\mathrm{d}_k$ that

$$\mathrm{d}_{k+1} S_k + S_{k+1} \mathrm{d}_k = 0.$$

This identity, for $k = n - 2$, i.e., $\mathrm{d}_{n-1} S_{n-2} + S_{n-1} \mathrm{d}_{n-2} = 0$ is the key identity in establishing stability of continuous and discrete variational formulations of elasticity with weak symmetry.

Note that this formula is much more complicated and also different in different dimensions when stated in terms of proxy fields (which are reasons why we have introduced differential forms). When $n = 2$ and $k = 0$, if we identify $\omega = (w_1, w_2)^T \in \Lambda^0(\Omega; \mathbb{V})$ with the vector W, then the formula $(\mathrm{d}_1 S_0 + S_1 \mathrm{d}_0)\omega = 0$ becomes

$$(\mathrm{div}\, W)\chi + 2\,\mathrm{skw}\, \mathbf{curl}\, W = 0.$$

When $n = 3$ and $k = 1$, if we identify $\omega \in \Lambda^1(\Omega; \mathbb{V})$ with the matrix W, then the formula $(\mathrm{d}_2 S_1 + S_2 \mathrm{d}_1)\omega = 0$ becomes

$$\mathrm{Skw}\, \mathrm{div}(\Xi W) - 2\,\mathrm{skw}\, \mathbf{curl}\, W = 0.$$

5 Mixed Formulation of the Equations of Elasticity with Weak Symmetry

In order to write (3) in the language of exterior calculus, we will use the spaces of vector-valued differential forms presented in the previous section. We assume that Ω is a contractible domain in $\mathbb{R}^n$, $\mathbb{V} = \mathbb{R}^n$, and $\mathbb{K}$ is again the space of skew-symmetric matrices. We showed in the last section that the operator $S = S_{n-1} : \Lambda^{n-1}(\Omega; \mathbb{V}) \to \Lambda^n(\Omega; \mathbb{K})$ corresponds (up to a factor of ± 2) to taking the skew-symmetric part of its argument. Setting $\mathrm{d}_{n-1} = \mathrm{d}$, the elasticity problem (3) becomes: Find $(\sigma, u, p) \in H\Lambda^{n-1}(\Omega; \mathbb{V}) \times L^2\Lambda^n(\Omega; \mathbb{V}) \times L^2\Lambda^n(\Omega; \mathbb{K})$ such that

$$\langle A\sigma, \tau \rangle + \langle \mathrm{d}\tau, u \rangle - \langle S\tau, p \rangle = 0, \quad \tau \in H\Lambda^{n-1}(\Omega; \mathbb{V}), \tag{11}$$
$$\langle \mathrm{d}\sigma, v \rangle = \langle f, v \rangle, \quad v \in L^2\Lambda^n(\Omega; \mathbb{V}), \qquad \langle S\sigma, q \rangle = 0, \quad q \in L^2\Lambda^n(\Omega; \mathbb{K}).$$

This problem is well-posed in the sense that, for each $f \in L^2\Lambda^n(\Omega; \mathbb{V})$, there exists a unique solution $(\sigma, u, p) \in H\Lambda^{n-1}(\Omega; \mathbb{V}) \times L^2\Lambda^n(\Omega; \mathbb{V}) \times L^2\Lambda^n(\Omega; \mathbb{K})$, and the solution operator is a bounded operator

$$L^2\Lambda^n(\Omega; \mathbb{V}) \to H\Lambda^{n-1}(\Omega; \mathbb{V}) \times L^2\Lambda^n(\Omega; \mathbb{V}) \times L^2\Lambda^n(\Omega; \mathbb{K}).$$

This will follow from the general theory of such saddle point problems [14] once we establish two conditions:

(W1) $\|\tau\|_{H\Lambda}^2 \le c_1 \langle A\tau, \tau \rangle$ whenever $\tau \in H\Lambda^{n-1}(\Omega; \mathbb{V})$ satisfies $\langle \mathrm{d}\tau, v \rangle = 0$ $\forall v \in L^2\Lambda^n(\Omega; \mathbb{V})$ and $\langle S\tau, q \rangle = 0\ \forall q \in L^2\Lambda^n(\Omega; \mathbb{K})$,

(W2) for all nonzero $(v, q) \in L^2\Lambda^n(\Omega; \mathbb{V}) \times L^2\Lambda^n(\Omega; \mathbb{K})$, there exists nonzero $\tau \in H\Lambda^{n-1}(\Omega; \mathbb{V})$ with $\langle \mathrm{d}\tau, v \rangle - \langle S\tau, q \rangle \ge c_2 \|\tau\|_{H\Lambda}(\|v\| + \|q\|)$,

for some positive constants c_1 and c_2. The first condition is obvious (and does not even utilize the orthogonality of $S\tau$). However, the second condition is more subtle. We will verify it in Theorem 7.2 in a subsequent section.

We next consider a finite element discretizations of (11). For this, we choose families of finite-dimensional subspaces

$$\Lambda_h^{n-1}(\mathbb{V}) \subset H\Lambda^{n-1}(\Omega; \mathbb{V}), \quad \Lambda_h^n(\mathbb{V}) \subset L^2\Lambda^n(\Omega; \mathbb{V}), \quad \Lambda_h^n(\mathbb{K}) \subset L^2\Lambda^n(\Omega; \mathbb{K}),$$

indexed by h, and seek the discrete solution $(\sigma_h, u_h, p_h) \in \Lambda_h^{n-1}(\mathbb{V}) \times \Lambda_h^n(\mathbb{V}) \times \Lambda_h^n(\mathbb{K})$ such that

$$\langle A\sigma_h, \tau \rangle + \langle \mathrm{d}\tau, u_h \rangle - \langle S\tau, p_h \rangle = 0, \quad \tau \in \Lambda_h^{n-1}(\mathbb{V}), \tag{12}$$
$$\langle \mathrm{d}\sigma_h, v \rangle = \langle f, v \rangle \quad v \in \Lambda_h^n(\mathbb{V}), \qquad \langle S\sigma_h, q \rangle = 0, \quad q \in \Lambda_h^n(\mathbb{K}).$$

In analogy with the well-posedness of the problem (11), the stability of the saddle point system (12) will be ensured by the Brezzi stability conditions:

(S1) $\|\tau\|_{H\Lambda}^2 \le c_1 (A\tau, \tau)$ whenever $\tau \in \Lambda_h^{n-1}(\mathbb{V})$ satisfies $\langle \mathrm{d}\tau, v \rangle = 0$ $\forall v \in \Lambda_h^n(\mathbb{V})$ and $\langle S\tau, q \rangle = 0\ \forall q \in \Lambda_h^n(\mathbb{K})$,

(S2) for all nonzero $(v,q) \in \Lambda_h^n(\mathbb{V}) \times \Lambda_h^n(\mathbb{K})$, there exists nonzero $\tau \in \Lambda_h^{n-1}(\mathbb{V})$ with $\langle \mathrm{d}\tau, v\rangle - \langle S\tau, q\rangle \geq c_2 \|\tau\|_{H\Lambda}(\|v\| + \|q\|)$,

where now the constants c_1 and c_2 must be independent of h. The difficulty is, of course, to design finite element spaces satisfying these conditions.

We have seen previously that there is a close relation between the construction of stable mixed finite element methods for the approximation of the equations of linear elasticity and discretization of the associated elasticity complex (6). This relationship extends an analogous relationship between the construction of stable mixed finite element methods for Poisson's equation and discretization of the de Rham complex. It turns out that there is also a close, but nonobvious, connection between the elasticity complex and the de Rham complex. This connection is described in [19] and is related to a general construction given in [13], called the BGG resolution (see also [16]).

The elasticity complex (6) is related to the formulation of the equations of elasticity with strong symmetry. It is also possible to derive an elasticity complex that is related to the equations of elasticity with weak symmetry, again starting from the de Rham complex. In [8] (two dimensions) and [6] (three dimensions), such an elasticity complex is derived and a discrete version of the BGG construction also developed. This was then used to derive stable mixed finite element methods for elasticity in a systematic manner based on the finite element versions of the de Rham sequence described earlier. The resulting elements in both two and three space dimensions are simpler than any derived previously. For example, the simple choice of $\mathcal{P}_1\Lambda^{n-1}(\mathcal{T}_h;\mathbb{V})$ for stress, $\mathcal{P}_0\Lambda^n(\mathcal{T}_h;\mathbb{V})$ for displacement, and $\mathcal{P}_0\Lambda^n(\mathcal{T}_h;\mathbb{K})$ for the multiplier results in a stable discretization of the problem (12). In Figure 4, this element is depicted in two dimensions. For stress, the degrees of freedom are the first two moments of its trace on the edges, and for the displacement and multiplier, their integrals on the triangle (two components for displacement, one for the multiplier). Moreover, this element is the lowest order of a family of stable elements in n dimensions utilizing $\mathcal{P}_r\Lambda^{n-1}(\mathcal{T}_h;\mathbb{V})$ for stress, $\mathcal{P}_{r-1}\Lambda^n(\mathcal{T}_h;\mathbb{V})$ for displacement, and $\mathcal{P}_{r-1}\Lambda^n(\mathcal{T}_h;\mathbb{K})$ for the multiplier. In fact, the lowest order element may be simplified further, so that only a subset of linear vectors is needed to approximate the stress. More details of this simplified element are presented in Sect. 11.

In the next section, we follow the approach in [9] and outline how an elasticity complex with weakly imposed symmetry can be derived from the de Rham complex. Since this derivation produces a sequence in the notation of differential forms, we then translate our results to the more classical notation for elasticity in two and three

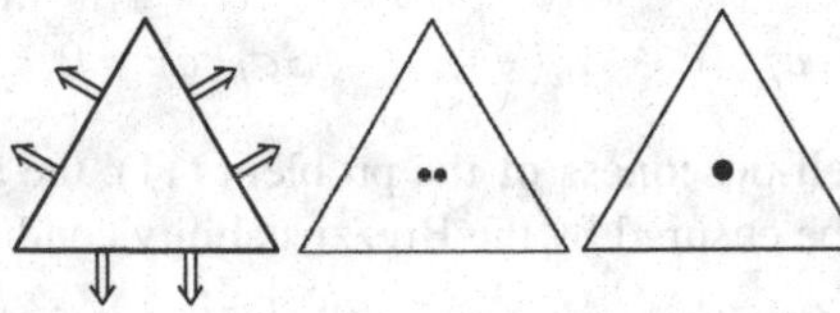

Fig. 4. Approximation of stress, displacement, and multiplier for the simplest element in two dimensions

dimensions. In Sect. 7, we give a proof of the well-posedness of the mixed formulation of elasticity with weak symmetry for the continuous problem, as a guide for establishing a similar result for the discrete problem. Based on this proof, we develop in Sect. 8 the conditions that we will need for stable approximation schemes. These results are then used to establish the main stability result for weakly symmetric mixed finite element approximations of the equations of elasticity in Sect. 9 and some more refined estimates in Sect. 10. The results presented in this paper are for the case of displacement boundary conditions. An extension to the equations of elasticity with traction boundary conditions can be found in [9].

6 From the de Rham Complex to an Elasticity Complex with Weak Symmetry

In this section, we discuss the connection of the elasticity complex in n dimensions with the de Rham complex. Details of the derivation can be found in [6] and [9] and follow the ideas in a derivation of elasticity from the de Rham sequence in the case of strongly imposed symmetry given in [19] in three dimensions.

We start with the two vector-valued de Rham sequences, one with values in $\mathbb{V}$ and one with values in $\mathbb{K}$, i.e.,

$$\Lambda^{n-2}(\Omega;\mathbb{K}) \xrightarrow{\mathrm{d}_{n-2}} \Lambda^{n-1}(\Omega;\mathbb{K}) \xrightarrow{\mathrm{d}_{n-1}} \Lambda^{n}(\mathbb{K}) \to 0,$$

$$\Lambda^{n-3}(\Omega;\mathbb{V}) \xrightarrow{\mathrm{d}_{n-3}} \Lambda^{n-2}(\Omega;\mathbb{V}) \xrightarrow{\mathrm{d}_{n-2}} \Lambda^{n-1}(\Omega;\mathbb{V}) \xrightarrow{\mathrm{d}_{n-1}} \Lambda^{n}(\mathbb{V}) \to 0.$$

Using the fact that these sequences are exact, one is able to show that the sequence

$$\Lambda^{n-3}(\mathbb{W}) \xrightarrow{(\mathrm{d}_{n-3},-S_{n-3})} \Lambda^{n-2}(\Omega;\mathbb{K}) \xrightarrow{\mathrm{d}_{n-2}\circ S_{n-2}^{-1}\circ \mathrm{d}_{n-2}} \Lambda^{n-1}(\Omega;\mathbb{V}) \xrightarrow{(-S_{n-1},\mathrm{d}_{n-1})^T} \Lambda^{n}(\mathbb{W}) \to 0 \quad (13)$$

is exact, where $\mathbb{W} = \mathbb{K} \times \mathbb{V}$. We refer to the sequence (13) as the *elasticity sequence with weak symmetry*. Crucial to this construction is the fact that the operator S_{n-2} : $H^1\Lambda^{n-2}(\Omega;\mathbb{V}) \to H^1\Lambda^{n-1}(\Omega;\mathbb{K})$ is an isomorphism.

We next interpret this sequence in the language of differential operators in two and three dimensions. When $n = 2$, we have the sequence

$$\Lambda^{0}(\Omega;\mathbb{K}) \xrightarrow{\mathrm{d}_0\circ S_0^{-1}\circ \mathrm{d}_0} \Lambda^{1}(\Omega;\mathbb{V}) \xrightarrow{(-S_1,\mathrm{d}_1)^T} \Lambda^{2}(\mathbb{W}) \to 0.$$

Hence, if we begin with an element $w\chi \in \Lambda^0(\Omega;\mathbb{K})$ that we identify with the scalar function w, then

$$\mathrm{d}_0(w\chi) = \frac{\partial w}{\partial x_1}\chi \mathrm{d}x_1 + \frac{\partial w}{\partial x_2}\chi \mathrm{d}x_2, \qquad S_0^{-1}[\mathrm{d}_0(w\chi)] = \left(\frac{\partial w}{\partial x_2}, -\frac{\partial w}{\partial x_1}\right)^T,$$

$$\mathrm{d}_0 S_0^{-1}[\mathrm{d}_0(w\chi)] = \begin{pmatrix} \partial^2 w/\partial x_1\partial x_2 \\ -\partial^2 w/\partial x_1^2 \end{pmatrix} \mathrm{d}x_1 + \begin{pmatrix} \partial^2 w/\partial x_2^2 \\ -\partial^2 w/\partial x_1\partial x_2 \end{pmatrix} \mathrm{d}x_2.$$

We then identity this vector-valued 1-form with the matrix

$$\begin{pmatrix} -\partial^2 w/\partial x_2^2 & \partial^2 w/\partial x_1 \partial x_2 \\ \partial^2 w/\partial x_1 \partial x_2 & -\partial^2 w/\partial x_1^2 \end{pmatrix} \equiv -Jw.$$

To translate the second part of the sequence, we begin with an element $\omega = \begin{pmatrix} w_{11} \\ w_{21} \end{pmatrix} \mathrm{d}x_1 + \begin{pmatrix} w_{12} \\ w_{22} \end{pmatrix} \mathrm{d}x_2 \in \Lambda^1(\mathbb{V};\mathbb{V})$ that we identify (as in (10)) with the matrix

$$W = \begin{pmatrix} W_{11} & W_{12} \\ W_{21} & W_{22} \end{pmatrix} = \begin{pmatrix} -w_{12} & w_{11} \\ -w_{22} & w_{21} \end{pmatrix}.$$

We have seen previously that $-S_1\omega$ corresponds to $-2\,\mathrm{skw}\, W$. Now

$$\mathrm{d}_1\omega = \begin{pmatrix} \partial w_{12}/\partial x_1 - \partial w_{11}/\partial x_2 \\ \partial w_{22}/\partial x_1 - \partial w_{21}/\partial x_2 \end{pmatrix} \mathrm{d}x_1 \wedge \mathrm{d}x_2 = -\operatorname{div} W \mathrm{d}x_1 \wedge \mathrm{d}x_2.$$

Hence, modulo some constants, we obtain the elasticity sequence

$$C^\infty(\Omega) \xrightarrow{J} C^\infty(\Omega;\mathbb{M}) \xrightarrow{(\mathrm{skw},\mathrm{div})^T} C^\infty(\Omega, \mathbb{K}\times\mathbb{V}) \to 0.$$

When $n = 3$, we have the sequence

$$\Lambda^0(\mathbb{W}) \xrightarrow{(\mathrm{d}_0,-S_0)} \Lambda^1(\Omega;\mathbb{K}) \xrightarrow{\mathrm{d}_1\circ S_1^{-1}\circ \mathrm{d}_1} \Lambda^2(\Omega;\mathbb{V}) \xrightarrow{(-S_2,\mathrm{d}_2)^T} \Lambda^3(\mathbb{W}) \to 0.$$

Hence, if we begin with a pair $(\mathrm{Skw}\, w, \mu) \in \Lambda^0(\mathbb{W}) = \Lambda^0(\Omega,\mathbb{K}) \times \Lambda^0(\Omega,\mathbb{V})$ that we identify with the pair $(w, \mathrm{Skw}\,\mu) \in C^\infty(\Omega,\mathbb{V})\times C^\infty(\Omega,\mathbb{K})$, then d_0 corresponds to the row-wise gradient and S_0 to the inclusion of $C^\infty(\Omega,\mathbb{K}) \to C^\infty(\Omega,\mathbb{M})$. We have discussed previously natural identifications of $\Lambda^1(\Omega;\mathbb{K})$ and $\Lambda^2(\Omega;\mathbb{V})$ with $C^\infty(\Omega;\mathbb{M})$. With these identifications, d_1 corresponds to the row-wise **curl** and S_1 to the operator Ξ. Finally, we have also seen how $-S_2$ corresponds to the operator $2\,\mathrm{skw}$. Since d_2 corresponds to the row-wise divergence, we obtain (modulo some unimportant constants), the elasticity sequence with weak symmetry

$$C^\infty(\mathbb{V}\times\mathbb{K}) \xrightarrow{(\mathbf{grad},I)} C^\infty(\mathbb{M}) \xrightarrow{\mathbf{curl}\,\Xi^{-1}\,\mathbf{curl}} C^\infty(\mathbb{M}) \xrightarrow{(\mathrm{skw},\mathrm{div})^T} C^\infty(\mathbb{K}\times\mathbb{V}) \to 0.$$

More details, and the extension of these ideas to more general domains, can be found in [9].

7 Well-Posedness of the Weak Symmetry Formulation of Elasticity

As discussed in Sect. 5, to establish well-posedness of the elasticity problem with weakly imposed symmetry (11), it suffices to verify condition (W2) of that section. This may be deduced from the following theorem, which says that the map

$$H\Lambda^{n-1}(\Omega;\mathbb{V}) \xrightarrow{(-S_{n-1},\mathrm{d}_{n-1})^T} H\Lambda^n(\Omega;\mathbb{K}) \times H\Lambda^n(\Omega;\mathbb{V})$$

is surjective. We present the proof in detail, since it will give us guidance as we construct stable discretizations. The proof will make use of the following well-known result from partial differential equations.

Lemma 7.1 *Let Ω be a bounded domain in $\mathbb{R}^n$ with a Lipschitz boundary. Then, for all $\mu \in L^2\Lambda^n(\Omega)$, there exists $\eta \in H^1\Lambda^{n-1}(\Omega)$ satisfying $\mathrm{d}_{n-1}\eta = \mu$. If, in addition, $\int_\Omega \mu = 0$, then we can choose $\eta \in \mathring{H}^1\Lambda^{n-1}(\Omega)$.*

Theorem 7.2 *Given $(\omega,\mu) \in L^2\Lambda^n(\Omega;\mathbb{K}) \times L^2\Lambda^n(\Omega;\mathbb{V})$, there exists $\sigma \in H\Lambda^{n-1}(\Omega;\mathbb{V})$ such that $\mathrm{d}_{n-1}\sigma = \mu$, $-S_{n-1}\sigma = \omega$. Moreover, we may choose σ so that*

$$\|\sigma\|_{H\Lambda} \le c(\|\omega\| + \|\mu\|),$$

for a fixed constant c.

Proof. The second sentence follows from the first Banach's theorem, (i.e., if a continuous linear operator between two Banach spaces has an inverse, then this inverse operator is continuous), so we need only prove the first.

1. By Lemma 7.1, we can find $\eta \in H^1\Lambda^{n-1}(\Omega;\mathbb{V})$ with $\mathrm{d}_{n-1}\eta = \mu$.
2. Since $\omega + S_{n-1}\eta \in H\Lambda^n(\Omega;K)$, we can apply Lemma 7.1 a second time to find $\tau \in H^1\Lambda^{n-1}(\Omega;\mathbb{K})$ with $\mathrm{d}_{n-1}\tau = \omega + S_{n-1}\eta$.
3. Since S_{n-2} is an isomorphism from $H^1\Lambda^{n-2}(\Omega;\mathbb{V})$ onto $H^1\Lambda^{n-1}(\Omega;\mathbb{K})$, we have $\varrho \in H^1\Lambda^{n-2}(\Omega;\mathbb{V})$ with $S_{n-2}\varrho = \tau$.
4. Define $\sigma = \mathrm{d}_{n-2}\varrho + \eta \in H\Lambda^{n-1}(\Omega;\mathbb{V})$.
5. From steps 1 and 4, it is immediate that $\mathrm{d}_{n-1}\sigma = \mu$.
6. From 4, $-S_{n-1}\sigma = S_{n-1}\mathrm{d}_{n-2}\varrho - S_{n-1}\eta$. But, since $\mathrm{d}_{n-1}S_{n-2} = -S_{n-1}\mathrm{d}_{n-2}$,

$$-S_{n-1}\mathrm{d}_{n-2}\varrho = \mathrm{d}_{n-1}S_{n-2}\varrho = \mathrm{d}_{n-1}\tau = \omega + S_{n-1}\eta,$$

so $-S_{n-1}\sigma = \omega$.

We note a few points from the proof.

(i) Although the elasticity problem (11) only involves the three spaces $H\Lambda^{n-1}(\Omega;\mathbb{V})$, $L^2\Lambda^n(\Omega;\mathbb{V})$, and $L^2\Lambda^n(\Omega;\mathbb{K})$, the proof brings in two additional spaces from the BGG construction: $H\Lambda^{n-2}(\Omega;\mathbb{V})$ and $H\Lambda^{n-1}(\Omega;\mathbb{K})$.

(ii) Although S_{n-1} is the only S operator arising in the formulation, S_{n-2} plays a role in the proof.

(iii) We do not fully use the fact that S_{n-2} is an isomorphism from $\Lambda^{n-2}(\mathbb{V};\mathbb{V})$ to $\Lambda^{n-1}(\mathbb{V};\mathbb{K})$, only the fact that it is a surjection. This will prove important in the next section, when we derive conditions for stable approximation schemes for elasticity.

(iv) Other slightly weaker conditions can be used in some places in the proof (a fact we also exploit in discrete versions for some choices of finite element spaces).

8 Conditions for Stable Approximation Schemes

To obtain stable approximation schemes, we now mimic the key structural elements present for the continuous problem. In particular, we see that to establish stability of the continuous problem, we do not use the complete exact sequences, but only the last two spaces in the top sequence and the last three spaces in the bottom sequence, connected by the operators S_{n-2} and S_{n-1}.

$$\begin{array}{ccccc} & & \Lambda^{n-1}(\mathbb{K}) & \xrightarrow{\mathrm{d}_{n-1}} & \Lambda^{n}(\mathbb{K}) \to 0 \\ & \nearrow S_{n-2} & & \nearrow S_{n-1} & \\ \Lambda^{n-2}(\mathbb{V}) & \xrightarrow{\mathrm{d}_{n-2}} & \Lambda^{n-1}(\mathbb{V}) & \xrightarrow{\mathrm{d}_{n-1}} & \Lambda^{n}(\mathbb{V}) \to 0. \end{array} \tag{14}$$

Thus, we look for five finite-dimensional spaces that are connected by a similar structure, i.e., in addition to the spaces $\Lambda_h^n(\mathbb{K}) \subset H\Lambda^n(\mathbb{K})$, $\Lambda_h^{n-1}(\mathbb{V}) \subset H\Lambda^{n-1}(\mathbb{V})$, and $\Lambda_h^n(\mathbb{V}) \subset H\Lambda^n(\mathbb{V})$ used in the finite element method, we also seek spaces $\Lambda_h^{n-1}(\mathbb{K}) \subset H\Lambda^{n-1}(\mathbb{K})$ and $\Lambda_h^{n-2}(\mathbb{V}) \subset H\Lambda^{n-2}(\mathbb{V})$.

To mimic the structure of the continuous problem, but taking into account the comments made following Theorem 7.2, we require that the finite element spaces are also connected by exact sequences, but where we introduce some additional flexibility by inserting the L^2 projection operator Π_h^n and using approximations of the operators S_{n-2} and S_{n-1}.

$$\begin{array}{ccccc} & & \Lambda_h^{n-1}(\mathbb{K}) & \xrightarrow{\Pi_h^n \mathrm{d}_{n-1}} & \Lambda_h^{n}(\mathbb{K}) \to 0 \\ & \nearrow S_{n-2,h} & & \nearrow S_{n-1,h} & \\ \Lambda_h^{n-2}(\mathbb{V}) & \xrightarrow{\mathrm{d}_{n-2}} & \Lambda_h^{n-1}(\mathbb{V}) & \xrightarrow{\mathrm{d}_{n-1}} & \Lambda_h^{n}(\mathbb{V}) \to 0. \end{array} \tag{15}$$

In anticipation of proving a stability result for the mixed finite element method for elasticity that mimics that proof used in the continuous case, we need to define interpolants into each of these finite element spaces that have appropriate properties. The reason for the choice of the specific properties will become apparent in the stability proof.

We first define Π_h^n and $\widetilde{\Pi}_h^n$ to be the L^2 projection operators into the spaces $\Lambda_h^n(\mathbb{K})$ and $\Lambda_h^n(\mathbb{V})$, respectively. We then define Π_h^{n-1} and $\widetilde{\Pi}_h^{n-1}$ to be interpolation operators mapping $H^1\Lambda^{n-1}(\mathbb{K})$ to $\Lambda_h^{n-1}(\mathbb{K})$ and $H^1\Lambda^{n-1}(\mathbb{V})$ to $\Lambda_h^{n-1}(\mathbb{V})$, respectively, and satisfying

$$\Pi_h^n \mathrm{d}_{n-1} \Pi_h^{n-1} \tau = \Pi_h^n \mathrm{d}_{n-1} \tau, \ \tau \in (\mathring{H}^1 + P^1)\Lambda^{n-1}(\mathbb{K}), \tag{16}$$

$$\mathrm{d}_{n-1} \widetilde{\Pi}_h^{n-1} \tau = \widetilde{\Pi}_h^n \mathrm{d}_{n-1} \tau, \ \tau \in H^1\Lambda^{n-1}(\mathbb{V}).$$

$$\|\Pi_h^{n-1} \tau\| \le C\|\tau\|_1, \ \tau \in (\mathring{H}^1 + P^1)\Lambda^{n-1}(\mathbb{K}), \tag{17}$$

$$\|\widetilde{\Pi}_h^{n-1} \tau\| \le C\|\tau\|_1, \ \tau \in H^1\Lambda^{n-1}(\mathbb{V}).$$

Next, we define $\widetilde{\Pi}_h^{n-2}$ mapping $H^1\Lambda^{n-2}(\mathbb{V})$ to $\Lambda_h^{n-2}(\mathbb{V})$ satisfying

$$\|\mathrm{d}_{n-2}\widetilde{\Pi}_h^{n-2}\varrho\| \le c\|\varrho\|_1, \quad \varrho \in H^1\Lambda^{n-2}. \tag{18}$$

(In (18), the exterior derivative d_{n-2} corresponds to the differential operator **curl**.) As we shall see in the examples, in some cases these will be the canonical interpolation operators we usually associate with standard finite element spaces, while in other cases, we will need to make some modifications so that the interpolation operators are defined on spaces of functions will less smoothness than we usually assume.

The key to the derivation of the formulation of elasticity with weak symmetry at the continuous level was the introduction of the operators $S = S_k : \Lambda^k(\mathbb{V}) \to \Lambda^{k+1}(\mathbb{K})$. In the reduced sequence (14), only the operators S_{n-2} and S_{n-1} will enter the analysis. One of the key properties of these operators was that

$$\mathrm{d}_{n-1}S_{n-2} = -S_{n-1}\mathrm{d}_{n-2}. \tag{19}$$

For the discrete version of this analysis, we will need to modify the definitions of S_{n-2} and S_{n-1} in a simple way. As a discrete analogue of the operator S_{n-1}, we define $S_{n-1,h} : \Lambda_h^{n-1}(\mathbb{V}) \to \Lambda_h^3(\mathbb{K})$ by $S_{n-1,h} = \Pi_h^n S_{n-1}$. As a discrete analogue of the operator S_{n-2}, we define $S_{n-2,h} : \Lambda_h^{n-2}(\mathbb{V}) \to \Lambda_h^2(\mathbb{K})$ by $S_{n-2,h} = \Pi_h^{n-1}S_{n-2}$. With these definitions, we establish the following discrete version of (19),

$$\Pi_h^n\mathrm{d}_{n-1}S_{n-2,h} = -S_{n-1,h}\mathrm{d}_{n-2}. \tag{20}$$

To see this, we observe that using (16) and (20),

$$\begin{aligned}\Pi_h^n\mathrm{d}_{n-1}S_{n-2,h} = \Pi_h^n\mathrm{d}_{n-1}\Pi_h^{n-1}S_{n-2} &= \Pi_h^n\mathrm{d}_{n-1}S_{n-2}\\ &= -\Pi_h^n S_{n-1}\mathrm{d}_{n-2} = -S_{n-1,h}\mathrm{d}_{n-2}.\end{aligned}$$

Another key property of the operator S_{n-2} was that it was invertible as a map from $H^1\Lambda^{n-2}(\mathbb{V})$ to $H^1\Lambda^{n-1}(\mathbb{K})$. This fact was used in the prove of stability of the weak symmetry formulation at the continuous level, although we observed that surjectivity of this map would be sufficient. We cannot expect invertibility of the map $S_{n-2,h}$. However, a key condition to prove stability of the finite element approximation to the weak symmetry formulation is that $S_{n-2,h}$ maps $\Lambda_h^{n-2}(\mathbb{V})$ onto $\Lambda_h^{n-1}(\mathbb{K})$. To ensure this condition, we will assume that $\Lambda_h^{n-2}(\mathbb{V})$ and $\Lambda_h^{n-1}(\mathbb{K})$ are related by the condition

$$S_{n-2,h}\widetilde{\Pi}_h^{n-2}\tau = \Pi_h^{n-1}S_{n-2}\tau, \quad \tau \in H^1\Lambda^{n-2}(\mathbb{V}). \tag{21}$$

To see that this condition ensures surjectivity, note that given a function $\sigma_h \in \Lambda_h^{n-1}(\mathbb{K})$, we can find $\sigma \in H^1\Lambda^{n-1}(\mathbb{K})$ (e.g., a continuous piecewise polynomial differential form), such that $\sigma_h = \Pi_h^{n-1}\sigma$. Defining $\tau = S_{n-2}^{-1}\sigma$ and $\tau_h = \widetilde{\Pi}_h^{n-2}\tau \in \Lambda_h^{n-2}(\mathbb{V})$, we find that

$$\sigma_h = \Pi_h^{n-1}\sigma = \Pi_h^{n-1}S_{n-2}\tau = S_{n-2,h}\widetilde{\Pi}_h^{n-2}\tau = S_{n-2,h}\tau_h.$$

To summarize the results of this section, we will develop stable mixed finite element approximation schemes by finding five finite element spaces. The three spaces $\Lambda_h^n(\mathbb{K}) \subset H\Lambda^n(\mathbb{K})$, $\Lambda_h^{n-1}(\mathbb{V}) \subset H\Lambda^{n-1}(\mathbb{V})$, $\Lambda_h^n(\mathbb{V}) \subset H\Lambda^n(\mathbb{V})$ are used in the method and the spaces $\Lambda_h^{n-1}(\mathbb{K}) \subset H\Lambda^{n-1}(\mathbb{K})$ and $\Lambda_h^{n-2}(\mathbb{V}) \subset H\Lambda^{n-2}(\mathbb{V})$ are auxiliary spaces crucial to the proof of stability. Associated with each of these spaces is an operator for which we need properties (16), (17), and (18). We further assume that the five spaces are connected by the exact sequences given in (15). Finally, we require (21), which ensures that $S_{n-2,h}$ maps $\Lambda_h^{n-2}(\mathbb{V})$ onto $\Lambda_h^{n-1}(\mathbb{K})$. Under these conditions, we can then prove the following stability result for the mixed finite element method for elasticity.

9 Stability of Finite Element Approximation Schemes

Theorem 9.1 *Assume that the finite element subspaces $\Lambda_h^k(\mathbb{K})$ and $\Lambda_h^k(\mathbb{V})$ are connected by the exact sequences given in* (15), *that there are operators associated with these subspaces satisfying conditions* (16), (17), (18), *and that condition* (21) *is satisfied. Then, given $(\omega, \mu) \in \Lambda_h^n(\mathbb{K}) \times \Lambda_h^n(\mathbb{V})$, there exists $\sigma \in \Lambda_h^{n-1}(\mathbb{V})$ such that* $\mathrm{d}_{n-1}\sigma = \mu$, $-S_{n-1,h}\sigma \equiv -\Pi_h^n S_{n-1}\sigma = \omega$, *and*

$$\|\sigma\|_{H\Lambda} \leq c(\|\omega\| + \|\mu\|), \tag{22}$$

where the constant c is independent of ω, μ and h.

Before proving this theorem, we note that condition (15) immediately implies that the first Brezzi condition (S1) is satisfied and that the second Brezzi condition (S2) easily follows from the conclusion of the theorem.

Proof.

1. By Lemma 7.1, we can find $\eta \in H^1\Lambda^{n-1}(\Omega;\mathbb{V})$ with $\mathrm{d}_{n-1}\eta = \mu$ and $\|\eta\|_1 \leq c\|\mu\|$.
2. Since $\omega + \Pi_h^n S_{n-1}\widetilde{\Pi}_h^{n-1}\eta \in H\Lambda^n(\Omega;K)$, we can apply Lemma 7.1 a second time to find $\tau \in H^1\Lambda^{n-1}(\Omega;\mathbb{K})$ with $\mathrm{d}_{n-1}\tau = \omega + \Pi_h^n S_{n-1}\widetilde{\Pi}_h^{n-1}\eta$ and $\|\tau\|_1 \leq c(\|\omega\| + \|\Pi_h^n S_{n-1}\widetilde{\Pi}_h^{n-1}\eta\|)$.
3. Since S_{n-2} is an isomorphism from $H^1\Lambda^{n-2}(\Omega;\mathbb{V})$ to $H^1\Lambda^{n-1}(\Omega;\mathbb{K})$, we have $\varrho \in H^1\Lambda^{n-2}(\Omega;\mathbb{V})$ with $S_{n-2}\varrho = \tau$, and $\|\varrho\|_1 \leq c\|\tau\|_1$.
4. Define $\sigma = \mathrm{d}_{n-2}\widetilde{\Pi}_h^{n-2}\varrho + \widetilde{\Pi}_h^{n-1}\eta \in \Lambda_h^{n-1}(\mathbb{V})$.
5. From step 4, (16), step 1, and the fact that $\widetilde{\Pi}_h^n$ is a projection, we have
$$\mathrm{d}_{n-1}\sigma = \mathrm{d}_{n-1}\widetilde{\Pi}_h^{n-1}\eta = \widetilde{\Pi}_h^n \mathrm{d}_{n-1}\eta = \widetilde{\Pi}_h^n\mu = \mu.$$
6. Also from step 4,
$$-S_{n-1,h}\sigma = -S_{n-1,h}\mathrm{d}_{n-2}\widetilde{\Pi}_h^{n-2}\varrho - S_{n-1,h}\widetilde{\Pi}_h^{n-1}\eta.$$

Applying, in order, (20), (21), step 3, (16), step 2, and the fact that Π_h^n is a projection, we obtain

$$
\begin{aligned}
S_{n-1,h} d_{n-2} \widetilde{\Pi}_h^{n-2} \varrho &= -\Pi_h^n d_{n-2} S_{n-2,h} \widetilde{\Pi}_h^{n-2} \varrho \\
&= -\Pi_h^n d_{n-2} \Pi_h^{n-1} S_{n-2} \varrho = -\Pi_h^n d_{n-1} \Pi_h^{n-1} \tau = -\Pi_h^n d_{n-1} \tau \\
&= -\Pi_h^n (\omega + \Pi_h^n S_{n-1} \widetilde{\Pi}_h^{n-1} \eta) = -\omega - S_{n-1,h} \widetilde{\Pi}_h^{n-1} \eta.
\end{aligned}
$$

Combining, we have $-\Pi_h^n S_{n-1} \equiv -S_{n-1,h}\sigma = \omega$.

7. Finally, we prove the norm bound. From the boundedness of S_{n-1} in L^2, (17), and step 1,

$$
\|\Pi_h^n S_{n-1} \widetilde{\Pi}_h^{n-1} \eta\| \le c \|S_{n-1} \widetilde{\Pi}_h^{n-1} \eta\| \le c \|\widetilde{\Pi}_h^{n-1} \eta\| \le c \|\eta\|_1 \le c \|\mu\|.
$$

Combining with the bounds in steps 3 and 2, this gives $\|\varrho\|_1 \le c(\|\omega\| + \|\mu\|)$. From (18), we then have $\|d_{n-2} \widetilde{\Pi}_h^{n-2} \varrho\| \le c(\|\omega\| + \|\mu\|)$. From (17) and the bound in step 1, $\|\widetilde{\Pi}_h^{n-1} \eta\| \le c\|\eta\|_1 \le c\|\mu\|$. In view of the definition of σ, these two last bounds imply that $\|\sigma\| \le c(\|\omega\| + \|\mu\|)$, while $\|d_{n-1}\sigma\| \le C\|\widetilde{\Pi}_h^n d_{n-1}\sigma\| = \|\mu\|$, and thus we have the desired bound (22).

We have thus verified the stability conditions (S1) and (S2), and so obtain the following quasi-optimal error estimate (see [14, 15]).

Theorem 9.2 *Suppose (σ, u, p) is the solution of the elasticity system* (11) *and (σ_h, u_h, p_h) is the solution of discrete system* (12)*, where the finite element spaces satisfy the hypotheses of Theorem 9.1. Then there is a constant C, independent of h, such that*

$$
\|\sigma - \sigma_h\|_{HA} + \|u - u_h\| + \|p - p_h\| \le C \inf(\|\sigma - \tau\|_{HA} + \|u - v\| + \|p - q\|),
$$

where the infimum is over all $\tau \in \Lambda_h^{n-1}(\mathbb{V})$, $v \in \Lambda_h^n(\mathbb{V})$, and $q \in \Lambda_h^n(\mathbb{K})$.

10 Refined Error Estimates

To see more precisely the contribution to the error from each of the approximating subspaces, we now follow the theory developed in [18] and [20] for error estimates for mixed finite element methods. Since the derivation is fairly simple and we are in an intermediate case to the general theory developed in the references above, we present the complete derivation for the problem we are considering.

Theorem 10.1 *Suppose (σ, u, p) is the solution of the elasticity system* (11) *and (σ_h, u_h, p_h) is the solution of discrete system* (12)*, where the finite element subspaces satisfy the hypotheses of Theorem 9.1. Then*

$$
\begin{aligned}
\|\sigma - \sigma_h\| + \|p - p_h\| + \|u_h - \widetilde{\Pi}_h^n u\| &\le C(\|\sigma - \widetilde{\Pi}_h^{n-1}\sigma\| + \|p - \Pi_h^n p\|), \\
\|u - u_h\| &\le C(\|\sigma - \widetilde{\Pi}_h^{n-1}\sigma\| + \|p - \Pi_h^n p\| + \|u - \widetilde{\Pi}_h^n u\|), \\
\|d_{n-1}(\sigma - \sigma_h)\| &= \|d_{n-1}\sigma - \widetilde{\Pi}_h^n d_{n-1}\sigma\|.
\end{aligned}
$$

Proof. Subtracting the equations in (12) from the corresponding equations in (11), and adding and subtracting appropriate interpolants, we get the error equations

$$\begin{aligned}
&\langle A(\sigma_h - \widetilde{\Pi}_h^{n-1}\sigma), \tau\rangle + \langle \mathrm{d}\tau, u_h - \widetilde{\Pi}_h^n u\rangle - \langle S\tau, p_h - \Pi_h^n p\rangle \\
&= \langle A(\sigma - \widetilde{\Pi}_h^{n-1}\sigma), \tau\rangle + \langle \mathrm{d}\tau, u - \widetilde{\Pi}_h^n u\rangle - \langle S\tau, p - \Pi_h^n p\rangle, \quad \tau \in \Lambda_h^{n-1}(\mathbb{V}), \\
&\langle \mathrm{d}(\sigma_h - \widetilde{\Pi}_h^{n-1}\sigma), v\rangle = \langle \mathrm{d}(\sigma - \widetilde{\Pi}_h^{n-1}\sigma), v\rangle, \quad v \in \Lambda_h^n(\mathbb{V}), \qquad (23) \\
&\langle S(\sigma_h - \widetilde{\Pi}_h^{n-1}\sigma), q\rangle = \langle S(\sigma - \widetilde{\Pi}_h^{n-1}\sigma), q\rangle, \quad q \in \Lambda_h^n(\mathbb{K}),
\end{aligned}$$

where we use d as an abbreviation for d_{n-1}. Now by (16), $\langle \mathrm{d}(\sigma - \widetilde{\Pi}_h^{n-1}\sigma), v\rangle = 0$ for $v \in \Lambda_h^n(\mathbb{V})$ and hence by (18), $\mathrm{d}(\sigma_h - \widetilde{\Pi}_h^{n-1}\sigma) = 0$. Setting

$$\tau = \sigma_h - \widetilde{\Pi}_h^{n-1}\sigma, \qquad v = u_h - \widetilde{\Pi}_h^n u, \qquad q = p_h - \Pi_h^n p,$$

and adding the equations, we get

$$\begin{aligned}
C\|\sigma_h - \widetilde{\Pi}_h^{n-1}\sigma\|^2 &\le \langle A(\sigma_h - \widetilde{\Pi}_h^{n-1}\sigma), \sigma_h - \widetilde{\Pi}_h^{n-1}\sigma\rangle \\
&= \langle A(\sigma - \widetilde{\Pi}_h^{n-1}\sigma), \sigma_h - \widetilde{\Pi}_h^{n-1}\sigma\rangle - \langle S(\sigma_h - \widetilde{\Pi}_h^{n-1}\sigma), p - \Pi_h^n p\rangle.
\end{aligned}$$

Applying standard estimates, we then obtain

$$\|\sigma_h - \widetilde{\Pi}_h^{n-1}\sigma\| \le C(\|\sigma - \widetilde{\Pi}_h^{n-1}\sigma\| + \|p - \Pi_h^n p\|), \qquad (24)$$

and hence,

$$\|\sigma - \sigma_h\| \le C(\|\sigma - \widetilde{\Pi}_h^{n-1}\sigma\| + \|p - \Pi_h^n p\|).$$

Next applying Theorem 9.1, with $\omega = p_h - \Pi_h^n p$ and $\mu = u_h - \widetilde{\Pi}_h^n u$, we can find $\tau \in \Lambda_h^{n-1}(\mathbb{V})$ such that

$$\begin{aligned}
\widetilde{\Pi}_h^n d\tau = u_h - \widetilde{\Pi}_h^n u, \qquad -S_{n-1,h}\tau \equiv -\Pi_h^n S_{n-1}\tau = p_h - \Pi_h^n p, \\
\|\tau\|_{H\Lambda} \le c(\|p_h - \Pi_h^n p\| + \|u_h - \widetilde{\Pi}_h^n u\|).
\end{aligned}$$

Making this choice of τ in (23), we get

$$\begin{aligned}
\|p_h - \Pi_h^n p\|^2 + \|u_h - \widetilde{\Pi}_h^n u\|^2 &= \langle A(\sigma - \widetilde{\Pi}_h^{n-1}\sigma), \tau\rangle - \langle A(\sigma_h - \widetilde{\Pi}_h^{n-1}\sigma), \tau\rangle \\
&\quad + \langle u_h - \widetilde{\Pi}_h^n u, u - \widetilde{\Pi}_h^n u\rangle + \langle p_h - \Pi_h^n p, p - \Pi_h^n p\rangle \\
&= \langle A(\sigma - \widetilde{\Pi}_h^{n-1}\sigma), \tau\rangle - \langle A(\sigma_h - \widetilde{\Pi}_h^{n-1}\sigma), \tau\rangle + \langle p_h - \Pi_h^n p, p - \Pi_h^n p\rangle.
\end{aligned}$$

Applying standard estimates and (24), we easily obtain

$$\begin{aligned}
\|p_h - \Pi_h^n p\| + \|u_h - \widetilde{\Pi}_h^n u\| &\le C(\|\sigma - \widetilde{\Pi}_h^{n-1}\sigma\| + \|\sigma_h - \widetilde{\Pi}_h^{n-1}\sigma\| + \|p - \Pi_h^n p\|) \\
&\le C(\|\sigma - \widetilde{\Pi}_h^{n-1}\sigma\| + \|p - \Pi_h^n p\|).
\end{aligned}$$

Hence,

$$\begin{aligned}
\|p - p_h\| &\le C(\|\sigma - \widetilde{\Pi}_h^{n-1}\sigma\| + \|p - \Pi_h^n p\|), \\
\|u - u_h\| &\le C(\|\sigma - \widetilde{\Pi}_h^{n-1}\sigma\| + \|p - \Pi_h^n p\| + \|u - \widetilde{\Pi}_h^n u\|).
\end{aligned}$$

Finally, since $\langle \mathrm{d}(\sigma - \sigma_h, v\rangle = 0$ for $v \in \Lambda_h^n(\mathbb{V})$, we get $\mathrm{d}\sigma_h = \widetilde{\Pi}_h^n \mathrm{d}\sigma$, which establishes the last estimate of the theorem.

11 Examples of Stable Finite Element Methods for the Weak Symmetry Formulation of Elasticity

The examples that follow are of two types. In the first two subsections, we present choices of finite element spaces for which diagram (15) is satisfied without the need for the additional projection Π_h^n in the top sequence. These methods make use of multiple copies of finite element spaces normally associated to the use of mixed methods for scalar second order elliptic problems. In the final three subsections, we consider methods which require the additional projection Π_h^n in the top sequence in diagram (15). This is because the two spaces in the top sequence are ones normally associated to stable pairs for the approximation of the stationary Stokes equations.

11.1 Arnold, Falk, Winther Families

In the approach of [8, 6, 9], the spaces are chosen for $r \geq 0$ to be:

$$\Lambda_h^{n-2}(\mathbb{V}) = \mathcal{P}_{r+2}^{-}\Lambda^{n-2}(\mathcal{T}_h), \quad \Lambda_h^{n-1}(\mathbb{V}) = \mathcal{P}_{r+1}\Lambda^{n-1}(\mathcal{T}_h;\mathbb{V}),$$
$$\Lambda_h^{n}(\mathbb{V}) = \mathcal{P}_{r}\Lambda^{n}(\mathcal{T}_h;\mathbb{V}),$$
$$\Lambda_h^{n-1}(\mathbb{K}) = \mathcal{P}_{r+1}^{-}\Lambda^{n-1}(\mathcal{T}_h;\mathbb{K}), \quad \Lambda_h^{n}(\mathbb{K}) = \mathcal{P}_{r}\Lambda^{n}(\mathcal{T}_h;\mathbb{K}).$$

The sequences

$$\mathcal{P}_{r+1}^{-}\Lambda^{n-1}(\mathcal{T}_h;\mathbb{K}) \xrightarrow{\mathrm{d}_{n-1}} \mathcal{P}_{r}\Lambda^{n}(\mathcal{T}_h;\mathbb{K}) \to 0$$

$$\mathcal{P}_{r+2}^{-}\Lambda^{n-2}(\mathcal{T}_h;\mathbb{V}) \xrightarrow{\mathrm{d}_{n-2}} \mathcal{P}_{r+1}\Lambda^{n-1}(\mathcal{T}_h;\mathbb{V}) \xrightarrow{\mathrm{d}_{n-1}} \mathcal{P}_{r}\Lambda^{n}(\mathcal{T}_h;\mathbb{V}) \to 0$$

are the final parts of longer exact sequences involving the $\mathcal{P}_r$ and $\mathcal{P}_r^{-}$ spaces. Hence, (15) is satisfied without the additional projection at the end of the first sequence. For these spaces, the canonical projection operators Π_h^{n-1}, Π_h^n, $\widetilde{\Pi}_h^{n-1}$, and $\widetilde{\Pi}_h^n$ satisfy conditions (16) and (17). Although the canonical projection operator $\widetilde{\Pi}_h^{n-2}$ does not satisfy (18), since this operator is not defined on functions in $H^1\Lambda^{n-2}(\mathbb{V})$, we can define a modification of this operator, $\widetilde{P}_h : \Lambda^{n-2}(\Omega;\mathbb{V}) \to \mathcal{P}_{r+2}^{-}\Lambda^{n-2}(\mathcal{T}_h;\mathbb{V})$ that does satisfy (18). The operator $\widetilde{P}_h\omega$ will have the same moments as ω on faces of codimension 0 and 1, but with moments of a smoothed approximation of ω on the faces of codimension 2. When $n = 2$, the issue is simply that the vertex values are not defined and this can be remedied by using the ideas of the interpolant of Clement. When $n = 3$, additional details are provided in [6]. Thus, to satisfy the hypotheses of Theorem 9.1, it remains to show that

$$\Pi_h^{n-1}S_{n-2}\widetilde{P}_h = \Pi_h^{n-1}S_{n-2}.$$

This is equivalent to showing that

$$\Pi_h^{n-1}S_{n-2}\omega = 0, \quad \forall \omega = (I - \widetilde{P}_h)\sigma, \quad \sigma \in \Lambda^{n-2}(\mathbb{V}).$$

Since $\widetilde{P}_h\omega = 0$, we have for $n - 1 \leq d \leq \min(n, r + n - 1)$,

$$\int_f \mathrm{Tr}_f\, \omega \wedge \zeta = 0, \quad \zeta \in \mathcal{P}_{r-d+n-1}\Lambda^{d-n+2}(f;\mathbb{V}), \quad f \in \Delta_d(\mathcal{T}_h). \tag{25}$$

Note that we have not included similar statements for the vertex degrees of freedom when $n = 2$ or the edge degrees of freedom when $n = 3$, since we will not need them here. We must show that (25) implies that for $n-1 \le d \le \min(n, r+n-1)$,

$$\int_f \mathrm{Tr}_f\, S_{n-2}\omega \wedge \mu = 0, \quad \mu \in \mathcal{P}_{r-d+n-1}\Lambda^{d-n+1}(f;\mathbb{K}), \quad f \in \Delta_d(\mathcal{T}_h).$$

The simplest case is when $r = 0$. When $n = 2$, (25) becomes

$$\int_f \mathrm{Tr}_f\, \omega \wedge \zeta = 0, \quad \zeta \in \mathcal{P}_0\Lambda^1(f;\mathbb{V}), \quad f \in \Delta_1(\mathcal{T}_h),$$

which for $\omega = (w_1, w_2)^T$, is simply the condition

$$\int_e w_i \, \mathrm{d}e = 0, \quad i = 1, 2, \quad e \in \Delta_1(\mathcal{T}_h). \tag{26}$$

We then require that

$$\int_e \mathrm{Tr}_e(-w_2\chi \mathrm{d}x_1 + w_1\chi \mathrm{d}x_2) = 0, \quad e \in \Delta_1(\mathcal{T}_h).$$

But if (t^1, t^2) is the unit tangent to e, then by (26),

$$\int_e \mathrm{Tr}_e(-w_2\chi \mathrm{d}x_1 + w_1\chi \mathrm{d}x_2) = \int_e (-w_2 t^1 + w_1 t^2)\chi \, \mathrm{d}e = 0.$$

An analogous argument works for general r when $n = 2$, and the basic outline of the proof is the same when $n = 3$, although in this case the operator S_1 is more complicated. The details can be found in [6].

Using Theorem 10.1, it is straightforward to derive the following error estimates, valid for $1 \le k \le r+1$, assuming that σ, p, and u are sufficiently smooth.

$$\|\sigma - \sigma_h\| + \|p - p_h\| + \|u_h - \tilde{\Pi}_h^n u\| \le Ch^k(\|\sigma\|_k + \|p\|_k),$$
$$\|u - u_h\| \le Ch^k(\|\sigma\|_k + \|p\|_k + \|u\|_k), \qquad \|\mathrm{d}_{n-1}(\sigma - \sigma_h)\| \le Ch^k\|\mathrm{d}_{n-1}\sigma\|_k.$$

11.2 Arnold, Falk, Winther Reduced Elements

In the reduced elements proposed in [8] (in two dimensions) and [6] (in three dimensions), the spaces $\Lambda_h^n(\mathbb{V})$, $\Lambda_h^{n-1}(\mathbb{K})$, and $\Lambda_h^n(\mathbb{K})$ remain as chosen above, while the spaces $\Lambda_h^{n-2}(\mathbb{V})$ and $\Lambda_h^{n-1}(\mathbb{V})$ are modified. Thus, the reduced elements have a somewhat simpler stress space than the methods described above. The basic idea is that in the verification of condition (21) in the last section, we did not use all the degrees of freedom of the space $\mathcal{P}_2^-\Lambda^0(\mathcal{T}_h)$, i.e., we did not use the vanishing of the edge integral of both components of ω, but only the combination

$-w_2 t^1 + w_1 t^2$ (the normal component). Hence, instead of the vector-valued quadratic space $\mathcal{P}_2\Lambda^0(\mathcal{T}_h, \mathbb{V})$, we can use the reduced space obtained from it by imposing the constraint that the tangential component on each edge vary only linearly on that edge. This space of vector fields, which we denote by $\mathcal{P}_{2-}\Lambda^0(\mathcal{T}_h, \mathbb{V})$ has been used previously to approximate the velocity field in the approximation of the stationary Stokes equations (cf. [23, p. 134 ff., 153 ff.]; see also the lecture on Stokes problem in the present volume). Together with piecewise constants, it gives a stable finite element approximation scheme for the Stokes equations. An element in this space is determined by its vertex values and the integral of its normal component on each edge. In order to complete the construction, we must provide a vector-valued discrete de Rham sequence in which the space of 0-forms is $\mathcal{P}_{2-}\Lambda^0(\mathcal{T}_h; \mathbb{R}^2)$. This will be the sequence

$$\mathcal{P}_{2-}\Lambda^0(\mathcal{T}_h; \mathbb{V}) \xrightarrow{\mathrm{d}_0} \mathcal{P}_{1-}\Lambda^1(\mathcal{T}_h; \mathbb{V}) \xrightarrow{\mathrm{d}_1} \mathcal{P}_0\Lambda^2(\mathcal{T}_h; \mathbb{V}) \to 0,$$

where it remains to define $\mathcal{P}_{1-}\Lambda^1(\mathcal{T}_h; \mathbb{V})$. This will be the set of $\tau \in \mathcal{P}_1\Lambda^1(\mathcal{T}_h; \mathbb{V})$ for which $\mathrm{Tr}_e(\tau) \cdot t$ is constant on any edge e with unit tangent t and unit normal n. More specifically, for $\tau \in \mathcal{P}_1\Lambda^1(\mathcal{T}_h; \mathbb{R}^2)$, $\mathrm{Tr}_e(\tau)$ is a vector-valued 1-form on e of the form $g\,\mathrm{d}s$ with $\mu : e \to \mathbb{R}^2$ linear and $\mathrm{d}s$ the volume form—i.e., length form—on e. If $\mu \cdot t$ is constant, then $\tau \in \mathcal{P}_1^-\Lambda^1(\mathcal{T}_h; \mathbb{V})$. The natural degrees of freedom for this space are the integral and first moment of $\mathrm{Tr}_e(\tau) \cdot n$ and the integral of $\mathrm{Tr}_e(\tau) \cdot t$. If we use (10) to identify vector-valued 1-forms and matrix fields, then the condition for a piecewise linear matrix field W to correspond to an element of $\mathcal{P}_1^-\Lambda^1(\mathcal{T}_h; \mathbb{R}^2)$ is that on each edge e with tangent t and normal n, $Wn \cdot t$ must be constant on e. This defines the reduced space Σ_h, with three degrees of freedom per edge. Together with piecewise constant for displacements and multipliers, this furnishes a stable choice of elements.

A three-dimensional simplified element can be constructed using a similar approach. We start from the space $\mathcal{P}_2^-\Lambda(\mathcal{T}_h; \mathbb{V})$ and see that we do not use all the degrees of freedom to satisfy condition (21). We thus define a reduced space $\mathcal{P}_{2-}^-\Lambda(\mathcal{T}_h; \mathbb{V})$ and a space $\mathcal{P}_{1-}\Lambda^2(\mathcal{T}_h; \mathbb{V})$ such that these spaces, together with $\mathcal{P}_0\Lambda^3(\mathcal{T}_h; \mathbb{V})$, form the exact sequence

$$\mathcal{P}_{2-}\Lambda^1(\mathcal{T}_h; \mathbb{V}) \xrightarrow{\mathrm{d}_1} \mathcal{P}_{1-}\Lambda^2(\mathcal{T}_h; \mathbb{V}) \xrightarrow{\mathrm{d}_2} \mathcal{P}_0\Lambda^3(\mathcal{T}_h; \mathbb{V}) \to 0.$$

We are then able to replace the space $\mathcal{P}_1\Lambda^1(\mathcal{T}_h; \mathbb{V})$, which has 36 degrees of freedom (9 per face), by the space $\mathcal{P}_{1-}\Lambda^2(\mathcal{T}_h; \mathbb{V})$, which has 24 degrees of freedom (6 per face). If we identify an element in our reduced space with a matrix W is the manner discussed previously, then we get on each face the six degrees of freedom:

$$\int_f Wn\,\mathrm{d}f, \quad \int_f (x \cdot t) n^T Wn\,\mathrm{d}f, \quad \int_f (x \cdot s) n^T Wn\,\mathrm{d}f, \quad \int_f [(x \cdot t)s^T - (x \cdot s)t^T] Wn\,\mathrm{d}f,$$

where s and t denote orthogonal unit tangent vectors on the face f. More details can be found in [6].

11.3 PEERS

In the PEERS method [3], $n = 2$ and we choose

$$\Lambda_h^1(\mathbb{V}) = \mathcal{P}_1^- \Lambda^1(\mathcal{T}_h;\mathbb{V}) + dB_3\Lambda^0(\mathcal{T}_h;\mathbb{V}), \quad \Lambda_h^2(\mathbb{V}) = \mathcal{P}_0\Lambda^2(\mathcal{T}_h;\mathbb{V}),$$
$$\Lambda_h^2(\mathbb{K}) = \mathcal{P}_1\Lambda^2(\mathcal{T}_h;\mathbb{K}) \cap H^1\Lambda^2(\mathbb{K}), \quad \text{which we denote by } \mathcal{P}_1^0\Lambda^2(\mathcal{T}_h;\mathbb{K}),$$

where B_3 denotes the space of cubic bubble functions. We then choose the two remaining spaces as

$$\Lambda_h^0(\mathbb{V}) = (\mathcal{P}_1 + B_3)\Lambda^0(\mathcal{T}_h;\mathbb{V}), \quad \Lambda_h^1(\mathbb{K}) = S_0\Lambda_h^0(\mathbb{V}).$$

It is easy to see that

$$\Lambda_h^1(\mathbb{K}) = (\mathcal{P}_1 + B_3)\Lambda^1(\mathcal{T}_h;\mathbb{K}) \cap H^1\Lambda^1(\mathbb{K}) \equiv (\mathcal{P}_1^0 + B_3)\Lambda^1(\mathcal{T}_h;\mathbb{K}).$$

Since the sequence

$$\mathcal{P}_1\Lambda^0(\mathcal{T}_h;\mathbb{V}) \xrightarrow{\mathrm{d}_0} \mathcal{P}_1^-\Lambda^1(\mathcal{T}_h;\mathbb{V}) \xrightarrow{\mathrm{d}_1} \mathcal{P}_0\Lambda^2(\mathcal{T}_h;\mathbb{V}) \to 0$$

is exact, so is the sequence

$$(\mathcal{P}_1 + B_3)\Lambda^0(\mathcal{T}_h;\mathbb{V}) \xrightarrow{\mathrm{d}_0} \mathcal{P}_1^-\Lambda^1(\mathcal{T}_h;\mathbb{V}) + \mathrm{d}_0B_3\Lambda^0(\mathcal{T}_h;\mathbb{V}) \xrightarrow{\mathrm{d}_1} \mathcal{P}_0\Lambda^2(\mathcal{T}_h;\mathbb{V}) \to 0.$$

For this choice of spaces, however, it is not true that $d\Lambda_h^1(\mathbb{K}) = \Lambda_h^2(\mathbb{K})$. Instead, we use the more general condition $\Pi_h^2 d\Lambda_h^1(\mathbb{K}) = \Lambda_h^2(\mathbb{K})$, which allows the use of stable Stokes elements. The proof that the combination $(\mathcal{P}_1^0 + B_3)\Lambda^1(\mathcal{T}_h;\mathbb{K})$ and $\mathcal{P}_1^0\Lambda^2(\mathcal{T}_h;\mathbb{K})$ is a stable Stokes pair (the Mini-element) involves construction of an interpolation operator $\Pi_h^1 : H^1\Lambda^1(\mathbb{K}) \mapsto (\mathcal{P}_1^0 + B_3)\Lambda^1(\mathcal{T}_h;\mathbb{K})$ satisfying

$$\langle \mathrm{d}_1(\tau - \Pi_h^1\tau), q_h\rangle = 0, \; q_h \in \Lambda_h^2(\mathbb{K}), \qquad \|\Pi_h^1\tau\|_1 \le C\|\tau\|_1, \; \tau \in H^1\Lambda^1(\mathbb{K}),$$

which gives properties (16) and (17) for the operators Π_h^1 and Π_h^2. Properties (16) and (17) for the operators $\widetilde{\Pi}_h^1$ and $\widetilde{\Pi}_h^2$ are satisfied by the Raviart–Thomas interpolant $\widetilde{\Pi}_h^1 : H^1\Lambda^1(\mathbb{V}) \mapsto \mathcal{P}_1^-\Lambda^1(\mathcal{T}_h;\mathbb{V})$. Finally, one can easily check that (18) and (21) are satisfied if we define

$$\widetilde{\Pi}_h^0 : H^1\Lambda^0(\mathbb{V}) \mapsto (\mathcal{P}_1 + B_3)\Lambda^0(\mathcal{T}_h;\mathbb{V})$$

by

$$\widetilde{\Pi}_h^0\tau = S_0^{-1}\Pi_h^1 S_0\tau.$$

Note that condition (21) is then trivial, since for $\tau \in H^1\Lambda^0(\mathbb{V})$,

$$S_{0,h}\widetilde{\Pi}_h^0\tau = \Pi_h^1 S_0 S_0^{-1}\Pi_h^1 S_0\tau = \Pi_h^1 S_0\tau.$$

Applying Theorem 10.1, and standard approximation and regularity results, we obtain the error estimates

$$\|\sigma - \sigma_h\|_0 + \|p - p_h\|_0 + \|u - u_h\|_0 \le Ch(\|\sigma\|_1 + \|p\|_1 + \|u\|_1) \le Ch\|f\|_0.$$

11.4 A PEERS-Like Method with Improved Stress Approximation

In this new method, we change one of the spaces used in the PEERS element and both of the auxiliary spaces used in the analysis, i.e., we choose

$$\Lambda_h^1(\mathbb{V}) = \mathcal{P}_1\Lambda^1(\mathcal{T}_h;\mathbb{V}), \quad \Lambda_h^2(\mathbb{V}) = \mathcal{P}_0\Lambda^2(\mathcal{T}_h;\mathbb{V}), \quad \Lambda_h^2(\mathbb{K}) = \mathcal{P}_1^0\Lambda^2(\mathcal{T}_h;\mathbb{K}),$$

and the two remaining spaces as

$$\Lambda_h^0(\mathbb{V}) = \mathcal{P}_2\Lambda^0(\mathcal{T}_h;\mathbb{V}), \quad \Lambda_h^1(\mathbb{K}) = S_0\Lambda_h^0(\mathbb{V}) \equiv \mathcal{P}_2\Lambda^1(\mathcal{T}_h;\mathbb{K}) \cap H^1\Lambda^1(\mathbb{K}).$$

The basic change from the analysis of the PEERS element is that we use the fact that the combination of $\mathcal{P}_2\Lambda^1(\mathcal{T}_h;\mathbb{K}) \cap H^1\Lambda^1(\mathbb{K})$ and $\mathcal{P}_1^0\Lambda^2(\mathcal{T}_h;\mathbb{K})$ is a stable pair of spaces for the Stokes problem (i.e., the Taylor–Hood element).

We may also view this new method as a modification of the lowest order Arnold–Falk–Winther method, where we are using the same stress and displacement spaces and lower exact sequence as in that method, but have changed the spaces with values in $\mathbb{K}$. The advantage of this modification is that it produces a higher order approximation to the stress variable. Looking at the error estimates given in Theorem 10.1, we see that the error estimate for $\|\sigma - \sigma_h\|_0$ depends both on $\|\sigma - \tilde{\Pi}_h^{n-1}\sigma\|_0$ and $\|p - \Pi_h^n p\|_0$. In the lowest order Arnold–Falk–Winther method, $\|\sigma - \tilde{\Pi}_h^{n-1}\sigma\|_0 \le Ch^2\|\sigma\|_2$, since we are using $\mathcal{P}_1$ elements to approximate σ. The fact that piecewise constants are used to approximate the multiplier results in only an $O(h)$ approximation for the second term. By using linear elements in the modified method, we recover second order convergence. Since we use only piecewise constants to approximate u, we can only obtain the estimate $\|u - u_h\|_0 \le Ch$. However, since the quantity $\|u_h - \Pi_h^n u\|_0$ is also $O(h^2)$, we might be able to obtain a better result by a post-processing procedure.

Remark 11.1. We note that some of these same ideas have been used to develop hybrid methods for the approximation of the elasticity equations. For example, see [21].

11.5 Methods of Stenberg

A family of methods proposed and analyzed by Stenberg [28] chooses for $r \ge 2$, $n = 2$ or $n = 3$,

$$\Lambda_h^{n-1}(\mathbb{V}) = \mathcal{P}_r\Lambda^{n-1}(\mathcal{T}_h;\mathbb{V}) + dB_{r+n}\Lambda^{n-2}(\mathcal{T}_h;\mathbb{V}), \quad \Lambda_h^n(\mathbb{V}) = \mathcal{P}_{r-1}\Lambda^n(\mathcal{T}_h;\mathbb{V}),$$
$$\Lambda_h^n(\mathbb{K}) = \mathcal{P}_r\Lambda^n(\mathcal{T}_h;\mathbb{K}),$$

where B_{r+n} denotes the space of functions which on each simplex T have the form $b_T\mathcal{P}_{r-1}$, where $b_T(x) = \prod_{i=1}^{n+1}\lambda_i(x)$, i.e., the space of bubbles of degree $r+n$. To fit our framework, we then choose the two remaining spaces as

$$\Lambda_h^{n-2}(\mathbb{V}) = (\mathcal{P}_{r+1} + B_{r+n})\Lambda^{n-2}(\mathcal{T}_h;\mathbb{V}),$$
$$\Lambda_h^{n-1}(\mathbb{K}) = (\mathcal{P}_{r+1} + B_{r+n})\Lambda^{n-1}(\mathcal{T}_h;\mathbb{K}) \cap H^1\Lambda^1(\mathbb{K}).$$

Since the sequence

$$\mathcal{P}_{r+1}\Lambda^{n-2}(\mathcal{T}_h;\mathbb{V}) \xrightarrow{\mathrm{d}_{n-2}} \mathcal{P}_r\Lambda^{n-1}(\mathcal{T}_h;\mathbb{V}) \xrightarrow{\mathrm{d}_{n-1}} \mathcal{P}_{r-1}\Lambda^n(\mathcal{T}_h;\mathbb{V}) \to 0$$

is exact, it is easy to see that the sequence

$$(\mathcal{P}_{r+1} + B_{r+n})\Lambda^{n-2}(\mathcal{T}_h;\mathbb{V}) \xrightarrow{\mathrm{d}_{n-2}} \mathcal{P}_r\Lambda^{n-1}(\mathcal{T}_h;\mathbb{V}) + \mathrm{d}_{n-2}B_{r+n}\Lambda^{n-2}(\mathcal{T}_h;\mathbb{V}) \\ \xrightarrow{\mathrm{d}_{n-1}} \mathcal{P}_{r-1}\Lambda^n(\mathcal{T}_h;\mathbb{V}) \to 0$$

will be exact. Again it is not true that $d\Lambda_h^{n-1}(\mathbb{K}) = \Lambda_h^n(\mathbb{K})$, and so we use the more general condition,

$$\Pi_h^n d\Lambda_h^{n-1}(\mathbb{K}) = \Lambda_h^n(\mathbb{K}),$$

which allows the use of stable Stokes spaces. From the definition of S_{n-2}, it it easy to see that when $n = 2$,

$$S_0\Lambda_h^0(\mathbb{V}) = (\mathcal{P}_{r+1} + B_{r+n})\Lambda^1(\mathcal{T}_h;\mathbb{K}) \cap H^1\Lambda^1(\mathbb{K}),$$

and when $n = 3$,

$$S_1[\Lambda_h^1(\mathbb{V}) \cap H^1\Lambda^2(\mathbb{V})] = (\mathcal{P}_{r+1} + B_{r+n})\Lambda^2(\mathcal{T}_h;\mathbb{K}) \cap H^1\Lambda^2(\mathbb{K}).$$

The proof that the combination $(\mathcal{P}_{r+1} + B_{r+n})\Lambda^{n-1}(\mathcal{T}_h;\mathbb{K}) \cap H^1\Lambda^{n-1}(\mathbb{K})$ and $\mathcal{P}_r\Lambda^n(\mathcal{T}_h;\mathbb{K})$ is a stable pair of Stokes elements (cf. [23, 15]) gives us precisely what we need to establish (16) and (17) for the operators Π_h^{n-1} and Π_h^n, i.e., the construction of an interpolation operator $\Pi_h^{n-1} : H^1\Lambda^{n-1}(\mathbb{K}) \mapsto (\mathcal{P}_{r+1} + B_{r+n})\Lambda^{n-1}(\mathcal{T}_h;\mathbb{K}) \cap H^1\Lambda^{n-1}(\mathbb{K})$ satisfying

$$\langle \mathrm{d}_{n-1}(\tau - \Pi_h^{n-1}\tau), q_h\rangle = 0, \; q_h \in \Lambda_h^n(\mathbb{K}),$$
$$\|\Pi_h^{n-1}\tau\|_1 \le C\|\tau\|_1, \; \tau \in H^1\Lambda^{n-1}(\mathbb{K}).$$

Properties (16) and (17) for the operators $\tilde{\Pi}_h^{n-1}$ and $\tilde{\Pi}_h^n$ are satisfied by the canonical interpolant $\tilde{\Pi}_h^{n-1} : H^1\Lambda^{n-1}(\mathbb{V}) \mapsto \mathcal{P}_r\Lambda^{n-1}(\mathcal{T}_h;\mathbb{V})$. Finally, it is easy to check that (18) and (21) are satisfied if we define

$$\tilde{\Pi}_h^{n-2} : H^1\Lambda^{n-2}(\mathbb{V}) \mapsto (\mathcal{P}_{r+1} + B_{r+n})\Lambda^{n-2}(\mathcal{T}_h;\mathbb{V}) \cap H^1\Lambda^{n-2}(\mathbb{V})$$

by

$$\tilde{\Pi}_h^{n-2}\tau = S_{n-2}^{-1}\Pi_h^{n-1}S_{n-2}\tau.$$

When $n = 2$, this same analysis also carries over to the case $r = 1$, since the combination $(\mathcal{P}_2 + B_3)\Lambda^1(\mathcal{T}_h;\mathbb{K}) \cap H^1\Lambda^1(\mathbb{K})$ and $\mathcal{P}_1\Lambda^2(\mathcal{T}_h;\mathbb{K})$ is a stable pair of Stokes elements. The situation is more complicated in three dimensions, since the analogous combination is not a stable pair of Stokes elements.

Using Theorem 10.1, it is straightforward to derive the following error estimates, assuming that σ, p, and u are sufficiently smooth.

$$\|\sigma - \sigma_h\| + \|p - p_h\| + \|u_h - \tilde{\Pi}_h^n u\| \le Ch^k(\|\sigma\|_k + \|p\|_k), \quad 1 \le k \le r+1,$$
$$\|u - u_h\| \le Ch^k(\|\sigma\|_k + \|p\|_k + \|u\|_k), \quad 1 \le k \le r,$$
$$\|\mathrm{d}_{n-1}(\sigma - \sigma_h)\| \le Ch^k\|\mathrm{d}_{n-1}\sigma\|_k, \quad 1 \le k \le r.$$

Acknowledgment

The work described here on finite element methods for the equations of elasticity with weak symmetry, which is the main focus of these notes, was done in collaboration with Douglas Arnold and Ragnar Winther.

References

1. S. Adams and B. Cockburn. A mixed finite element method for elasticity in three dimensions. *J. Sci. Comput.*, 25:515–521, 2005.
2. M. Amara and J. M. Thomas. Equilibrium finite elements for the linear elastic problem. *Numer. Math.*, 33:367–383, 1979.
3. D. N. Arnold, F. Brezzi, and J. Douglas, Jr. PEERS: a new mixed finite element for plane elasticity. *Jpn. J. Appl. Math.*, 1:347–367, 1984.
4. D. N. Arnold, J. Douglas, Jr., and C. P. Gupta. A family of higher order mixed finite element methods for plane elasticity. *Numer. Math.*, 45:1–22, 1984.
5. D. N. Arnold and R. S. Falk. A new mixed formulation for elasticity. *Numer. Math.*, 53:13–30, 1988.
6. D. N. Arnold, R. S. Falk, and R. Winther. Mixed finite element methods for linear elasticity with weakly imposed symmetry. *Math. Comput.*, 76(260):1699–1723, 2007.
7. D. N. Arnold, R. S. Falk, and R. Winther. Differential complexes and stability of finite element methods I. The de Rham complex. In *Compatible Spatial Discretizations, The IMA Volumes in Mathematics and its Applications*, vol. 142, pp. 23–46. Springer, Berlin Heidelberg New York, 2006.
8. D. N. Arnold, R. S. Falk, and R. Winther. Differential complexes and stability of finite element methods II: The elasticity complex. In *Compatible Spatial Discretizations, The IMA Volumes in Mathematics and its Applications*, vol. 142, pp. 47–68. Springer, Berlin Heidelberg New York, 2006.
9. D. N. Arnold, R. S. Falk, and R. Winther. Finite element exterior calculus, homological techniques, and applications. *Acta Numer.*, vol. 15, pp. 1–155, 2006.
10. D. N. Arnold and R. Winther. Mixed finite elements for elasticity. *Numer. Math.*, 92:401–419, 2002.
11. D. N. Arnold and R. Winther. Mixed finite elements for elasticity in the stress-displacement formulation. In *Current trends in scientific computing (Xi'an, 2002), Contemp. Math.*, vol. 329, pp. 33–42. American Mathematical Society, Providence, RI, 2003.
12. D. N. Arnold and R. Winther. Nonconforming mixed elements for elasticity. *Math. Models Methods Appl. Sci.*, 13(3):295–307, 2003. Dedicated to Jim Douglas, Jr. on the occasion of his 75th birthday.
13. I. N. Bernšteĭn, I. M. Gel′fand, and S. I. Gel′fand. Differential operators on the base affine space and a study of $\mathfrak{g}$-modules. In *Lie Groups and Their Representations* (Proc. Summer School, Bolyai János Math. Soc., Budapest, 1971), pp. 21–64. Halsted, New York, 1975.
14. F. Brezzi. On the existence, uniqueness and approximation of saddle-point problems arising from Lagrangian multipliers. *Rev. Française Automat. Informat. Recherche Opérationnelle Sér. Rouge*, 8:129–151, 1974.
15. F. Brezzi and M. Fortin. *Mixed and Hybrid Finite Element Methods, Springer Series in Computational Mathematics*, vol. 15. Springer, Berlin Heidelberg New York, 1991.

16. A. Čap, J. Slovák, and V. Souček. Bernstein–Gelfand–Gelfand sequences. *Ann. Math. (2)*, 154:97–113, 2001.
17. J. Douglas, Jr., T. Dupont, P. Percell, and L. R. Scott. A family of C^1 finite elements with optimal approximation properties for various Galerkin methods for 2nd and 4th order problems. *RAIRO Anal. Numér.*, 13(3):227–255, 1979.
18. J. Douglas, Jr. and J. E. Roberts. Global estimates for mixed methods for second order elliptic equations. *Math. Comp.*, 44:39–52, 1985.
19. M. Eastwood. A complex from linear elasticity. *Rend. Circ. Mat. Palermo (2) Suppl.*, (63):23–29, 2000.
20. R. S. Falk and J. E. Osborn. Error estimates for mixed methods. *RAIRO Anal. Numér.*, 14:249–277, 1980.
21. M. Farhloul and M. Fortin. Dual hybrid methods for the elasticity and the Stokes problems: a unified approach. *Numer. Math.*, 76:419–440, 1997.
22. B. M. Fraeijs de Veubeke. Stress function approach. In *Proc. of the World Congress on Finite Element Methods in Structural Mechanics*, vol. 1, pp. J.1–J.51. Bournemouth, Dorset, England, 1975.
23. V. Girault and P.-A. Raviart. *Finite element methods for Navier-Stokes equations*, *Springer Series in Computational Mathematics*, vol. 5. Springer, Berlin Heidelberg New York, 1986. Theory and Algorithms.
24. C. Johnson and B. Mercier. Some equilibrium finite element methods for two-dimensional elasticity problems. *Numer. Math.*, 30:103–116, 1978.
25. M. E. Morley. A family of mixed finite elements for linear elasticity. *Numer. Math.*, 55:633–666, 1989.
26. E. Stein and R. Rolfes. Mechanical conditions for stability and optimal convergence of mixed finite elements for linear plane elasticity. *Comput. Methods Appl. Mech. Eng.*, 84:77–95, 1990.
27. R. Stenberg. On the construction of optimal mixed finite element methods for the linear elasticity problem. *Numer. Math.*, 48:447–462, 1986.
28. R. Stenberg. A family of mixed finite elements for the elasticity problem. *Numer. Math.*, 53:513–538, 1988.
29. R. Stenberg. Two low-order mixed methods for the elasticity problem. In *The Mathematics of Finite Elements and Applications, VI* (Uxbridge, 1987), pp. 271–280. Academic Press, London, 1988.
30. V. B. Watwood, Jr. and B. J. Hartz. An equilibrium stress field model for finite element solution of two-dimensional elastostatic problems. *Int. J. Solids Struct.*, 4:857–873, 1968.

Finite Elements for the Reissner–Mindlin Plate

Richard S. Falk*

Department of Mathematics - Hill Center, Rutgers, The State University of New Jersey, 110 Frelinghuysen Rd., Piscataway, NJ 08854-8019, USA
falk@math.rutgers.edu

1 Introduction

In this paper, we consider the approximation of the equations of linear elasticity in the case when the body is an isotropic, homogeneous, linearly elastic plate. To describe the geometry of the plate, it will be convenient to consider the plate as occupying the region $P_t = \Omega \times (-t/2, t/2)$, where Ω is a bounded domain in $\mathbb{R}^2$ and $t \in (0, 1]$. We are interested in the case when the plate is thin, so that the thickness t will be small. We denote the union of the top and bottom surfaces of the plate by $\partial P_t^{\pm} = \Omega \times \{-t/2, t/2\}$ and the lateral boundary by $\partial P_t^{\mathrm{L}} = \partial\Omega \times (-t/2, t/2)$ (see Fig. 1). We suppose that the plate is loaded by a surface force density $\underline{g}\colon \partial P_t^{\pm} \to \mathbb{R}^3$ and a volume force density $\underline{f}\colon P_t \to \mathbb{R}^3$, and is clamped along its lateral boundary. The resulting stress $\underline{\underline{\sigma}}^*\colon P_t \to \mathbb{R}^{3\times 3}_{\mathrm{sym}}$ and displacement $\underline{u}^*\colon P_t \to \mathbb{R}^3$ then satisfy the boundary-value problem

$$\begin{aligned} \mathcal{A}\underline{\underline{\sigma}}^* &= \underline{\underline{\varepsilon}}(\underline{u}^*), \quad -\underline{\operatorname{div}}\,\underline{\underline{\sigma}}^* = \underline{f} \text{ in } P_t, \\ \underline{\underline{\sigma}}^*\underline{n} &= \underline{g} \text{ on } \partial P_t^{\pm}, \quad \underline{u}^* = 0 \text{ on } \partial P_t^{\mathrm{L}}. \end{aligned} \tag{1}$$

Here $\underline{\underline{\varepsilon}}(\underline{u}^*)$ denotes the infinitesimal strain tensor associated to the displacement vector $\underline{u}^*$, namely the symmetric part of its gradient, and $\underline{\operatorname{div}}\,\underline{\underline{\sigma}}$ denotes the vector divergence of the symmetric matrix $\underline{\underline{\sigma}}$ taken by rows. The compliance tensor $\mathcal{A}$ is given by $\mathcal{A}\underline{\underline{\tau}} = (1+\nu)\underline{\underline{\tau}}/E - \nu\,\mathrm{tr}(\underline{\underline{\tau}})\underline{\underline{\delta}}/E$, with $E > 0$ Young's modulus, $\nu \in [0, 1/2)$ Poisson's ratio, and $\underline{\underline{\delta}}$ the 3×3 identity matrix.

A plate model seeks to approximate the solution of the elasticity problem (1) in terms of the solution of a system of partial differential equations on the two-dimensional domain Ω without requiring the solution of a three-dimensional problem. The passage from the 3-D problem to a plate model is known as *dimensional reduction.*

By taking odd and even parts with respect to the variable x_3, the three-dimensional plate problem splits into two decoupled problems which correspond to *stretching* and *bending* of the plate. The most common plate stretching

* This work supported by NSF grants DMS03-08347 and DMS06-09755. 9/8/07.

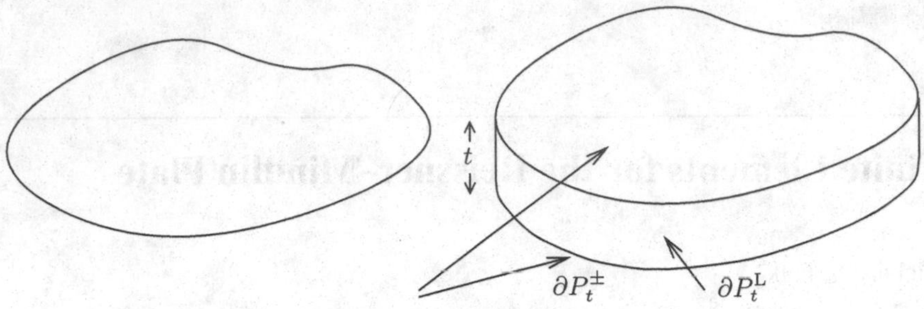

Fig. 1. The two-dimensional domain Ω and plate domain P_t

models are variants of the equations of generalized plane stress. The most common plate bending models are variants of the Kirchhoff–Love biharmonic plate model or of the Reissner–Mindlin plate model. We speak of variants here, because the specification of the forcing functions for the 2-D differential equations in terms of the 3-D loads $\underset{\sim}{g}$ and $\underline{f}$ differs for different models to be found in the literature, as does the specification of the approximate 3-D stresses and displacements in terms of the solutions of the 2-D boundary-value problems. Moreover, there is a coefficient in the Reissner–Mindlin model, the so-called shear correction factor, which is given different values in the literature. So there is no universally accepted basic two-dimensional model of plate stretching or bending.

2 A Variational Approach to Dimensional Reduction

The Hellinger–Reissner principle gives a variational characterization of the solution to the three-dimensional problem (1). We will consider two forms of this principle.

2.1 The First Variational Approach

To state the first form of the Hellinger–Reissner principle, which we label HR, we define

$$\underline{\underline{\Sigma}}^{\bullet} = \underline{\underline{L}}^2(P_t), \quad \underline{V}^{\bullet} = \{\underline{v} \in \underline{H}^1(P_t) : \underline{v} = 0 \text{ on } \partial P_t^{\mathrm{L}}\}.$$

Then HR characterizes $(\underline{\underline{\sigma}}^*, \underline{u}^*)$ as the unique critical point (namely a saddle point) of the HR functional

$$J(\underline{\underline{\tau}}, \underline{v}) = \frac{1}{2}\int_{P_t} \mathcal{A}\underline{\underline{\tau}} : \underline{\underline{\tau}}\, d\underline{x} - \int_{P_t} \underline{\underline{\tau}} : \underline{\underline{\varepsilon}}(\underline{v})\, d\underline{x} + \int_{P_t} \underline{f} \cdot \underline{v}\, d\underline{x} + \int_{\partial P_t^{\pm}} \underline{g} \cdot \underline{v}\, d\underset{\sim}{x}$$

on $\underline{\underline{\Sigma}}^{\bullet} \times \underline{V}^{\bullet}$. Equivalently, $(\underline{\underline{\sigma}}^*, \underline{u}^*)$ is the unique element of $\underline{\underline{\Sigma}}^{\bullet} \times \underline{V}^{\bullet}$ satisfying the weak equations

$$\int_{P_t} \mathcal{A}\underline{\underline{\sigma}}^* : \underline{\underline{\tau}}\, d\underline{x} - \int_{P_t} \underline{\underline{\varepsilon}}(\underline{u}) : \underline{\underline{\tau}}\, d\underline{x} = 0 \quad \text{for all } \underline{\underline{\tau}} \in \underline{\underline{\Sigma}}^{\bullet}, \tag{2}$$

$$\int_{P_t} \underline{\underline{\sigma}} : \underline{\underline{\varepsilon}}(\underline{v})\, d\underline{x} = \int_{P_t} \underline{f} \cdot \underline{v}\, d\underline{x} + \int_{\partial P_t^{\pm}} \underline{g} \cdot \underline{v}\, d\underset{\sim}{x} \quad \text{for all } \underline{v} \in \underline{V}^{\bullet}. \tag{3}$$

Plate models may be derived by replacing $\underline{\underline{\Sigma}}^\bullet$ and $\underline{V}^\bullet$ in HR with subspaces $\underline{\underline{\Sigma}}$ and $\underline{V}$ which admit only a specified polynomial dependence on x_3 and then defining $(\underline{\underline{\sigma}}, \underline{u})$ as the unique critical point of J over $\underline{\underline{\Sigma}} \times \underline{V}$. This is equivalent to restricting the trial and test spaces in the weak formulation to $\underline{\underline{\Sigma}} \times \underline{V}$. We insure a unique solution by requiring that $\underline{\underline{\varepsilon}}(\underline{V}) \subset \underline{\underline{\Sigma}}$. Here we shall consider only one of these models, which we denote HR(1). Define the two-dimensional analogue of the compliance tensor by $\underset{\approx\approx}{A}\underset{\approx}{\tau} = (1+\nu)\underset{\approx}{\tau}/E - \nu\,\mathrm{tr}(\underset{\approx}{\tau})\underset{\approx}{\delta}/E$. It can be shown that the HR(1) solution is given by

$$\underline{u}(\underline{x}) = \begin{pmatrix} \underset{\sim}{\eta}(\underset{\sim}{x}) \\ 0 \end{pmatrix} + \begin{pmatrix} -\underset{\sim}{\phi}(\underset{\sim}{x})x_3 \\ \omega(\underset{\sim}{x}) \end{pmatrix},$$

$$\underline{\underline{\sigma}}(\underline{x}) = \begin{pmatrix} \underset{\approx\approx}{A}^{-1}\underset{\approx}{\varepsilon}(\underset{\sim}{\eta}) & 0 \\ 0 & 0 \end{pmatrix} + \begin{pmatrix} -\underset{\approx\approx}{A}^{-1}\underset{\approx}{\varepsilon}(\underset{\sim}{\phi})x_3 & \frac{E}{2(1+\nu)}(\underset{\sim}{\nabla}\omega - \underset{\sim}{\phi}) \\ \frac{E}{2(1+\nu)}(\underset{\sim}{\nabla}\omega - \underset{\sim}{\phi})^T & 0 \end{pmatrix},$$

where $\underset{\sim}{\eta}$ is determined by a classical generalized plane stress problem and $\underset{\sim}{\phi}$ and ω by a Reissner–Mindlin problem. Specifically,

$$-t\,\mathrm{div}\,\underset{\approx\approx}{A}^{-1}\underset{\approx}{\varepsilon}(\underset{\sim}{\eta}) = 2\underset{\sim}{g}^0 + \underset{\sim}{f}^0 \text{ in } \Omega, \qquad \underset{\sim}{\eta} = 0 \text{ on } \partial\Omega, \tag{4}$$

$$\frac{t^3}{12}\,\mathrm{div}\,\underset{\approx\approx}{A}^{-1}\underset{\approx}{\varepsilon}(\underset{\sim}{\phi}) + t\frac{E}{2(1+\nu)}(\underset{\sim}{\phi} - \underset{\sim}{\nabla}\omega) = -t(\underset{\sim}{g}^1 + \underset{\sim}{f}^1) \text{ in } \Omega, \tag{5}$$

$$t\frac{E}{2(1+\nu)}\mathrm{div}(\underset{\sim}{\phi} - \underset{\sim}{\nabla}\omega) = 2g_3^0 + f_3^0 \text{ in } \Omega, \tag{6}$$

$$\underset{\sim}{\phi} = 0, \quad \omega = 0 \text{ on } \partial\Omega. \tag{7}$$

In the above (and in this section only for clarity), we use $\sim$ and $\approx$ to denote two-dimensional vectors and 2×2 matrices and $\_$ and $=$ to denote three-dimensional vectors and 3×3 matrices, respectively. We also define

$$g_3^0(\underset{\sim}{x}) = \frac{1}{2}[g_3(\underset{\sim}{x}, t/2) + g_3(\underset{\sim}{x}, -t/2)], \quad g_3^1(\underset{\sim}{x}) = \frac{1}{2}[g_3(\underset{\sim}{x}, t/2) - g_3(\underset{\sim}{x}, -t/2)],$$

$$f_3^0(\underset{\sim}{x}) = \int_{-t/2}^{t/2} f_3(\underset{\sim}{x}, x_3)\,\mathrm{d}x_3, \quad f_3^1(\underset{\sim}{x}) = \int_{-t/2}^{t/2} f_3(\underset{\sim}{x}, x_3)\frac{x_3}{t}\,\mathrm{d}x_3,$$

with $\underset{\sim}{g}^0$, $\underset{\sim}{g}^1$, $\underset{\sim}{f}^0$, and $\underset{\sim}{f}^1$ defined analogously. The verification of these equations is straightforward, but tedious.

In the case of a purely transverse bending load, the system (5)–(7) is the classical Reissner–Mindlin system with shear correction factor 1. When the bending is also affected by nonzero $\underset{\sim}{g}^1$ or $\underset{\sim}{f}^1$, then these appear as an applied couple in the Reissner–Mindlin system. Thus we see that the HR(1) method is a simple approach to deriving generalized plane stress and Reissner–Mindlin type models. There is an alternative approach, however, that produces models that are both more accurate and more amenable to rigorous justification than the methods based on HR discussed above. We discuss this approach below.

2.2 An Alternative Variational Approach

A second form of the Hellinger–Reissner principle, which we shall call HR′, leads to somewhat different plate models. For HR′ we define

$$\underset{\approx}{\Sigma}_g^* = \{\underset{\approx}{\sigma} \in \underset{\approx}{H}(\operatorname{div}, P_t) \mid \underset{\approx}{\sigma}\underset{\sim}{n} = \underset{\sim}{g} \text{ on } \partial P_t^{\pm}\}, \quad \underset{\sim}{V}^* = \underset{\sim}{L}^2(P).$$

Then HR′ characterizes $(\underset{\approx}{\sigma}^*, \underset{\sim}{u}^*)$ as the unique critical point (again a saddle point) of the HR′ functional

$$J'(\underset{\approx}{\tau}, \underset{\sim}{v}) = \frac{1}{2}\int_{P_t} \mathcal{A}\underset{\approx}{\tau} : \underset{\approx}{\tau}\, d\underset{\sim}{x} + \int_{P_t} \operatorname{div} \underset{\approx}{\tau} \cdot \underset{\sim}{v}\, d\underset{\sim}{x} + \int_{P_t} \underset{\sim}{f} \cdot \underset{\sim}{v}\, d\underset{\sim}{x}$$

on $\underset{\approx}{\Sigma}_g^* \times \underset{\sim}{V}^*$. Equivalently, $(\underset{\approx}{\sigma}^*, \underset{\sim}{u}^*)$ is the unique element of $\underset{\approx}{\Sigma}_g^* \times \underset{\sim}{V}^*$ satisfying the weak equations

$$\int_{P_t} \mathcal{A}\underset{\approx}{\sigma}^* : \underset{\approx}{\tau}\, d\underset{\sim}{x} + \int_{P_t} \underset{\sim}{u} \cdot \operatorname{div} \underset{\approx}{\tau}\, d\underset{\sim}{x} = 0 \quad \text{for all } \underset{\approx}{\tau} \in \underset{\approx}{\Sigma}_0^*,$$

$$\int_{P_t} \operatorname{div} \underset{\approx}{\sigma} \cdot \underset{\sim}{v}\, d\underset{\sim}{x} = -\int_{P_t} \underset{\sim}{f} \cdot \underset{\sim}{v}\, d\underset{\sim}{x} \quad \text{for all } \underset{\sim}{v} \in \underset{\sim}{V}^*.$$

Here $\underset{\approx}{\Sigma}_0^* = \{\underset{\approx}{\sigma} \in \underset{\approx}{H}(\operatorname{div}, P_t) \mid \underset{\approx}{\sigma}\underset{\sim}{n} = 0 \text{ on } \partial P_t^{\pm}\}$. Note that the displacement boundary conditions, which were essential to the first form of the Hellinger–Reissner principle, are natural in this setting, while the reverse situation holds for the traction boundary conditions.

By restricting J' to subspaces of $\underset{\approx}{\Sigma}_g^*$ and $\underset{\sim}{V}^*$ with a specified polynomial dependence on x_3, we also obtain a variety of plate models. Here we shall consider only one of these, which we denote HR′(1). The HR′(1) solution is:

$$\underset{\sim}{u}(\underset{\sim}{x}) = \begin{pmatrix} \underset{\sim}{\eta}(\underset{\sim}{x}) \\ \varrho(\underset{\sim}{x})x_3 \end{pmatrix} + \begin{pmatrix} -\underset{\sim}{\phi}(\underset{\sim}{x})x_3 \\ \omega(\underset{\sim}{x}) + \omega_2(\underset{\sim}{x})r(x_3) \end{pmatrix},$$

$$\underset{\approx}{\sigma}(\underset{\sim}{x}) = \begin{pmatrix} \underset{\approx}{\sigma}^0(\underset{\sim}{x}) & \frac{2x_3}{t}\underset{\sim}{g}^0(\underset{\sim}{x}) \\ \frac{2x_3}{t}\underset{\sim}{g}^0(\underset{\sim}{x})^T & g_3^1(\underset{\sim}{x}) + \sigma_{33}^0(\underset{\sim}{x})q(x_3) \end{pmatrix}$$

$$+ \begin{pmatrix} \underset{\approx}{\sigma}^1(\underset{\sim}{x})\frac{x_3}{t} & \underset{\sim}{g}^1(\underset{\sim}{x}) + \underset{\sim}{\varsigma}^0(\underset{\sim}{x})q(x_3) \\ \underset{\sim}{g}^1(\underset{\sim}{x})^T + \underset{\sim}{\varsigma}^0(\underset{\sim}{x})^T q(x_3) & g_3^0(\underset{\sim}{x})\frac{2x_3}{t} + \sigma_{33}^1(x_3)s(x_3) \end{pmatrix},$$

where the coefficient functions $\underset{\sim}{\eta}$, ϱ, $\underset{\sim}{\phi}$, ω, ω_2, $\underset{\approx}{\sigma}^0$, σ_{33}^0, $\underset{\approx}{\sigma}^1$, $\underset{\sim}{\varsigma}^0$, and σ_{33}^1 are functions of $\underset{\sim}{x}$ which we shall describe, and the polynomials q, r, and s are given by $q(z) = 3/2 - 6z^2/t^2$, $r(z) = 6z^2/t^2 - 3/10$, and $s(z) = (5/2)z/t - 10z^3/t^3$.

The stretching portion of the solution is determined by the solution to the boundary-value problem

$$-t \operatorname{div}_{\sim} \underset{\approx}{A}^{-1} \underset{\approx}{\varepsilon}(\underset{\sim}{\eta}) = \underset{\sim}{l}_1 + t\frac{\nu}{1-\nu}\underset{\sim}{\nabla} l_2 \text{ in } \Omega, \qquad \underset{\sim}{\eta} = 0 \text{ on } \partial\Omega, \tag{8}$$

$$\text{where} \quad \underset{\sim}{l}_1 = 2\underset{\sim}{g}^0 + \underset{\sim}{f}^0, \quad l_2 = g_3^1 + \frac{t}{6}\operatorname{div} \underset{\sim}{g}^0 + f_3^1.$$

With $\underset{\sim}{\eta}$ uniquely determined by (8), the remaining solution quantities are

$$\underset{\approx}{\sigma}^0 = \underset{\approx}{A}^{-1}\underset{\approx}{\varepsilon}(\underset{\sim}{\eta}) + \frac{\nu}{1-\nu} l_2 \underset{\approx}{\delta}, \qquad \sigma_{33}^0 = \frac{t}{6}\operatorname{div}\underset{\sim}{g}^0 + f_3^1,$$
$$\varrho = \frac{1}{E}\big[-\nu\operatorname{tr}(\underset{\approx}{\sigma}^0) + \frac{6}{5}\sigma_{33}^0 + g_3^1\big].$$

The bending portion of the solution is determined by the solution to the boundary-value problem

$$-\frac{t^3}{12}\operatorname{div}\underset{\approx}{A}^{-1}\underset{\approx}{\varepsilon}(\underset{\sim}{\phi}) + t\frac{5}{6}\frac{E}{2(1+\nu)}(\underset{\sim}{\phi} - \underset{\sim}{\nabla}\omega) = t\underset{\sim}{k}_1 - \frac{t^2}{12}\underset{\sim}{\nabla}k_2 \text{ in } \Omega,$$

$$t\frac{5}{6}\frac{E}{2(1+\nu)}\operatorname{div}(\underset{\sim}{\phi} - \underset{\sim}{\nabla}\omega) = k_3 \text{ in } \Omega, \qquad \underset{\sim}{\phi} = 0, \quad \omega = 0 \text{ on } \partial\Omega, \tag{9}$$

where $\underset{\sim}{k}_1 = -\frac{5}{6}\underset{\sim}{g}^1 - \underset{\sim}{f}^1, \quad k_2 = \frac{\nu}{1-\nu}\left[\frac{t}{5}\operatorname{div}\underset{\sim}{g}^1 + \frac{12}{5}g_3^0 + f_3^2\right],$

$$f_3^2(\underset{\sim}{x}) = \int_{-t/2}^{t/2} f_3(\underset{\sim}{x})r(x_3)\,\mathrm{d}x_3, \qquad k_3 = \frac{t}{6}\operatorname{div}\underset{\sim}{g}^1 + 2g_3^0 + f_3^0.$$

The boundary-value problem (9) determining the bending solution is a somewhat different version of the Reissner–Mindlin equations than (5)–(7), which arose from the HR(1) model. Not only are the formulas for the applied load and couple more involved, but a shear correction factor of $5/6$ has been introduced. With $\underset{\sim}{\phi}$ and ω determined by (9), we find

$$\underset{\approx}{\sigma}^1 = -t\underset{\approx}{A}^{-1}\underset{\approx}{\varepsilon}(\underset{\sim}{\phi}) + k_2\underset{\approx}{\delta}, \qquad \underset{\sim}{\sigma}^0 = \frac{5}{6}\left[\frac{E}{2(1+\nu)}(-\underset{\sim}{\phi} + \underset{\sim}{\nabla}\omega) - \underset{\sim}{g}^1\right],$$
$$\sigma_{33}^1 = \frac{t}{5}\operatorname{div}\underset{\sim}{g}^1 + \frac{2}{5}g_3^0 + f_3^2, \qquad \omega_2 = \frac{t}{E}\left[\frac{1}{6}g_3^0 + \frac{5}{42}\sigma_{33}^1 - \frac{\nu}{12}\operatorname{tr}(\underset{\approx}{\sigma}^1)\right].$$

For this model, it is possible to use the "two-energies principle" to derive rigorous error estimates between the solution of the three-dimensional model and the two-dimensional reduced model as a function of the plate thickness (see [1] for details).

3 The Reissner–Mindlin Model

From the previous section, we see that if we introduce the tensor $\mathfrak{C} = \underset{\approx}{A}^{-1}$ and scale the right hand side, then the Reissner–Mindlin equations may be written in the form

$$-\operatorname{\mathbf{div}}\mathfrak{C}\,\mathcal{E}(\boldsymbol{\theta}) - \lambda t^{-2}(\operatorname{\mathbf{grad}} w - \boldsymbol{\theta}) = -\boldsymbol{f},$$
$$-\operatorname{div}(\operatorname{\mathbf{grad}} w - \boldsymbol{\theta}) = \lambda^{-1}t^2 g,$$

with λ a constant depending on the particular version of the model that is chosen. We also have a Reissner–Mindlin energy

$$J(\boldsymbol{\theta}, w) = \frac{1}{2}\int_\Omega \mathfrak{C}\varepsilon(\boldsymbol{\theta}) : \varepsilon(\boldsymbol{\theta}) + \frac{1}{2}\lambda t^{-2}\int_\Omega |\,\mathbf{grad}\, w - \boldsymbol{\theta}|^2 - \int_\Omega gw + \int_\Omega \boldsymbol{f}\cdot\boldsymbol{\theta}, \quad (10)$$

for which the above equations are the Euler equations. As both a theoretical and computational tool, it is useful to introduce the shear stress $\boldsymbol{\gamma} = \lambda t^{-2}(\mathbf{grad}\, w - \boldsymbol{\theta})$. Then we have the equivalent Reissner–Mindlin system

$$-\,\mathbf{div}\,\mathfrak{C}\,\mathcal{E}(\boldsymbol{\theta}) - \boldsymbol{\gamma} = -\boldsymbol{f}, \quad (11)$$

$$-\,\mathrm{div}\,\boldsymbol{\gamma} = g, \quad (12)$$

$$\mathbf{grad}\, w - \boldsymbol{\theta} - \lambda^{-1}t^2\boldsymbol{\gamma} = 0, \quad (13)$$

For simplicity we restrict our attention to the case of a clamped plate, i.e., we consider the boundary conditions $\boldsymbol{\theta} = 0$ and $w = 0$ on the boundary $\partial\Omega$. A weak formulation of the Reissner–Mindlin model is then given by:

Find $\boldsymbol{\theta} \in \mathring{\boldsymbol{H}}^1(\Omega)$, $w \in \mathring{H}^1(\Omega)$, $\boldsymbol{\gamma} \in \boldsymbol{L}^2(\Omega)$ such that

$$a(\boldsymbol{\theta}, \boldsymbol{\phi}) + (\boldsymbol{\gamma}, \mathbf{grad}\, v - \boldsymbol{\phi}) = (g, v) - (\boldsymbol{f}, \boldsymbol{\phi}), \quad \boldsymbol{\phi} \in \mathring{\boldsymbol{H}}^1(\Omega), v \in \mathring{H}^1(\Omega), \quad (14)$$

$$(\mathbf{grad}\, w - \boldsymbol{\theta}, \boldsymbol{\eta}) - \lambda^{-1}t^2(\boldsymbol{\gamma}, \boldsymbol{\eta}) = 0, \quad \boldsymbol{\eta} \in \boldsymbol{L}^2(\Omega), \quad (15)$$

where $a(\boldsymbol{\theta}, \boldsymbol{\phi}) = (\mathfrak{C}\,\mathcal{E}(\boldsymbol{\theta}), \mathcal{E}(\boldsymbol{\phi}))$.

4 Properties of the Solution

As $t \to 0$, $\boldsymbol{\theta} \to \boldsymbol{\theta}^0$ and $w \to w^0$, where $\boldsymbol{\theta}^0 = \mathbf{grad}\, w^0$. One can then show that w^0 satisfies the limit problem: Find $w^0 \in \mathring{H}^2(\Omega) = \{v \in H^2(\Omega) : v = \partial v/\partial n = 0 \text{ on } \partial\Omega\}$ such that

$$a(\mathbf{grad}\, w^0, \mathbf{grad}\, v) = (g, v) - (\boldsymbol{f}, \mathbf{grad}\, v), \qquad v \in \mathring{H}^2(\Omega).$$

This is the weak form of the equation: $\mathrm{div}\,\mathbf{div}\,\mathfrak{C}\,\mathcal{E}(\mathbf{grad}\, w^0) = g + \mathrm{div}\,\boldsymbol{f}$, which after the application of some calculus identities becomes:

$$D\,\Delta^2 w^0 = g + \mathrm{div}\,\boldsymbol{f}, \qquad D = \frac{E}{12(1-\nu^2)}. \quad (16)$$

Hence, the limiting problem is the biharmonic problem.

To understand this limiting behavior and also to derive the regularity results presented in the next section, it is useful to introduce the Helmholtz decomposition, and rewrite the Reissner–Mindlin system as a perturbed Stokes equation. For some $r \in \mathring{H}^1(\Omega)$ and $p \in \hat{H}^1(\Omega)$, we can write

$$\boldsymbol{\gamma} = \lambda t^{-2}(\mathbf{grad}\, w - \boldsymbol{\theta}) = \mathbf{grad}\, r + \mathbf{curl}\, p.$$

Then it is easy to check that problem (14) and (15) is equivalent to the system:

Find $(r, \boldsymbol{\theta}, p, w) \in \mathring{H}^1(\Omega) \times \mathring{\boldsymbol{H}}^1(\Omega) \times \hat{H}^1(\Omega) \times \mathring{H}^1(\Omega)$ such that

$$(\mathbf{grad}\, r, \mathbf{grad}\, \mu) = (g, \mu), \quad \mu \in \mathring{H}^1(\Omega), \tag{17}$$

$$a(\boldsymbol{\theta}, \boldsymbol{\phi}) - (\mathbf{curl}\, p, \boldsymbol{\phi}) = (\mathbf{grad}\, r, \boldsymbol{\phi}) - (\boldsymbol{f}, \boldsymbol{\phi}), \quad \boldsymbol{\phi} \in \mathring{\boldsymbol{H}}^1(\Omega), \tag{18}$$

$$-(\boldsymbol{\theta}, \mathbf{curl}\, q) - \lambda^{-1} t^2 (\mathbf{curl}\, p, \mathbf{curl}\, q) = 0, \quad q \in \hat{H}^1(\Omega), \tag{19}$$

$$(\mathbf{grad}\, w, \mathbf{grad}\, s) = (\boldsymbol{\theta} + \lambda^{-1} t^2\, \mathbf{grad}\, r, \mathbf{grad}\, s), \quad s \in \mathring{H}^1(\Omega). \tag{20}$$

We then define $(\boldsymbol{\theta}^0, p^0, w^0) \in \mathring{\boldsymbol{H}}^1(\Omega) \times \hat{H}^1(\Omega) \times \mathring{H}^1(\Omega)$ as the solution of (17)–(20) with $t = 0$. Note that for r known and $t = 0$, (18) and (19) is the ordinary Stokes system for $(\theta_2^0, -\theta_1^0, p^0)$.

5 Regularity Results

A key issue in the approximation of the Reissner–Mindlin plate problem is the regularity of the solution and especially its dependence on the plate thickness t. For this problem, there is a boundary layer, whose strength depends on the particular boundary condition. There are a number of physically interesting boundary conditions:

$$\begin{aligned}
\boldsymbol{\theta} \cdot \boldsymbol{n} = \boldsymbol{\theta} \cdot \boldsymbol{s} = w = 0 &\quad \text{hard clamped (hc)},\\
\boldsymbol{\theta} \cdot \boldsymbol{n} = M_s(\boldsymbol{\theta}) \cdot \boldsymbol{s} = w = 0 &\quad \text{soft clamped (sc)},\\
M_n(\boldsymbol{\theta}) = \boldsymbol{\theta} \cdot \boldsymbol{s} = w = 0 &\quad \text{hard simply supported (hss)},\\
M_n(\boldsymbol{\theta}) = M_s(\boldsymbol{\theta}) = w = 0 &\quad \text{soft simply supported (sss)},\\
M_n(\boldsymbol{\theta}) = M_s(\boldsymbol{\theta}) = \partial w/\partial n - \boldsymbol{\theta} \cdot \boldsymbol{n} = 0 &\quad \text{free (f)},
\end{aligned}$$

where $\boldsymbol{n}$ and $\boldsymbol{s}$ denote the unit normal and counterclockwise unit tangent vectors, respectively, and $M_n(\boldsymbol{\theta}) = \boldsymbol{n} \cdot \mathfrak{C}\varepsilon(\boldsymbol{\theta})\boldsymbol{n}$, $M_s(\boldsymbol{\theta}) = \boldsymbol{s} \cdot \mathfrak{C}\varepsilon(\boldsymbol{\theta})\boldsymbol{n}$. In the case of a domain with smooth boundary, it is shown in [12, 13] that for all boundary conditions, the transverse displacement and all its derivatives are bounded uniformly in t, i.e., $\|w\|_s \le C$, $s \in \mathbb{R}$. Estimates showing the boundary layers, ordered from weakest to strongest, are given below.

$$\begin{aligned}
\|\boldsymbol{\theta}\|_s &\le Ct^{\min(0,7/2-s)}, &\quad \|\gamma\|_s &\le Ct^{\min(0,3/2-s)}, \quad s \in \mathbb{R}, \quad \text{(sc)}\\
\|\boldsymbol{\theta}\|_s &\le Ct^{\min(0,5/2-s)}, &\quad \|\gamma\|_s &\le Ct^{\min(0,1/2-s)}, \quad s \in \mathbb{R}, \quad \text{(hc), (hss)},\\
\|\boldsymbol{\theta}\|_s &\le Ct^{\min(0,3/2-s)}, &\quad \|\gamma\|_s &\le Ct^{\min(0,-1/2-s)}, \quad s \in \mathbb{R}, \quad \text{(sss), (f)}.
\end{aligned}$$

Additional results can be found in [10].

We will also need estimates that show the precise dependence on the data of the problem and which are valid when Ω is a convex polygon, the case we consider in the derivation of error estimates for finite element approximation schemes. We establish such estimates below for the case of the clamped plate.

Theorem 5.1 *Let Ω be a convex polygon or a smoothly bounded domain in the plane. For any $t \in (0,1]$, $\boldsymbol{f} \in \boldsymbol{H}^{-1}(\Omega)$, and $g \in H^{-1}(\Omega)$, there exists a unique solution $(r, \boldsymbol{\theta}, p, w) \in \mathring{H}^1(\Omega) \times \mathring{\boldsymbol{H}}^1(\Omega) \times \hat{H}^1(\Omega) \times \mathring{H}^1(\Omega)$ satisfying* (17)–(20). *Moreover, if $\boldsymbol{f} \in \boldsymbol{L}^2(\Omega)$, then $\boldsymbol{\theta} \in \boldsymbol{H}^2(\Omega)$ and there exists a constant C independent of t, $\boldsymbol{f}$, and g, such that*

$$\|\boldsymbol{\theta}\|_2 + \|r\|_1 + \|p\|_1 + t\|p\|_2 + \|w\|_1 + \|\gamma\|_0 \le C(\|\boldsymbol{f}\|_0 + \|g\|_{-1}), \tag{21}$$

If, in addition, $g \in L^2(\Omega)$, then r and $w \in H^2(\Omega)$ and

$$\|r\|_2 + \|w\|_2 + t\|\gamma\|_1 + \|\operatorname{div}\gamma\|_0 \le C(\|g\|_0 + \|\boldsymbol{f}\|_0). \tag{22}$$

Finally, if $(\boldsymbol{\theta}^0, w^0)$ denotes the solution of (17)–(20) *with $t = 0$, then*

$$\|\boldsymbol{\theta} - \boldsymbol{\theta}^0\|_1 \le Ct(\|\boldsymbol{f}\|_0 + \|g\|_{-1}), \qquad \|w - w^0\|_2 \le Ct(\|\boldsymbol{f}\|_0 + \|g\|_{-1} + t\|g\|_0),$$
$$\|w^0\|_3 \le C(\|\boldsymbol{f}\|_0 + \|g\|_{-1}). \tag{23}$$

Proof. Existence and uniqueness are easy to establish using the equivalence of this system to (14) and (15) and standard results, so we concentrate on the regularity estimates. We first observe that standard regularity results for Poisson's equation gives

$$\|r\|_1 \le C\|g\|_{-1}, \qquad \|r\|_2 \le \|g\|_0.$$

We next recall a regularity result for the Stokes system, valid both for the case of a domain with smooth boundary and for a convex polygon.

$$\|\boldsymbol{\theta}^0\|_2 + \|p^0\|_1 \le C(\|\boldsymbol{f}\|_0 + \|r\|_1) \le C(\|\boldsymbol{f}\|_0 + \|g\|_{-1}).$$

Now from (18) and (19), and the corresponding equations for $\boldsymbol{\theta}^0$ and p^0, we get

$$a(\boldsymbol{\theta}-\boldsymbol{\theta}^0, \boldsymbol{\phi}) - (\mathbf{curl}(p-p^0), \boldsymbol{\phi}) + (\boldsymbol{\theta}-\boldsymbol{\theta}^0, \mathbf{curl}\, q) + \lambda^{-1}t^2(\mathbf{curl}(p-p^0), \mathbf{curl}\, q)$$
$$= -\lambda^{-1}t^2(\mathbf{curl}\, p^0, \mathbf{curl}\, q), \quad (\boldsymbol{\phi}, q) \in \mathring{\boldsymbol{H}}^1(\Omega) \times \hat{H}^1(\Omega).$$

Choosing $\boldsymbol{\phi} = \boldsymbol{\theta} - \boldsymbol{\theta}^0$ and $q = p - p^0$, we obtain

$$\|\boldsymbol{\theta} - \boldsymbol{\theta}^0\|_1^2 + t^2\|\mathbf{curl}(p - p^0)\|_0^2 \le Ct^2\|p^0\|_1\|\mathbf{curl}(p - p^0)\|_0.$$

It easily follows that

$$\|\boldsymbol{\theta} - \boldsymbol{\theta}^0\|_1 + t\|p - p^0\|_1 \le Ct\|p^0\|_1 \le Ct(\|\boldsymbol{f}\|_0 + \|g\|_{-1}), \tag{24}$$

which establishes the first estimate in (23). We also get that

$$\|p\|_1 \le C(\|\boldsymbol{f}\|_0 + \|g\|_{-1}).$$

Applying standard estimates for second order elliptic problems to (18), we further obtain

$$\|\boldsymbol{\theta}\|_2 \le C(\|p\|_1 + \|r\|_1 + \|\boldsymbol{f}\|_0) \le C(\|\boldsymbol{f}\|_0 + \|g\|_{-1}).$$

Now from (19) and the definition of $\boldsymbol{\theta}^0$, we get

$$\lambda^{-1}t^2(\mathbf{curl}\, p, \mathbf{curl}\, q) = -(\boldsymbol{\theta}, \mathbf{curl}\, q) = (\boldsymbol{\theta}^0 - \boldsymbol{\theta}, \mathbf{curl}\, q), \quad q \in \hat{H}^1(\Omega).$$

Thus p is the weak solution of the boundary-value problem

$$-\Delta p = \lambda t^{-2} \operatorname{rot}(\boldsymbol{\theta}^0 - \boldsymbol{\theta}) \ \text{ in } \Omega, \qquad \partial p/\partial n = 0 \text{ on } \partial\Omega.$$

Applying elliptic regularity and (24), we get

$$\|p\|_2 \le Ct^{-2}\|\boldsymbol{\theta}^0 - \boldsymbol{\theta}\|_1 \le Ct^{-1}(\|\boldsymbol{f}\|_0 + \|g\|_{-1}).$$

The estimate for w in (21) now follows directly from (20) and the estimate for γ in (21) follows immediately from its definition and the estimates for r and p. The estimate (22) follows directly from the regularity result for r, the definition of γ, elliptic regularity of w, and the previous results. Finally, it remains to establish the last two estimates in (23). Subtracting the analogue of (20) from (20), we get that

$$(\mathbf{grad}(w - w^0), \mathbf{grad}\, s) = (\boldsymbol{\theta} - \boldsymbol{\theta}^0 + \lambda^{-1}t^2\, \mathbf{grad}\, r, \mathbf{grad}\, s), \quad s \in \mathring{H}^1(\Omega).$$

This is the weak form of the equation

$$-\Delta(w - w^0) = -\operatorname{div}(\boldsymbol{\theta} - \boldsymbol{\theta}^0) - \lambda^{-1}t^2 \Delta r.$$

Combining standard regularity estimates for Poisson's equation with our previous results, we get

$$\|w - w^0\|_2 \le C(\|\boldsymbol{\theta} - \boldsymbol{\theta}^0\|_1 + t^2\|r\|_2) \le Ct(\|\boldsymbol{f}\| + \|g\|_{-1} + t\|g\|_0).$$

Finally, using the fact that w^0 satisfies the biharmonic equation (16), together with the boundary conditions $w^0 = \partial w^0/\partial n = 0$, we get the estimate

$$\|w^0\|_3 \le C\|g + \operatorname{div} \boldsymbol{f}\|_{-1} \le C(\|g\|_{-1} + \|\boldsymbol{f}\|_0).$$

6 Finite Element Discretizations

The challenge in devising finite element approximation schemes for the Reissner–Mindlin plate model is to find schemes whose approximation accuracy does not deteriorate as the plate thickness becomes very small. For example, if one minimizes the Reissner–Mindlin energy over subspaces consisting of low order finite elements, then the resulting approximation suffers from the problem of "locking." This problem is most easily described by recalling that as $t \to 0$, the minimizer $(\boldsymbol{\theta}, w)$ of (10) approaches $(\boldsymbol{\theta}^0, w^0)$, where $\boldsymbol{\theta}^0 = \mathbf{grad}\, w^0$. If we discretize the problem directly by seeking $\boldsymbol{\theta}_h \in \boldsymbol{\Theta}_h$ and $w_h \in W_h$ minimizing $J(\boldsymbol{\theta}, w)$ over $\boldsymbol{\Theta}_h \times W_h$, then as $t \to 0$ we will have $(\boldsymbol{\theta}_h, w_h) \to (\boldsymbol{\theta}_h^0, w_h^0)$ where, again, $\boldsymbol{\theta}_h^0 = \mathbf{grad}\, w_h^0$. The locking problem occurs because, for low order finite element spaces, this last condition is too restrictive to allow for good approximations of smooth functions. In particular, if continuous piecewise linear functions are chosen to approximate both variables, then $\boldsymbol{\theta}_h^0 \equiv \mathbf{grad}\, w_h^0$ would be continuous *and* piecewise constant, with zero boundary conditions: Only the choice $\boldsymbol{\theta}_h^0 = 0$ can satisfy all these conditions. Hence, unless the combination of finite element spaces is chosen carefully, this problem is likely to occur.

Many of the successful locking-free finite element schemes have taken the following approach. Let $\boldsymbol{\Theta}_h \subset \mathring{\boldsymbol{H}}^1(\Omega)$, $W_h \subset \mathring{H}^1(\Omega)$, $\boldsymbol{\Gamma}_h \subset \boldsymbol{L}^2(\Omega)$, where $\mathbf{grad}\, W_h \subset \boldsymbol{\Gamma}_h$. Let $\boldsymbol{\Pi}^{\boldsymbol{\Gamma}}$ be an interpolation operator mapping $\mathring{\boldsymbol{H}}^1(\Omega)$ to $\boldsymbol{\Gamma}_h$. Then consider finite element approximation schemes of the form:

Find $\boldsymbol{\theta}_h \in \boldsymbol{\Theta}_h$, $w_h \in W_h$, $\gamma_h \in \boldsymbol{\Gamma}_h$ such that

$$a(\boldsymbol{\theta}_h, \boldsymbol{\phi}) + (\gamma_h, \mathbf{grad}\, v - \boldsymbol{\Pi}^{\boldsymbol{\Gamma}}\boldsymbol{\phi}) = (g, v) - (\boldsymbol{f}, \boldsymbol{\phi}), \qquad \boldsymbol{\phi} \in \boldsymbol{\Theta}_h, v \in W_h,$$
$$(\mathbf{grad}\, w_h - \boldsymbol{\Pi}^{\boldsymbol{\Gamma}}\boldsymbol{\theta}_h, \boldsymbol{\eta}) - \lambda^{-1}t^2(\gamma_h, \boldsymbol{\eta}) = 0, \qquad \boldsymbol{\eta} \in \boldsymbol{\Gamma}_h. \tag{25}$$

The point of introducing the operator $\boldsymbol{\Pi}^{\boldsymbol{\Gamma}}$ is that now, as $t \to 0$, we will get that $\mathbf{grad}\, w_{h,0} \to \boldsymbol{\Pi}^{\boldsymbol{\Gamma}}\boldsymbol{\theta}_{h,0}$. If $\boldsymbol{\Pi}^{\boldsymbol{\Gamma}}$ is chosen properly, this condition may be much easier to satisfy, while still maintaining good approximation properties of each subspace.

We will also consider some nonconforming discretizations in which either space $\boldsymbol{\Theta}_h$ or W_h consists of functions which belong to H^1 on each triangle, but not globally. In the first case, the operator $\mathcal{E}$ entering into the definition of the bilinear form a must be replaced with $\mathcal{E}_h$, the operator obtained by applying $\mathcal{E}$ piecewise on each triangle. Similarly, in the second case, the operator $\mathbf{grad}$ must be replaced by its piecewise counterpart, $\mathbf{grad}_h$.

7 Abstract Error Analysis

In order to analyze approximation schemes using a common framework, we first prove several abstract approximation results. These results will make use of the following assumptions about the approximation properties of the finite-dimensional subspaces and the operator $\boldsymbol{\Pi}^{\boldsymbol{\Gamma}}$ that define the various methods.

$$\mathbf{grad}\, W_h \subset \boldsymbol{\Gamma}_h, \tag{26}$$

$$\|\boldsymbol{\eta} - \boldsymbol{\Pi}^{\boldsymbol{\Gamma}}\eta\| \le ch\|\boldsymbol{\eta}\|_1, \qquad \boldsymbol{\eta} \in \boldsymbol{H}^1(\Omega), \tag{27}$$

for some constant c independent of h. Letting $\boldsymbol{M}_r$ denote the space of discontinuous piecewise polynomials of degree $\le r$, we also define $r_0 \ge -1$ as the greatest integer r for which

$$(\boldsymbol{\eta} - \boldsymbol{\Pi}^{\boldsymbol{\Gamma}}\eta, \boldsymbol{\zeta}) = 0, \qquad \boldsymbol{\zeta} \in \boldsymbol{M}_r. \tag{28}$$

Of course this relation trivially holds for $r = -1$. We then let $\boldsymbol{\Pi}^0$ denote the $\boldsymbol{L}^2$ projection into $\boldsymbol{M}_{r_0}$.

The following basic result is close to Lemma 3.1 of Durán and Liberman [33].

Theorem 7.1 *Let* $\boldsymbol{\theta}^I \in \boldsymbol{\Theta}_h$, $w^I \in W_h$ *be arbitrary, and define* $\boldsymbol{\gamma}^I = \lambda t^{-2}(\mathbf{grad}\, w^I - \boldsymbol{\Pi}^{\boldsymbol{\Gamma}}\boldsymbol{\theta}^I) \in \boldsymbol{\Gamma}_h$. *Then*

$$\|\boldsymbol{\theta} - \boldsymbol{\theta}_h\|_1 + t\|\boldsymbol{\gamma} - \boldsymbol{\gamma}_h\|_0 \le C(\|\boldsymbol{\theta} - \boldsymbol{\theta}^I\|_1 + t\|\boldsymbol{\gamma} - \boldsymbol{\gamma}^I\|_0 + h\|\boldsymbol{\gamma} - \boldsymbol{\Pi}^0\boldsymbol{\gamma}\|_0).$$

Proof. Clearly

$$a(\boldsymbol{\theta} - \boldsymbol{\theta}_h, \boldsymbol{\phi}) + (\boldsymbol{\gamma} - \boldsymbol{\gamma}_h, \mathbf{grad}\, v - \boldsymbol{\Pi}^{\boldsymbol{\Gamma}}\boldsymbol{\phi}) = (\boldsymbol{\gamma}, [\boldsymbol{I} - \boldsymbol{\Pi}^{\boldsymbol{\Gamma}}]\boldsymbol{\phi}), \tag{29}$$

for all $\boldsymbol{\phi} \in \boldsymbol{\Theta}_h$ and $v \in W_h$, so

$$a(\boldsymbol{\theta}^I - \boldsymbol{\theta}_h, \boldsymbol{\phi}) + (\boldsymbol{\gamma}^I - \boldsymbol{\gamma}_h, \mathbf{grad}\, v - \boldsymbol{\Pi}^{\boldsymbol{\Gamma}}\boldsymbol{\phi}) = a(\boldsymbol{\theta}^I - \boldsymbol{\theta}, \boldsymbol{\phi}) + (\boldsymbol{\gamma}^I - \boldsymbol{\gamma}, \mathbf{grad}\, v - \boldsymbol{\Pi}^{\boldsymbol{\Gamma}}\boldsymbol{\phi}) + (\boldsymbol{\gamma}, [\boldsymbol{I} - \boldsymbol{\Pi}^{\boldsymbol{\Gamma}}]\boldsymbol{\phi}).$$

Taking $\boldsymbol{\phi} = \boldsymbol{\phi}^I - \boldsymbol{\phi}_h$ and $v = w^I - w_h$, noting that $\mathbf{grad}\, w^I - \boldsymbol{\Pi}^{\boldsymbol{\Gamma}}\boldsymbol{\theta}^I = \lambda^{-1}t^2\boldsymbol{\gamma}^I$ and $\mathbf{grad}\, w_h - \boldsymbol{\Pi}^{\boldsymbol{\Gamma}}\boldsymbol{\theta}_h = \lambda^{-1}t^2\boldsymbol{\gamma}_h$, and using (28), we get the identity

$$a(\boldsymbol{\theta}^I - \boldsymbol{\theta}_h, \boldsymbol{\theta}^I - \boldsymbol{\theta}_h) + \lambda^{-1}t^2(\boldsymbol{\gamma}^I - \boldsymbol{\gamma}_h, \boldsymbol{\gamma}^I - \boldsymbol{\gamma}_h) = a(\boldsymbol{\theta}^I - \boldsymbol{\theta}, \boldsymbol{\theta}^I - \boldsymbol{\theta}_h) + \lambda^{-1}t^2(\boldsymbol{\gamma}^I - \boldsymbol{\gamma}, \boldsymbol{\gamma}^I - \boldsymbol{\gamma}_h) + (\boldsymbol{\gamma}, [\boldsymbol{I} - \boldsymbol{\Pi}^{\boldsymbol{\Gamma}}][\boldsymbol{\theta}^I - \boldsymbol{\theta}_h]).$$

Using Schwarz's inequality, and (27) and (28), we can bound the last term:

$$|(\boldsymbol{\gamma}, [\boldsymbol{I} - \boldsymbol{\Pi}^{\boldsymbol{\Gamma}}][\boldsymbol{\theta}^I - \boldsymbol{\theta}_h])| \le Ch\|\boldsymbol{\gamma} - \boldsymbol{\Pi}^0\boldsymbol{\gamma}\|_0\|\boldsymbol{\theta}^I - \boldsymbol{\theta}_h\|_1.$$

The theorem then follows easily.

Note that if we apply this theorem in a naive way, then the error estimates we obtain will blow up as $t \to 0$. More specifically, if we use the simple estimate

$$t\|\boldsymbol{\gamma} - \boldsymbol{\gamma}^I\| = \lambda t^{-1}\|\,\mathbf{grad}(w - w^I) - (\boldsymbol{\theta} - \Pi^{\boldsymbol{\Gamma}}\boldsymbol{\theta}^I)\| \le \lambda t^{-1}(\|\,\mathbf{grad}(w - w^I)\| + \|\boldsymbol{\theta} - \Pi^{\boldsymbol{\Gamma}}\boldsymbol{\theta}^I\|),$$

and use approximation theory to bound each of the terms on the right separately, then the bound will contain the term t^{-1}.

The key idea to using this theorem to obtain error estimates that are independent of the plate thickness t is to find functions $\boldsymbol{\theta}^I \in \boldsymbol{\Theta}_h$ and $w^I \in W_h$ that satisfy

$$\boldsymbol{\gamma}^I = \lambda t^{-2}(\mathbf{grad}\, w^I - \boldsymbol{\Pi}^{\boldsymbol{\Gamma}}\boldsymbol{\theta}^I) = \boldsymbol{\Pi}^{\boldsymbol{\Gamma}}\boldsymbol{\gamma}. \tag{30}$$

We then have the following corollary.

Corollary 7.2 *If* $\boldsymbol{\theta}^I \in \boldsymbol{\Theta}_h$ *and* $w^I \in W_h$ *satisfy* (30), *then*

$$\|\boldsymbol{\theta} - \boldsymbol{\theta}_h\|_1 + t\|\boldsymbol{\gamma} - \boldsymbol{\gamma}_h\|_0 \le C(\|\boldsymbol{\theta} - \boldsymbol{\theta}^I\|_1 + t\|\boldsymbol{\gamma} - \boldsymbol{\Pi}^{\boldsymbol{\Gamma}}\boldsymbol{\gamma}\|_0 + h\|\boldsymbol{\gamma} - \boldsymbol{\Pi}^0\boldsymbol{\gamma}\|_0).$$

If we also make assumptions about the approximation properties of the functions $\boldsymbol{\theta}^I$, w^I, and $\boldsymbol{\Pi}^{\boldsymbol{\Gamma}}\boldsymbol{\gamma}$, we immediately obtain order of convergence estimates. One such result is the following.

Theorem 7.3 *Let $n \geq 1$ and assume for each $\boldsymbol{\theta} \in \boldsymbol{H}^{n+1}(\Omega) \cap \mathring{\boldsymbol{H}}^1(\Omega)$ and $w \in H^{n+2}(\Omega) \cap \mathring{H}^1(\Omega)$, there exists $\boldsymbol{\theta}^I \in \boldsymbol{\Theta}_h$ and $w^I \in W_h$ satisfying* (30). *If for $1 \leq r \leq n$,*

$$\|\boldsymbol{\theta} - \boldsymbol{\theta}^I\|_1 \leq Ch^r \|\boldsymbol{\theta}\|_{r+1}, \tag{31}$$

$$\|\boldsymbol{\gamma} - \boldsymbol{\Pi}^{\boldsymbol{\Gamma}} \boldsymbol{\gamma}\|_0 \leq Ch^r \|\boldsymbol{\gamma}\|_r, \tag{32}$$

then

$$\|\boldsymbol{\theta} - \boldsymbol{\theta}_h\|_1 + t\|\boldsymbol{\gamma} - \boldsymbol{\gamma}_h\|_0 \leq C\left(h^r\|\boldsymbol{\theta}\|_{r+1} + h^r t\|\boldsymbol{\gamma}\|_r + h^{r_0+2}\|\boldsymbol{\gamma}\|_{r_0+1}\right).$$

Proof. The proof follows immediately from the hypotheses of the theorem and standard approximation properties of $\boldsymbol{\Pi}^0$.

We now state and prove an abstract estimate for the L^2 errors for the rotation and the transverse displacement. To do so, we first define an appropriate dual problem. Given $\boldsymbol{F} \in \boldsymbol{L}^2(\Omega)$ and $G \in L^2(\Omega)$, define $\boldsymbol{\psi}$, u, and $\boldsymbol{\zeta}$ to be the solution to the auxiliary problem

$$a(\boldsymbol{\phi}, \boldsymbol{\psi}) + (\mathbf{grad}\, v - \boldsymbol{\phi}, \boldsymbol{\zeta}) = (\boldsymbol{\phi}, \boldsymbol{F}) + (v, G), \quad \boldsymbol{\phi} \in \mathring{\boldsymbol{H}}^1, v \in \mathring{H}^1(\Omega), \tag{33}$$

$$(\boldsymbol{\eta}, \mathbf{grad}\, u - \boldsymbol{\psi}) - \lambda^{-1} t^2 (\boldsymbol{\eta}, \boldsymbol{\zeta}) = 0, \quad \boldsymbol{\eta} \in \boldsymbol{L}^2(\Omega). \tag{34}$$

Then by the regularity results (21) and (22),

$$\|\boldsymbol{\psi}\|_2 + \|u\|_2 + \|\boldsymbol{\zeta}\| + t\|\boldsymbol{\zeta}\|_1 + \|\operatorname{div} \boldsymbol{\zeta}\|_0 \leq c(\|\boldsymbol{F}\|_0 + \|G\|_0). \tag{35}$$

With these definitions we have the following estimate.

Theorem 7.4 *If the hypotheses of Theorems 7.1 and 7.3 are satisfied, then*

$$\begin{aligned}\|\boldsymbol{\theta} - \boldsymbol{\theta}_h\|^2/2 + \|w - w_h\|_0^2/2 \leq{} & Ch^2(\|\boldsymbol{\theta} - \boldsymbol{\theta}_h\|_1^2 + t^2\|\boldsymbol{\gamma} - \boldsymbol{\gamma}_h\|_0^2) \\ & + ([\boldsymbol{I} - \boldsymbol{\Pi}^{\boldsymbol{\Gamma}}]\boldsymbol{\theta}_h, \boldsymbol{\zeta}) + (\boldsymbol{\gamma}, [\boldsymbol{I} - \boldsymbol{\Pi}^{\boldsymbol{\Gamma}}]\boldsymbol{\psi}^I).\end{aligned} \tag{36}$$

Proof. Let $\boldsymbol{F} = \boldsymbol{\theta} - \boldsymbol{\theta}_h$ and $G = (w - w_h)$. Then, setting $\boldsymbol{\phi} = \boldsymbol{\theta} - \boldsymbol{\theta}_h$, $v = w - w_h$ in (34) and using the definitions of $\boldsymbol{\gamma}$ and $\boldsymbol{\gamma}_h$ we get

$$\|\boldsymbol{\theta} - \boldsymbol{\theta}_h\|_0^2 + \|w - w_h\|_0^2 = a(\boldsymbol{\theta} - \boldsymbol{\theta}_h, \boldsymbol{\psi}) + \lambda^{-1}t^2(\boldsymbol{\gamma} - \boldsymbol{\gamma}_h, \boldsymbol{\zeta}) + ([\boldsymbol{I} - \boldsymbol{\Pi}^{\boldsymbol{\Gamma}}]\boldsymbol{\theta}_h, \boldsymbol{\zeta}). \tag{37}$$

Now, the error equation (29) gives

$$a(\boldsymbol{\theta} - \boldsymbol{\theta}_h, \boldsymbol{\psi}^I) + \lambda^{-1}t^2(\boldsymbol{\gamma} - \boldsymbol{\gamma}_h, \boldsymbol{\zeta}^I) = (\boldsymbol{\gamma}, [\boldsymbol{I} - \boldsymbol{\Pi}^{\boldsymbol{\Gamma}}]\bar{\boldsymbol{\psi}})$$

where $\boldsymbol{\zeta}^I = \lambda t^{-2}(\mathbf{grad}\, u^I - \boldsymbol{\Pi}^{\boldsymbol{\Gamma}}\boldsymbol{\psi}^I)$, so (37) becomes

$$\begin{aligned}\|\boldsymbol{\theta} - \boldsymbol{\theta}_h\|^2 + \|w - w_h\|_0^2 = {} & a(\boldsymbol{\theta} - \boldsymbol{\theta}_h, \boldsymbol{\psi} - \boldsymbol{\psi}^I) \\ & + \lambda^{-1}t^2(\boldsymbol{\gamma} - \boldsymbol{\gamma}_h, \boldsymbol{\zeta} - \boldsymbol{\zeta}^I) + ([\boldsymbol{I} - \boldsymbol{\Pi}^{\boldsymbol{\Gamma}}]\boldsymbol{\theta}_h, \boldsymbol{\zeta}) + (\boldsymbol{\gamma}, [\boldsymbol{I} - \boldsymbol{\Pi}^{\boldsymbol{\Gamma}}]\boldsymbol{\psi}^I).\end{aligned} \tag{38}$$

The first two terms on the right side of (38) are easily bounded by

$$\begin{aligned} C(\|\boldsymbol{\theta}-\boldsymbol{\theta}_h\|_1\|\boldsymbol{\psi}-\boldsymbol{\psi}^I\|_1+t^2\|\boldsymbol{\gamma}-\boldsymbol{\gamma}_h\|_0\|\boldsymbol{\zeta}-\boldsymbol{\zeta}^I\|_0)\\ \le Ch(\|\boldsymbol{\theta}-\boldsymbol{\theta}_h\|_1+t\|\boldsymbol{\gamma}-\boldsymbol{\gamma}_h\|_0)(\|\boldsymbol{\psi}\|_2+t\|\boldsymbol{\zeta}\|_1)\\ \le Ch(\|\boldsymbol{\theta}-\boldsymbol{\theta}_h\|_1+t\|\boldsymbol{\gamma}-\boldsymbol{\gamma}_h\|_0)(\|\boldsymbol{\theta}-\boldsymbol{\theta}_h\|_0+\|w-w_h\|_0). \qquad (39) \end{aligned}$$

Application of the arithmetic–geometric mean inequality establishes the result.

Remark 7.1. Bounds on the last two terms will depend on the particular method being analyzed.

Next, we establish an abstract estimate for the approximation of the derivatives of the transverse displacement.

Theorem 7.5 *For all $w_I \in W_h$, we have*

$$\begin{aligned} &\|\,\mathbf{grad}[w-w_h]\|_0\\ &\quad\le C(\|\,\mathbf{grad}[w-w_I]\|_0+\|[\boldsymbol{I}-\boldsymbol{\Pi}^\Gamma]\boldsymbol{\theta}\|_0+h\|\boldsymbol{\theta}-\boldsymbol{\theta}_h\|_1+\|\boldsymbol{\theta}-\boldsymbol{\theta}_h\|_0). \end{aligned}$$

Proof. Choosing $\boldsymbol{\eta} = \mathbf{grad}\, v_h$, $v_h \in W_h$, we get for all $w_I \in W_h$,

$$(\mathbf{grad}[w_I-w_h],\mathbf{grad}\,v_h)=(\mathbf{grad}[w_I-w],\mathbf{grad}\,v_h)+(\boldsymbol{\theta}-\boldsymbol{\Pi}^\Gamma\boldsymbol{\theta}_h,\mathbf{grad}\,v_h).$$

Then choosing $v_h = w_h - w_I$, it easily follows that

$$\begin{aligned} &\|\,\mathbf{grad}[w_I-w_h]\|_0 \le \|\,\mathbf{grad}[w_I-w]\|_0+\|\boldsymbol{\theta}-\boldsymbol{\Pi}^\Gamma\boldsymbol{\theta}_h\|_0\\ &\le \|\,\mathbf{grad}[w_I-w]\|_0+\|[\boldsymbol{I}-\boldsymbol{\Pi}^\Gamma]\boldsymbol{\theta}\|_0+\|[\boldsymbol{I}-\boldsymbol{\Pi}^\Gamma][\boldsymbol{\theta}_h-\boldsymbol{\theta}]\|_0+\|\boldsymbol{\theta}-\boldsymbol{\theta}_h\|_0\\ &\le \|\,\mathbf{grad}[w_I-w]\|_0+\|[\boldsymbol{I}-\boldsymbol{\Pi}^\Gamma]\boldsymbol{\theta}\|_0+Ch\|\boldsymbol{\theta}-\boldsymbol{\theta}_h\|_1+\|\boldsymbol{\theta}-\boldsymbol{\theta}_h\|_0. \end{aligned}$$

The result follows from the triangle inequality.

In some cases, it is also possible to establish improved estimates for the shear stress $\boldsymbol{\gamma}$ in negative norms. We will not derive such estimates here, but will state known results in some cases.

8 Applications of the Abstract Error Estimates

Most of our discussion will be centered on triangular elements. We will henceforth assume that Ω is a convex polygonal domain in the plane, and we let $\mathcal{T}_h$ denote a triangulation of Ω. Let **V** and **E** denote the set of vertices and edges, respectively in the mesh $\mathcal{T}_h$. We will use the following finite element spaces based on the mesh $\mathcal{T}_h$.

$M_k(\mathcal{T}_h)$:	arbitrary piecewise polynomials of degree $\leq k$,
$M_k^l(\mathcal{T}_h)$:	$M_k \cap C^l(\Omega)$,
$M_k^*(\mathcal{T}_h)$:	elements of M_k continuous at k Gauss-points of each interelement edge,
$B_k(\mathcal{T}_h)$:	elements of M_k^0 which vanish on interelement edges,
$\boldsymbol{RT}_k^\perp(\mathcal{T}_h)$:	Raviart–Thomas discretization of order k to $\boldsymbol{H}(\mathrm{rot}, \Omega)$,
$\boldsymbol{BDM}_k^\perp(\mathcal{T}_h)$:	Brezzi–Douglas–Marini discretization of order k to $\boldsymbol{H}(\mathrm{rot}, \Omega)$,
$\boldsymbol{BDFM}_k^\perp(\mathcal{T}_h)$:	Brezzi–Douglas–Fortin–Marini discretization of order k to $\boldsymbol{H}(\mathrm{rot}, \Omega)$.

When there is no risk of confusion, we write M_k for $M_k(\mathcal{T}_h)$, etc. For the scalar-valued function spaces in this list, we have vector-valued analogues in the obvious way. For example, $\boldsymbol{M}_k := M_k \times M_k$. Note that $B_k = 0$ for $k < 3$. For convenience, we interpret M_{-1} as the zero space.

The degrees of freedom for each space determine an interpolation operator from $C^\infty(\Omega)$ or $\boldsymbol{C}^\infty(\Omega)$ into the corresponding space. We denote these operators Π^{M_k}, etc. The operators Π^{M_k} and Π^{B_k} extend boundedly to L^2; the operators $\Pi^{M_k^0}$ extend boundedly to $W_p^1(\Omega)$ for any $p > 2$; the other interpolation operators extend boundedly to H^1 or $\boldsymbol{H}^1$ (these are not the largest possible domain spaces). With each space we have a corresponding space in which all degrees of freedom associated with edges or vertices contained in the boundary are set equal to zero. Thus $\mathring{M}_k = M_k \cap \mathring{H}^1$.

We will now consider some specific choices of the subspaces in the general method (25).

8.1 The Durán–Liberman Element [33]

See also [25, p. 145]. This element corresponds to the choices

$$\boldsymbol{\Theta}_h = \{\boldsymbol{\phi} \in \mathring{\boldsymbol{M}}_2^0 \,|\, \boldsymbol{\phi} \cdot \boldsymbol{n} \in P_1(e), e \in \mathbf{E}\}, \quad W_h = \mathring{M}_1^0, \quad \boldsymbol{\Gamma}_h = \boldsymbol{RT}_0^\perp,$$

depicted in the element diagram below. We then take $\boldsymbol{\Pi}^{\boldsymbol{\Gamma}}$ to the usual interpolant into $\boldsymbol{RT}_0^\perp$ defined for $\gamma \in \boldsymbol{H}^1(\Omega)$ by

$$\int_e \boldsymbol{\Pi}^{\boldsymbol{\Gamma}} \gamma \cdot \boldsymbol{s} \, \mathrm{d}s = \int_e \gamma \cdot \boldsymbol{s} \, \mathrm{d}s, \qquad e \in \mathbf{E}.$$

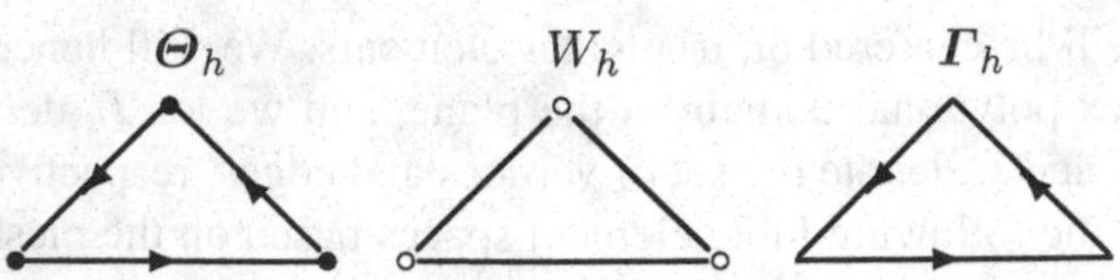

Durán–Liberman

We then get the following error estimate.

Theorem 8.1

$$\|\boldsymbol{\theta} - \boldsymbol{\theta}_h\|_1 + t\|\gamma - \gamma_h\|_0 + \|w - w_h\|_1 \leq Ch(\|\boldsymbol{f}\|_0 + \|g\|_0).$$

Proof. Using standard approximation properties of the space $\boldsymbol{\Theta}_h$, we may find a function $\boldsymbol{\theta}^I$ satisfying (31) with $n = 1$ and the condition $\int_e \boldsymbol{\theta}^I \cdot \boldsymbol{s}\,\mathrm{d}s = \int_e \boldsymbol{\theta} \cdot \boldsymbol{s}\,\mathrm{d}s$ on each edge e. Then

$$\int_e \boldsymbol{\Pi}^{\Gamma} \boldsymbol{\theta}^I \cdot \boldsymbol{s}\,\mathrm{d}s = \int_e \boldsymbol{\theta}^I \cdot \boldsymbol{s}\,\mathrm{d}s = \int_e \boldsymbol{\theta} \cdot \boldsymbol{s}\,\mathrm{d}s = \int_e \boldsymbol{\Pi}^{\Gamma} \boldsymbol{\theta} \cdot \boldsymbol{s}\,\mathrm{d}s,$$

so

$$\boldsymbol{\Pi}^{\Gamma} \boldsymbol{\theta}^I = \boldsymbol{\Pi}^{\Gamma} \boldsymbol{\theta}.$$

Next observe that if $\Pi^W w$ is the standard piecewise linear interpolant of w, and e is the edge joining vertices v_a and v_b, then

$$\begin{aligned}\int_e \mathbf{grad}\, \Pi^W w \cdot \boldsymbol{s}\,\mathrm{d}s &= \int_e \partial \Pi^W w / \partial s\,\mathrm{d}s = \Pi^W w(v_b) - \Pi^W w(v_a) \\ &= w(v_b) - w(v_a) = \int_e \partial w/\partial s\,\mathrm{d}s = \int_e \mathbf{grad}\, w \cdot \boldsymbol{s}\,\mathrm{d}s, \qquad (40)\end{aligned}$$

so

$$\mathbf{grad}\, \Pi^W w = \boldsymbol{\Pi}^{\Gamma}\, \mathbf{grad}\, w.$$

If we choose $w^I = \Pi^W w$, then $\gamma^I = \boldsymbol{\Pi}^{\Gamma} \gamma$, so (30) is satisfied and Theorem 7.3 is satisfied with $n = 1$. Since (28) is satisfied with $r_0 = -1$, the first two estimates of the theorem follow directly from Theorem 7.3 and the a priori estimate (21). The final estimate is an easy consequence of Theorem 7.5.

To obtain L^2 estimates, we apply Theorem 7.4. In this regard, the following technical lemma will be useful.

Lemma 8.2 *(cf. [32]) For $\psi \in \mathring{H}^1(\Omega)$, denote by ψ_c a piecewise linear approximation to ψ satisfying*

$$\|\psi_c\|_1 \leq C\|\psi\|_1, \qquad \|\psi - \psi_c\|_1 \leq Ch\|\psi\|_2.$$

Then for all $\boldsymbol{\zeta} \in \boldsymbol{H}(\mathrm{div}, \Omega)$

$$|(\boldsymbol{\zeta}, \psi_c - \boldsymbol{\Pi}^{\Gamma} \psi_c)| \leq Ch^2 \|\operatorname{div} \boldsymbol{\zeta}\|_0 \|\psi\|_1.$$

Theorem 8.3

$$\|\boldsymbol{\theta} - \boldsymbol{\theta}_h\|_0 + \|w - w_h\|_0 \leq Ch^2(\|\boldsymbol{f}\|_0 + \|g\|_0).$$

Proof. Estimates for the first two terms on the right side of (36) are given by Theorem 8.1. For the third term in (36), let $\boldsymbol{\theta}_c$ be an approximation to $\boldsymbol{\theta}$ satisfying the hypotheses of Lemma 8.2 and write

$$([\boldsymbol{I} - \boldsymbol{\Pi}^{\Gamma}]\boldsymbol{\theta}_h, \boldsymbol{\zeta}) = ([\boldsymbol{I} - \boldsymbol{\Pi}^{\Gamma}][\boldsymbol{\theta}_h - \boldsymbol{\theta}_c], \boldsymbol{\zeta}) + ([\boldsymbol{I} - \boldsymbol{\Pi}^{\Gamma}]\boldsymbol{\theta}_c, \boldsymbol{\zeta}).$$

From Lemma 8.2 we have

$$([\boldsymbol{I} - \boldsymbol{\Pi}^{\Gamma}]\boldsymbol{\theta}_c, \boldsymbol{\zeta}) \leq Ch^2 \|\operatorname{div} \boldsymbol{\zeta}\|_0 \|\boldsymbol{\theta}\|_1,$$

and using Lemma 8.2 and Theorem 8.1, we have

$$\begin{aligned}([\boldsymbol{I} - \boldsymbol{\Pi}^{\Gamma}][\boldsymbol{\theta}_h - \boldsymbol{\theta}_c], \boldsymbol{\zeta}) &\leq Ch\|\boldsymbol{\zeta}\|_0 \|\boldsymbol{\theta}_h - \boldsymbol{\theta}_c\|_1 \\ &\leq Ch\|\boldsymbol{\zeta}\|_0 (\|\boldsymbol{\theta}_h - \boldsymbol{\theta}\|_1 + \|\boldsymbol{\theta} - \boldsymbol{\theta}_c\|_1) \leq Ch^2(\|\boldsymbol{\theta}\|_2 + \|\boldsymbol{f}\|_0 + \|g\|_0)\|\boldsymbol{\zeta}\|_0.\end{aligned}$$

Combining these results and applying (21) and (35), we get

$$|([\boldsymbol{I} - \boldsymbol{\Pi}^{\Gamma}]\boldsymbol{\theta}_h, \boldsymbol{\zeta})| \leq Ch^2(\|\boldsymbol{f}\|_0 + \|g\|_0)\|\boldsymbol{\theta} - \boldsymbol{\theta}_h\|_0.$$

We bound the last term in (36) in an analogous manner, obtaining

$$|(\boldsymbol{\gamma}, [\boldsymbol{I} - \boldsymbol{\Pi}^{\Gamma}]\boldsymbol{\psi}^I)| \leq Ch^2(\|\boldsymbol{f}\|_0 + \|g\|_0)\|\boldsymbol{\theta} - \boldsymbol{\theta}_h\|_0.$$

The theorem follows directly by combining these results.

We note that it is also possible to show that

$$\|\boldsymbol{\gamma} - \boldsymbol{\gamma}_h\|_{-1} \leq Ch(\|\boldsymbol{f}\|_0 + \|g\|_0).$$

8.2 The MITC Triangular Families

See [23], [25], and [44] for analysis of these methods and [19] for some experimental results. There are three triangular families considered in [25], defined for integer $k \geq 2$. For each of these families, the space $\boldsymbol{\Theta}_h$ is chosen to be

$$\boldsymbol{\Theta}_h = \mathring{\boldsymbol{M}}_k^0 + \boldsymbol{B}_{k+1}, \quad k = 2, 3, \qquad \boldsymbol{\Theta}_h = \mathring{\boldsymbol{M}}_k^0, \quad k \geq 4.$$

We then define

$$\begin{aligned}&\text{Family I:} && W_h = \mathring{M}_k^0, && \boldsymbol{\Gamma}_h = \boldsymbol{RT}_{k-1}^{\perp}, \\ &\text{Family II:} && W_h = \mathring{M}_k^0 + B_{k+1}, && \boldsymbol{\Gamma}_h = \boldsymbol{BDFM}_k^{\perp}, \\ &\text{Family III:} && W_h = \mathring{M}_{k+1}^0 && \boldsymbol{\Gamma}_h = \boldsymbol{BDM}_k^{\perp},\end{aligned}$$

and choose $\boldsymbol{\Pi}^{\Gamma}$ to be the usual interpolant into each $\boldsymbol{\Gamma}_h$ space.

The MITC elements are based on a common idea expressed in [23], i.e., "to combine in a proper way some known results on the approximation of Stokes problems with other known results on the approximation of linear elliptic problems." This combination is summarized in a list of five properties relating the spaces $\boldsymbol{\Theta}_h$, W_h, $\boldsymbol{\Gamma}_h$, and an auxiliary space Q_h (not part of the method). These properties are:

P1 $\mathbf{grad}\, W_h \subset \boldsymbol{\Gamma}_h$.
P2 $\operatorname{rot} \boldsymbol{\Gamma}_h \subset Q_h$.

P3 $\operatorname{rot} \boldsymbol{\Pi}^{\Gamma} \boldsymbol{\phi} = \Pi^0 \operatorname{rot} \boldsymbol{\phi}$, for $\boldsymbol{\phi} \in \mathring{\boldsymbol{H}}^1(\Omega)$, with $\Pi^0 : L^2_0(\Omega) \mapsto Q_h$ denoting the L^2-projection ($L^2_0(\Omega)$ denotes functions in $L^2(\Omega)$ with mean value zero.)
P4 If $\eta \in \boldsymbol{\Gamma}_h$ satisfies $\operatorname{rot} \boldsymbol{\eta} = 0$, then $\boldsymbol{\eta} = \mathbf{grad}\, v$ for some $v \in W_h$.
P5 $(\boldsymbol{\Theta}_h^{\perp}, Q_h)$ is a stable pair for the Stokes problem, i.e.,

$$\sup_{0 \neq \boldsymbol{\phi} \in \boldsymbol{\Theta}_h} \frac{(\operatorname{rot} \boldsymbol{\phi}, q)}{\|\boldsymbol{\phi}\|_1} \geq C\|q\|_0, \qquad q \in Q_h.$$

For each of the three families described above, we define the space

$$Q_h = \{q \in L^2_0(\Omega) : q_T \in P_{k-1}(T),\ T \in \mathcal{T}_h\}.$$

For this choice, the fact that the pair of spaces $(\boldsymbol{\Theta}_h, Q_h)$ satisfies **P5** follows from the corresponding results known for the Stokes equation.

Although these families are only defined for $k \geq 2$, it is interesting to see what the difficulties are in extending them to the case $k = 1$. Most obvious is that $\boldsymbol{B}_{k+1}$ is only defined for $k \geq 2$, so this space must be replaced. A suitable replacement space for $\boldsymbol{\Theta}_h$ in Family I is the one chosen in the Durán–Liberman element. With this choice, the Durán–Liberman element also fits this general framework, with $k = 1$. For Family II, a similar problem occurs for the choice of W_h and in addition $\boldsymbol{BDFM}_1^{\perp} = \boldsymbol{RT}_0^{\perp}$, so the method needs substantial change and does not give anything new. For Family III, the choices $W_h = \mathring{M}_2^0$ and $\boldsymbol{\Gamma}_h = \boldsymbol{BDM}_1^{\perp}$ make sense, and one can choose $\boldsymbol{\Theta}_h = \mathring{\boldsymbol{M}}_2^0$. This would correspond to the choice of piecewise constants for Q_h and the $\boldsymbol{P}_2 - P_0$ Stokes element. An element of this type is mentioned in [23] (page 1798). This element, which we label MITC6 is depicted below along with MITC7, the $k = 2$ element of Family II.

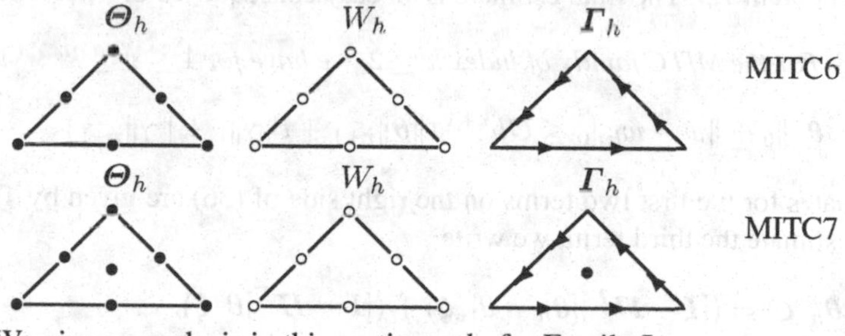

We give an analysis in this section only for Family I:

$$\boldsymbol{\Theta}_h = \begin{cases} \mathring{\boldsymbol{M}}_k^0 + \boldsymbol{B}_{k+1} & k = 2, 3 \\ \mathring{\boldsymbol{M}}_k^0 & k \geq 4 \end{cases}, \qquad W_h = \mathring{M}_k^0, \qquad \boldsymbol{\Gamma}_h = \boldsymbol{RT}_{k-1}^{\perp}.$$

The analysis of the other two families can be done in a similar manner.

Theorem 8.4 *For the MITC family of index $k \geq 2$, we have for $1 \leq r \leq k$*

$$\|\boldsymbol{\theta} - \boldsymbol{\theta}_h\|_1 + t\|\boldsymbol{\gamma} - \boldsymbol{\gamma}_h\|_0 + \|w - w_h\|_1 \leq Ch^r \left(\|\boldsymbol{\theta}\|_{r+1} + t\|\boldsymbol{\gamma}\|_r + \|\boldsymbol{\gamma}\|_{r-1}\right).$$

Proof. Using standard results about stable Stokes elements, we can find an interpolant $\boldsymbol{\theta}^I$ of $\boldsymbol{\theta} \in \boldsymbol{\Theta}_h$ satisfying (31) with $n = k$ and

$$\int_\Omega \operatorname{rot}(\boldsymbol{\theta} - \boldsymbol{\theta}^I)\, q \,\mathrm{d}x = 0, \quad \forall q \in M_{k-1}^{-1}.$$

By the definition of $\boldsymbol{\Pi}^{\boldsymbol{\Gamma}}$, we have $\forall q \in M_{k-1}^{-1}$

$$0 = \int_\Omega \operatorname{rot}(\boldsymbol{\theta} - \boldsymbol{\theta}^I)\, q \,\mathrm{d}x = \int_\Omega \operatorname{rot} \boldsymbol{\Pi}^{\boldsymbol{\Gamma}}(\boldsymbol{\theta} - \boldsymbol{\theta}^I)\, q \,\mathrm{d}x.$$

Choosing $q = \operatorname{rot} \boldsymbol{\Pi}^{\boldsymbol{\Gamma}}(\boldsymbol{\theta} - \boldsymbol{\theta}^I)$ implies $\operatorname{rot} \boldsymbol{\Pi}^{\boldsymbol{\Gamma}}(\boldsymbol{\theta} - \boldsymbol{\theta}^I) = 0$. Hence,

$$\boldsymbol{\Pi}^{\boldsymbol{\Gamma}}(\boldsymbol{\theta} - \boldsymbol{\theta}^I) = \mathbf{grad}\, v^I, \qquad \text{for some } v^I \in W_h.$$

Let $\Pi^W w \in M_0^k$ be the interpolant of w defined for each vertex x, edge e and triangle T by

$$\Pi^W w(x) = w(x), \qquad \int_e \Pi^W w\, p \,\mathrm{d}s = \int_e w\, p \,\mathrm{d}s, \quad \text{forall } p \in P_{k-2}(e), \tag{41}$$

$$\int_T \Pi^W w p \,\mathrm{d}x = \int_T w p \,\mathrm{d}x, \quad \text{forall } p \in P_{k-3}(T). \tag{42}$$

It is easy to check that $\boldsymbol{\Pi}^{\boldsymbol{\Gamma}}(\mathbf{grad}\, w) = \mathbf{grad}\, \Pi^W w$. Hence, (30) is satisfied with $w^I = \Pi^W w - v^I$. By the definition of the space $\boldsymbol{\Gamma}_h$, (32) is satisfied with $n = k$ and (28) is satisfied with $r_0 = k - 2$. The estimate for the first two terms follows directly from Theorem 7.3. The final estimate is an easy consequence of Theorem 7.5.

Theorem 8.5 *For the MITC family of index $k \geq 2$, we have for $1 \leq r \leq k$*

$$\|\boldsymbol{\theta} - \boldsymbol{\theta}_h\|_0 + \|w - w_h\|_0 \leq C h^{r+1} \left(\|\boldsymbol{\theta}\|_{r+1} + t\|\gamma\|_r + \|\gamma\|_{r-1} \right).$$

Proof. Estimates for the first two terms on the right side of (36) are given by Theorem 8.4. To estimate the third term, we write

$$\begin{aligned}
([\boldsymbol{I} - \boldsymbol{\Pi}^{\boldsymbol{\Gamma}}]\boldsymbol{\theta}_h, \boldsymbol{\zeta}) &= ([\boldsymbol{I} - \boldsymbol{\Pi}^{\boldsymbol{\Gamma}}][\boldsymbol{\theta}_h - \boldsymbol{\theta}], \boldsymbol{\zeta}) + ([\boldsymbol{I} - \boldsymbol{\Pi}^{\boldsymbol{\Gamma}}]\boldsymbol{\theta}, \boldsymbol{\zeta}). \\
&= ([\boldsymbol{I} - \boldsymbol{\Pi}^{\boldsymbol{\Gamma}}][\boldsymbol{\theta}_h - \boldsymbol{\theta}], \boldsymbol{\zeta}) - \lambda^{-1} t^2 (\boldsymbol{I} - \boldsymbol{\Pi}^{\boldsymbol{\Gamma}})\gamma, \zeta) + ([\boldsymbol{I} - \boldsymbol{\Pi}^{\boldsymbol{\Gamma}}]\, \mathbf{grad}\, w, \zeta) \\
&= ([\boldsymbol{I} - \boldsymbol{\Pi}^{\boldsymbol{\Gamma}}][\boldsymbol{\theta}_h - \boldsymbol{\theta}], \boldsymbol{\zeta}) - \lambda^{-1} t^2 ([\boldsymbol{I} - \boldsymbol{\Pi}^{\boldsymbol{\Gamma}}]\gamma, \zeta) \\
&\quad + (\mathbf{grad}\, w - \mathbf{grad}\, \Pi^W w, \zeta) \\
&= ([\boldsymbol{I} - \boldsymbol{\Pi}^{\boldsymbol{\Gamma}}][\boldsymbol{\theta}_h - \boldsymbol{\theta}], \boldsymbol{\zeta}) - \lambda^{-1} t^2 ([\boldsymbol{I} - \boldsymbol{\Pi}^{\boldsymbol{\Gamma}}]\gamma, [\boldsymbol{I} - \boldsymbol{\Pi}^{M^0}]\zeta) \\
&\quad - (w - \Pi^W w, \operatorname{div} \boldsymbol{\zeta}).
\end{aligned}$$

Hence,

$$\begin{aligned}
|([\boldsymbol{I} - \boldsymbol{\Pi}^{\boldsymbol{\Gamma}}]\boldsymbol{\theta}_h, \boldsymbol{\zeta})| &\le \|[\boldsymbol{I} - \boldsymbol{\Pi}^{\boldsymbol{\Gamma}}][\boldsymbol{\theta}_h - \boldsymbol{\theta}]\|_0 \|\boldsymbol{\zeta}\|_0 \\
&\quad + \lambda^{-1} t^2 \|[\boldsymbol{I} - \boldsymbol{\Pi}^{\boldsymbol{\Gamma}}]\boldsymbol{\gamma}\|_0 \|[\boldsymbol{I} - \boldsymbol{\Pi}^{M^0}]\boldsymbol{\zeta}\|_0 + \|w - \Pi^W w\|_0 \|\operatorname{div} \boldsymbol{\zeta}\|_0 \\
&\le Ch \Big(\|\boldsymbol{\theta}_h - \boldsymbol{\theta}\|_1 \|\boldsymbol{\zeta}\|_0 + t \|[\boldsymbol{I} - \boldsymbol{\Pi}^{\boldsymbol{\Gamma}}]\boldsymbol{\gamma}\|_0 t \|\boldsymbol{\zeta}\|_1 \\
&\quad + h^{-1} \|w - \Pi^W w\|_0 \|\operatorname{div} \boldsymbol{\zeta}\|_0 \Big).
\end{aligned}$$

To estimate the final term, we write

$$\begin{aligned}
(\boldsymbol{\gamma}, [\boldsymbol{I} - \boldsymbol{\Pi}^{\boldsymbol{\Gamma}}]\boldsymbol{\psi}^I) &= ([\boldsymbol{I} - \boldsymbol{\Pi}^0]\boldsymbol{\gamma}, [\boldsymbol{I} - \boldsymbol{\Pi}^{\boldsymbol{\Gamma}}]\boldsymbol{\psi}^I). \\
&= ([\boldsymbol{I} - \boldsymbol{\Pi}^0]\boldsymbol{\gamma}, [\boldsymbol{I} - \boldsymbol{\Pi}^{\boldsymbol{\Gamma}}][\boldsymbol{\psi}^I - \boldsymbol{\psi}]) + ([\boldsymbol{I} - \boldsymbol{\Pi}^0]\boldsymbol{\gamma}, [\boldsymbol{I} - \boldsymbol{\Pi}^{\boldsymbol{\Gamma}}]\boldsymbol{\psi}]).
\end{aligned}$$

Hence,

$$\begin{aligned}
&|(\boldsymbol{\gamma}, [\boldsymbol{I} - \boldsymbol{\Pi}^{\boldsymbol{\Gamma}}]\boldsymbol{\psi}^I)| \\
&\quad \le C \|[\boldsymbol{I} - \boldsymbol{\Pi}^0]\boldsymbol{\gamma}\|_0 (\|[\boldsymbol{I} - \boldsymbol{\Pi}^{\boldsymbol{\Gamma}}][\boldsymbol{\psi}^I - \boldsymbol{\psi}]\|_0 + \|[\boldsymbol{I} - \boldsymbol{\Pi}^{\boldsymbol{\Gamma}}]\boldsymbol{\psi}]\|_0) \\
&\quad \le Ch^2 \|[\boldsymbol{I} - \boldsymbol{\Pi}^0]\boldsymbol{\gamma}\|_0 \|\boldsymbol{\psi}\|_2.
\end{aligned}$$

The theorem now follows by combining these results and applying (35) and standard estimates.

8.3 The Falk–Tu Elements With Discontinuous Shear Stresses [35]

For $k = 2, 3, \ldots$ we choose

$$\boldsymbol{\Theta}_h = \mathring{\boldsymbol{M}}_{k-1}^0 + \boldsymbol{B}_{k+2}, \quad W_h = \mathring{M}_k^0, \quad \boldsymbol{\Gamma}_h = \boldsymbol{M}_{k-1},$$

and $\boldsymbol{\Pi}^{\boldsymbol{\Gamma}}$ to be the $\boldsymbol{L}^2$ projection into $\boldsymbol{\Gamma}_h$. See also the related element of Zienkiewicz–Lefebvre [52]. The element diagram for the lowest order Falk–Tu element ($k = 2$) is depicted below.

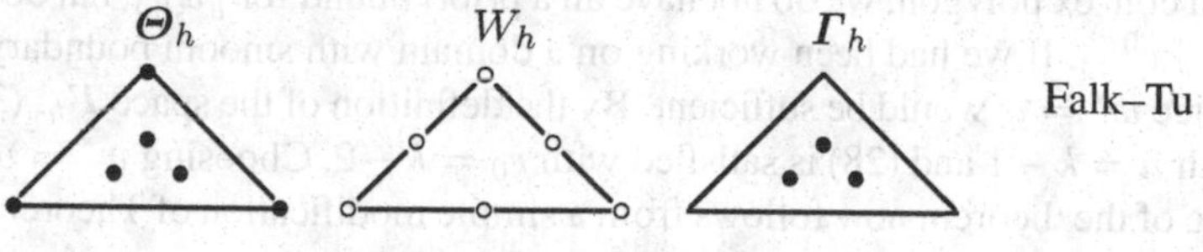

Theorem 8.6 *For the discontinuous shear stress family of index $k \ge 2$, we have for $1 \le r \le k - 1$*

$$\|\boldsymbol{\theta} - \boldsymbol{\theta}_h\|_1 + t\|\boldsymbol{\gamma} - \boldsymbol{\gamma}_h\|_0 \le Ch^r (\|\boldsymbol{\theta}\|_{r+1} + \|w\|_{r+2} + t\|\boldsymbol{\gamma}\|_r + \|\boldsymbol{\gamma}\|_{r-1}).$$

For $k = 2$ and $r = 1$, we also have the estimate

$$\begin{aligned}
&\|\boldsymbol{\theta} - \boldsymbol{\theta}_h\|_1 + t\|\boldsymbol{\gamma} - \boldsymbol{\gamma}_h\|_0 \\
&\quad \le Ch (\|\boldsymbol{\theta}\|_2 + \|w^0\|_3 + \|\boldsymbol{\gamma}\|_0 + t\|\boldsymbol{\gamma}\|_1 + t^{-1}\|w - w^0\|_2) \\
&\quad \le Ch (\|\boldsymbol{f}\|_0 + \|g\|_0).
\end{aligned}$$

Proof. For $1 \le r \le k-1$, let $\Pi^W w$ be a standard interpolant of w satisfying

$$\|w - \Pi^W w\|_0 + h\|w - \Pi^W w\|_1 \le Ch^{r+2}\|w\|_{r+2}$$

and $\boldsymbol{\Pi}^M \boldsymbol{\theta} \in \mathring{\boldsymbol{M}}_0^{k-1}$ a standard interpolant of $\boldsymbol{\theta}$ satisfying

$$\|\boldsymbol{\theta} - \boldsymbol{\Pi}^M \boldsymbol{\theta}\|_0 + h\|\boldsymbol{\theta} - \boldsymbol{\Pi}^M \boldsymbol{\theta}\|_1 \le Ch^{r+1}\|\boldsymbol{\theta}\|_{r+1}.$$

Define $\boldsymbol{\Pi}^B(\boldsymbol{\theta}, w^*) \in \boldsymbol{B}^{k+3}$ by

$$\boldsymbol{\Pi}^{\boldsymbol{\Gamma}} \boldsymbol{\Pi}^B(\boldsymbol{\theta}, w^*) = \boldsymbol{\Pi}^{\boldsymbol{\Gamma}} \boldsymbol{\theta} - \boldsymbol{\Pi}^{\boldsymbol{\Gamma}} \boldsymbol{\Pi}^M \boldsymbol{\theta} - \boldsymbol{\Pi}^{\boldsymbol{\Gamma}} \,\mathbf{grad}\, w^* + \mathbf{grad}\, \Pi^W w^*,$$

where w^* shall be chosen as either w or w^0, the limiting transverse displacement obtained from the Reissner–Mindlin system when $t = 0$. We then set $w^I = \Pi^W w$ and $\boldsymbol{\theta}^I = \boldsymbol{\Pi}^M \boldsymbol{\theta} + \boldsymbol{\Pi}^B(\boldsymbol{\theta}, w^*)$. In this case, $\boldsymbol{\theta}^I$ is not an interpolant of $\boldsymbol{\theta}$, since it depends on w^* also. Hence, (31) does not hold. However, we will show (the proof is postponed until after the completion of the proof of the theorem) that for $1 \le r \le k-1$,

$$\|\boldsymbol{\theta} - \boldsymbol{\theta}^I\|_1 \le Ch^r\big(\|\boldsymbol{\theta}\|_{r+1} + \|w^*\|_{r+2}\big). \tag{43}$$

Using the definitions given above, we also get

$$\begin{aligned}
\boldsymbol{\gamma}^I &= \lambda t^{-2}(\mathbf{grad}\, w^I - \boldsymbol{\Pi}^{\boldsymbol{\Gamma}} \boldsymbol{\theta}^I) \\
&= \lambda t^{-2}(\mathbf{grad}\, w^I - \boldsymbol{\Pi}^{\boldsymbol{\Gamma}} \boldsymbol{\Pi}^M \boldsymbol{\theta} - \boldsymbol{\Pi}^{\boldsymbol{\Gamma}} \boldsymbol{\Pi}^B(\boldsymbol{\theta}, w^*)) \\
&= \lambda t^{-2}(\mathbf{grad}\, w^I - \boldsymbol{\Pi}^{\boldsymbol{\Gamma}} \boldsymbol{\theta} + \boldsymbol{\Pi}^{\boldsymbol{\Gamma}} \,\mathbf{grad}\, w^* - \mathbf{grad}\, \Pi^W w^*) \\
&= \lambda t^{-2}[\boldsymbol{\Pi}^{\boldsymbol{\Gamma}}(\mathbf{grad}\, w - \boldsymbol{\theta}) + \boldsymbol{\Pi}^{\boldsymbol{\Gamma}} \,\mathbf{grad}(w^* - w) - \mathbf{grad}\, \Pi^W(w^* - w)] \\
&= \boldsymbol{\Pi}^{\boldsymbol{\Gamma}} \boldsymbol{\gamma} + \lambda t^{-2} \boldsymbol{\Pi}^{\boldsymbol{\Gamma}} \,\mathbf{grad}([I - \Pi^W][w^* - w]).
\end{aligned}$$

Note that if we choose $w^* = w$, then (30) will be satisfied, while the choice $w^* = w^0$ does not satisfy (30). The need for the second choice is a technical one, namely the fact that on a convex polygon, we do not have an a priori bound for $\|w\|_3$, but do have a bound for $\|w^0\|_3$. If we had been working on a domain with smooth boundary, the simpler choice $w^* = w$ would be sufficient. By the definition of the space $\boldsymbol{\Gamma}_h$, (32) is satisfied with $n = k-1$ and (28) is satisfied with $r_0 = k-2$. Choosing $w^* = w$, the first estimate of the theorem now follows from a simple modification of Theorem 7.3, in which we replace (31) by (43).

To establish the second estimate in the theorem, we choose $k = 2$, $r = 1$, and $w^* = w^0$, and first apply Theorem 5.1 to obtain

$$\|\boldsymbol{\theta} - \boldsymbol{\theta}^I\|_1 \le Ch(\|\boldsymbol{\theta}\|_2 + \|w^0\|_3) \le Ch(\|\boldsymbol{f}\|_0 + \|g\|_0). \tag{44}$$

Since (30) does not hold in this case, we cannot obtain an error estimate by the same simple modification of Theorem 7.3 used above. Instead, we return to Theorem 7.1 and estimate each of the terms. From our approximability assumption on the space W_h and Theorem 5.1, we get that

$$\begin{aligned} t\|\gamma-\gamma^I\| &\le t\|\gamma-\boldsymbol{\Pi}^{\boldsymbol{\Gamma}}\gamma\|+\lambda t^{-1}\|\boldsymbol{\Pi}^{\boldsymbol{\Gamma}}\,\mathbf{grad}([I-\Pi^W][w^0-w]]\| \\ &\le t\|\gamma-\boldsymbol{\Pi}^{\boldsymbol{\Gamma}}\gamma\|+Ct^{-1}\|\,\mathbf{grad}([I-\Pi^W][w^0-w]]\| \\ &\le Ch(t\|\gamma\|_1+t^{-1}\|w^0-w\|_2)\le Ch(\|\boldsymbol{f}\|_0+\|g\|_0). \end{aligned}$$

The estimate of the final term is straightforward, i.e.,

$$\|\gamma-\boldsymbol{\Pi}^0\gamma\|\le\|\gamma\|\le C(\|\boldsymbol{f}\|_0+\|g\|_{-1}).$$

Inserting the above estimates into Theorem 7.1, we obtain the second estimate of the theorem.

Finally, it remains to prove (43).

Lemma 8.7 *For* $1\le r\le k-1$,

$$\|\boldsymbol{\theta}-\boldsymbol{\theta}^I\|_1\le Ch^r\big(\|\boldsymbol{\theta}\|_{r+1}+\|w^*\|_{r+2}\big).$$

Proof. We first note that it is easy to show that if $\boldsymbol{\psi}\in\boldsymbol{B}_{k+2}$ and $\boldsymbol{\Pi}^{\boldsymbol{\Gamma}}$ denotes the L^2 projection into $\boldsymbol{M}_{k-1}$, then

$$\|\boldsymbol{\psi}\|_0\le C\|\boldsymbol{\Pi}^{\boldsymbol{\Gamma}}\boldsymbol{\psi}\|_0. \tag{45}$$

Hence, we have

$$\begin{aligned} \|\boldsymbol{\Pi}^{\boldsymbol{\Gamma}}\boldsymbol{\Pi}^B(\boldsymbol{\theta},w^*)\|_0 &= \|\boldsymbol{\Pi}^{\boldsymbol{\Gamma}}\boldsymbol{\theta}-\boldsymbol{\Pi}^{\boldsymbol{\Gamma}}\boldsymbol{\Pi}^M\boldsymbol{\theta}-\boldsymbol{\Pi}^{\boldsymbol{\Gamma}}\,\mathbf{grad}\,w^*+\mathbf{grad}\,\Pi^W w^*\|_0 \\ &\le \|\boldsymbol{\Pi}^{\boldsymbol{\Gamma}}(\boldsymbol{\theta}-\boldsymbol{\Pi}^M\boldsymbol{\theta})\|_0+\|(\boldsymbol{\Pi}^{\boldsymbol{\Gamma}}-\boldsymbol{I})\,\mathbf{grad}\,w^*\|_0+\|\,\mathbf{grad}(w^*-\Pi^W w^*)\|_0 \\ &\le C(\|\boldsymbol{\theta}-\boldsymbol{\Pi}^M\boldsymbol{\theta}\|_0+\|(\boldsymbol{\Pi}^{\boldsymbol{\Gamma}}-\boldsymbol{I})\,\mathbf{grad}\,w^*\|_0+\|\,\mathbf{grad}(w^*-\Pi^W w^*)\|_0. \end{aligned} \tag{46}$$

Now by the triangle inequality, standard approximation theory, (45), and (46):

$$\begin{aligned} \|\boldsymbol{\theta}-\boldsymbol{\theta}^I\|_1 = \|\boldsymbol{\theta}-\boldsymbol{\Pi}^M\boldsymbol{\theta}-\boldsymbol{\Pi}^B(\boldsymbol{\theta},w^*)\|_1 &\le \|\boldsymbol{\theta}-\boldsymbol{\Pi}^M\boldsymbol{\theta}\|_1+\|\boldsymbol{\Pi}^B(\boldsymbol{\theta},w^*)\|_1 \\ &\le \|\boldsymbol{\theta}-\boldsymbol{\Pi}^M\boldsymbol{\theta}\|_1+Ch^{-1}\|\boldsymbol{\Pi}^B(\boldsymbol{\theta},w^*)\|_0 \\ &\le \|\boldsymbol{\theta}-\boldsymbol{\Pi}^M\boldsymbol{\theta}\|_1+Ch^{-1}\|\boldsymbol{\Pi}^{\boldsymbol{\Gamma}}\boldsymbol{\Pi}^B(\boldsymbol{\theta},w^*)\|_0 \\ &\le C\Big[\|\boldsymbol{\theta}-\boldsymbol{\Pi}^M\boldsymbol{\theta}\|_1+h^{-1}(\|\boldsymbol{\theta}-\boldsymbol{\Pi}^M\boldsymbol{\theta}\|_0 \\ &\quad +\|(\boldsymbol{\Pi}^{\boldsymbol{\Gamma}}-\boldsymbol{I})\,\mathbf{grad}\,w^*\|_0+\|\,\mathbf{grad}(w^*-\Pi^W w^*)\|_0)\Big]. \end{aligned}$$

Applying our approximation theory results, we get for $1\le r\le k-1$

$$\|\boldsymbol{\theta}-\boldsymbol{\theta}^I\|_1\le Ch^r\big(\|\boldsymbol{\theta}\|_{r+1}+\|w^*\|_{r+2}\big).$$

Using a slightly modified version of Theorem 7.4, (due to the fact that $\boldsymbol{\theta}^I$ depends on both $\boldsymbol{\theta}$ and w^*), one can derive L^2 error estimates for $\boldsymbol{\theta}-\boldsymbol{\theta}_h$ and then error estimates for $w-w_h$. We state the results below.

Theorem 8.8 *For the discontinuous shear stress family of index $k \geq 2$, we have for $1 \leq r \leq k-1$*

$$\|\boldsymbol{\theta} - \boldsymbol{\theta}_h\|_0 + \|w - w_h\|_1 \leq Ch^{r+1}\left(\|\boldsymbol{\theta}\|_{r+1} + \|w\|_{r+2} + t\|\gamma\|_r + \|\gamma\|_{r-1}\right).$$

For $k = 2$ and $r = 1$, we also have the estimate

$$\|\boldsymbol{\theta} - \boldsymbol{\theta}_h\|_0 + \|w - w_h\|_1 \leq Ch^2\left(\|\boldsymbol{f}\|_0 + \|g\|_0\right).$$

We note that we do not obtain a higher order of convergence for $\|w - w_h\|_0$.

8.4 Linked Interpolation Methods

There are a number of formulations of the linked interpolation method. One approach is to use the mixed formulation (25), but replace the space $\boldsymbol{\Theta}_h \times W_h$ by a space $\boldsymbol{V}_h$ in which the two spaces are linked by a constraint. The simplest example of such a method is the one introduced by Xu [51] and Auricchio and Taylor [16, 49], and analyzed in [41, 39, 15]. In this method, we choose

$$\boldsymbol{\Theta}_h = \mathring{\boldsymbol{M}}_1^0 + \boldsymbol{B}_3, \quad W_h = \mathring{M}_1^0, \quad \boldsymbol{\Gamma}_h = \boldsymbol{M}_0,$$
$$\boldsymbol{V}_h = \{(\boldsymbol{\phi}, v + L\boldsymbol{\phi}) : \boldsymbol{\phi} \in \boldsymbol{\Theta}_h, v \in W_h\},$$

where following [41], we may define $L_T = L|_T$ as a mapping from $\boldsymbol{H}^1(T)$ onto $P_{2,-}(T)$ by

$$\int_e [(\mathbf{grad}\, L_T\boldsymbol{\phi} - \boldsymbol{\phi}) \cdot \boldsymbol{s}] \frac{\partial v}{\partial s}\, ds = 0, \quad v \in P_{2,-}(T), \tag{47}$$

for every edge e of T, where $P_{2,-}(T)$ is the space of piecewise quadratics which vanish at the vertices of T.

We then seek an approximation $(\boldsymbol{\theta}_h, w_h^*; \gamma_h) \in \boldsymbol{V}_h \times \boldsymbol{\Gamma}_h$ such that (25) holds for all $(\boldsymbol{\phi}, v^*; \boldsymbol{\eta}) \in \boldsymbol{V}_h \times \boldsymbol{\Gamma}_h$. Equivalently, we can write this method in terms of the usual spaces, but with a modified bilinear form, i.e., we seek $(\boldsymbol{\theta}_h, w_h, \gamma_h) \in \boldsymbol{\Theta}_h \times W_h \times \boldsymbol{\Gamma}_h$ such that

$$a(\boldsymbol{\theta}_h, \boldsymbol{\phi}) + \lambda^{-1}t^2(\gamma_h, \mathbf{grad}(v + L\boldsymbol{\phi}) - \boldsymbol{\phi}) = (g, v + L\boldsymbol{\phi}) - (\boldsymbol{f}, \boldsymbol{\phi}),$$
$$\boldsymbol{\phi} \in \boldsymbol{\Theta}_h, v \in W_h,$$
$$(\mathbf{grad}(w_h + L\boldsymbol{\theta}_h) - \boldsymbol{\theta}_h, \boldsymbol{\eta}) - \lambda^{-1}t^2(\gamma_h, \boldsymbol{\eta}) = 0, \quad \boldsymbol{\eta} \in \boldsymbol{\Gamma}_h.$$

Note that we can write this discrete variational formulation as a slight perturbation of the formulation (25), by defining $\boldsymbol{\Pi}^{\Gamma} = \boldsymbol{\Pi}^0(I - \mathbf{grad}\, L)$ (where $\boldsymbol{\Pi}^0$ denotes the $\boldsymbol{L}^2$ projection onto $\boldsymbol{\Gamma}_h$), and replacing the term (g, v) by $(g, v + L\boldsymbol{\phi})$. We omit the element diagram for this method, since depicting only the three basic spaces, without the additional space $P_{2,-}(T)$, is somewhat misleading.

We shall analyze this method using the usual spaces and the interpolation operator $\boldsymbol{\Pi}^{\Gamma}$ defined above. We first observe that from [41],

$$|(g, L\phi)|_T \leq \|g\|_{0,T}\|L_T\phi\|_{0,T} \leq Ch_T\|g\|_{0,T}\|\nabla L_T\phi\|_{0,T} \leq Ch_T^2\|g\|_{0,T}\|\phi\|_{1,T},$$

so this term is a high order perturbation and may be dropped. To apply our previous error estimates, we first define $w^I = \Pi^W w$, the continuous piecewise linear interpolant of w, and $\boldsymbol{\theta}^I = \boldsymbol{\Pi}^M\boldsymbol{\theta} + \boldsymbol{\Pi}^B\boldsymbol{\theta}$, where $\boldsymbol{\Pi}^M\boldsymbol{\theta}$ denotes an interpolant of $\boldsymbol{\theta}$ satisfying

$$\|\boldsymbol{\theta} - \boldsymbol{\Pi}^M\boldsymbol{\theta}\|_0 + \|\boldsymbol{\theta} - \boldsymbol{\Pi}^M\boldsymbol{\theta}\|_1 \leq Ch^s\|\boldsymbol{\theta}\|_s, \quad s = 1, 2,$$

and $\boldsymbol{\Pi}^B\boldsymbol{\theta} \in \boldsymbol{B}_3$ is defined by:

$$\boldsymbol{\Pi}^0\boldsymbol{\Pi}^B\boldsymbol{\theta} = \boldsymbol{\Pi}^0[(\boldsymbol{I} - \mathbf{grad}\, L)(\boldsymbol{\theta} - \boldsymbol{\Pi}^M\boldsymbol{\theta})]. \tag{48}$$

We note that

$$\begin{aligned}\|\boldsymbol{\Pi}^0\boldsymbol{\Pi}^B\boldsymbol{\theta}\|_0 &\leq \|(\boldsymbol{I} - \mathbf{grad}\, L)(\boldsymbol{\theta} - \boldsymbol{\Pi}^M\boldsymbol{\theta})\|_0 \leq \|\boldsymbol{\theta} - \boldsymbol{\Pi}^M\boldsymbol{\theta}\|_0 \\ &\quad + \|\mathbf{grad}\, L(\boldsymbol{\theta} - \boldsymbol{\Pi}^M\boldsymbol{\theta})\|_0 \leq \|\boldsymbol{\theta} - \boldsymbol{\Pi}^M\boldsymbol{\theta}\|_0 + Ch\|\boldsymbol{\theta} - \boldsymbol{\Pi}^M\boldsymbol{\theta}\|_1.\end{aligned}$$

Since $\|\boldsymbol{\Pi}^B\boldsymbol{\theta}\|_0 \leq C\|\boldsymbol{\Pi}^0\boldsymbol{\Pi}^B\boldsymbol{\theta}\|_0$, we easily obtain for $s = 1, 2$ that

$$\|\boldsymbol{\theta} - \boldsymbol{\theta}^I\|_0 \leq C(\|\boldsymbol{\theta} - \boldsymbol{\Pi}^M\boldsymbol{\theta}\|_0 + h\|\boldsymbol{\theta} - \boldsymbol{\Pi}^M\boldsymbol{\theta}\|_1 \leq Ch^s\|\boldsymbol{\theta}\|_s.$$

Using the inverse inequality $\|\boldsymbol{\Pi}^B\boldsymbol{\theta}\|_1 \leq Ch^{-1}\|\boldsymbol{\Pi}^B\boldsymbol{\theta}\|_0$, we then obtain

$$\|\boldsymbol{\theta} - \boldsymbol{\theta}^I\|_1 \leq C(\|\boldsymbol{\theta} - \boldsymbol{\Pi}^M\boldsymbol{\theta}\|_1 + h^{-1}\|\boldsymbol{\theta} - \boldsymbol{\Pi}^M\boldsymbol{\theta}\|_0) \leq Ch\|\boldsymbol{\theta}\|_2.$$

Hence, hypotheses (31) and (32) of Theorem 7.3 are satisfied with $r = 1$ and $r_0 = -1$. Thus, it only remains to show that (30) is satisfied. Applying (47) with $\phi = \mathbf{grad}(w - w^I)$, and noting that $(L_T\,\mathbf{grad} - I)(w - w^I) = 0$ at the vertices of T, we get

$$\begin{aligned}0 &= \int_e [(\mathbf{grad}\, L_T - \boldsymbol{I})\,\mathbf{grad}(w - w^I)] \cdot \boldsymbol{s}\frac{\mathrm{d}v}{\mathrm{d}s}\,\mathrm{d}s \\ &= \int_e \frac{d}{ds}\left[(L_T\,\mathbf{grad} - I)(w - w^I)\right]\frac{\mathrm{d}v}{\mathrm{d}s}\,\mathrm{d}s \\ &= -\int_e (L_T\,\mathbf{grad} - I)(w - w^I)\frac{\mathrm{d}^2v}{\mathrm{d}s^2}\,\mathrm{d}s, \quad v \in P_{2,-}(T).\end{aligned}$$

Since $\mathrm{d}^2v/\mathrm{d}s^2$ is a constant on the edge e, we get for all $\boldsymbol{q} \in \boldsymbol{P}_0(T)$,

$$\begin{aligned}\int_T (\mathbf{grad}\, L - I)\,\mathbf{grad}(w - w^I) \cdot \boldsymbol{q}\,\mathrm{d}x &= \int_T \mathbf{grad}(L\,\mathbf{grad} - I)(w - w^I) \cdot \boldsymbol{q}\,\mathrm{d}x \\ &= -\int_{\partial T} (L\,\mathbf{grad} - I)(w - w^I)\boldsymbol{q} \cdot \boldsymbol{n}\,\mathrm{d}s = 0,\end{aligned}$$

and so

$$\boldsymbol{\Pi}^{\Gamma}\,\mathbf{grad}(w - w^{I}) = \boldsymbol{\Pi}^{0}(\mathbf{grad}\,L_T - \boldsymbol{I})\,\mathbf{grad}(w - w^{I}) = 0.$$

Finally, from (48) and the fact that $L\boldsymbol{\Pi}^{B}\boldsymbol{\theta} = 0$, we get

$$\boldsymbol{\Pi}^{\Gamma}(\boldsymbol{\theta} - \boldsymbol{\theta}^{I}) = \boldsymbol{\Pi}^{0}(\mathbf{grad}\,L_T - \boldsymbol{I})(\boldsymbol{\theta} - \boldsymbol{\Pi}^{M}\boldsymbol{\theta} - \boldsymbol{\Pi}^{B}\boldsymbol{\theta}) = 0.$$

If we drop the term $(g, L\boldsymbol{\phi})$ from the right hand side of the method, then we get immediately from Theorems 7.3 and 7.5 the following estimate:

$$\begin{aligned}\|\boldsymbol{\theta} - \boldsymbol{\theta}_h\|_1 + t\|\boldsymbol{\gamma} - \boldsymbol{\gamma}_h\|_0 + \|w - w_h\|_1 \\ \le Ch(\|\boldsymbol{\theta}\|_2 + t\|\boldsymbol{\gamma}\|_1 + \|\boldsymbol{\gamma}_0 + \|w\|_2) \le Ch(\|g\|_0 + \|\boldsymbol{f}\|_0).\end{aligned}$$

A simple extension of this argument gives the same final result with this term included (the term $h^2\|g\|_0$ would need to be added to the intermediate result).

We note that the method of [53] analyzed in [34] is also of this type. The analysis given in [34] proceeds by comparing the method to the Durán–Liberman element described above. The two methods have the same choices for the spaces $\boldsymbol{\Theta}_h$ and $\boldsymbol{\Gamma}_h$,

8.5 The Nonconforming Element of Arnold and Falk [11]

See also [29].

$$\boldsymbol{\Theta}_h = \mathring{\boldsymbol{M}}_1^0 + \boldsymbol{B}_3, \quad W_h = \mathring{M}_1^{*}, \quad \boldsymbol{\Gamma}_h = \boldsymbol{M}_0$$

where $\boldsymbol{\Pi}^{\Gamma}$ is the $\boldsymbol{L}^2$ projection into $\boldsymbol{\Gamma}_h$.

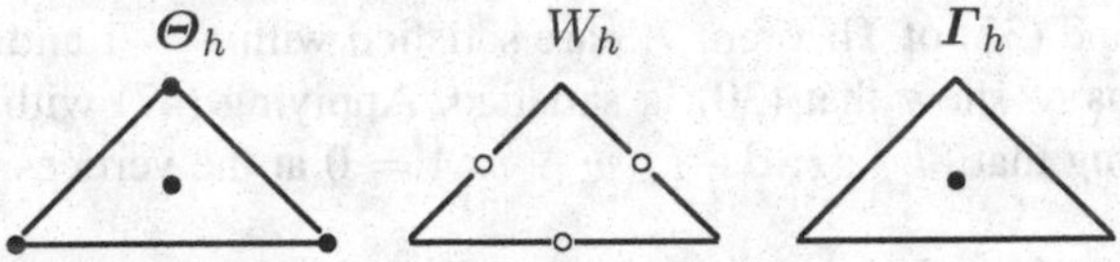

Arnold–Falk

Since the space W_h is not contained in $\mathring{H}^1(\Omega)$, $\mathbf{grad}$ must be replaced by $\mathbf{grad}_h$ and some modifications need to be made in the basic error estimates proved earlier. Rather than prove a general abstract version of these results taking into account several types of nonconformity, we simply modify the proofs for the particular method being analyzed. We begin by first stating a standard result basic to the analysis of nonconforming methods.

Lemma 8.9 (cf. [28]) *Let $\boldsymbol{\phi} \in \boldsymbol{H}^1(\Omega)$ and $v \in W_h$. Then*

$$\left|\sum_{T\in\mathcal{T}} \int_{\partial T} v\boldsymbol{\phi}\cdot\boldsymbol{n}_T\right| \le Ch\|\boldsymbol{\phi}\|_1\|\,\mathbf{grad}_h\, v\|_0.$$

Using this result, we can derive the following energy norm error estimate.

Theorem 8.10

$$\|\boldsymbol{\theta} - \boldsymbol{\theta}_h\|_1 + t\|\boldsymbol{\gamma} - \boldsymbol{\gamma}_h\|_0 + \|\,\mathbf{grad}_h[w - w_h]\|_0 \le Ch(\|\boldsymbol{f}\|_0 + \|g\|_0).$$

Proof. Since $W_h \notin \mathring{H}^1_0(\Omega)$, we cannot apply Theorems 7.1 and 7.3 directly. In particular, the error equation (29) must be replaced by a modified equation which contains an additional term for the consistency error.

$$a(\boldsymbol{\theta}-\boldsymbol{\theta}_h,\boldsymbol{\phi})+(\boldsymbol{\gamma}-\boldsymbol{\gamma}_h,\mathbf{grad}_h\,v-\boldsymbol{\Pi}^{\boldsymbol{\Gamma}}\boldsymbol{\phi})=(\boldsymbol{\gamma},[\boldsymbol{I}-\boldsymbol{\Pi}^{\boldsymbol{\Gamma}}]\boldsymbol{\phi})+\sum_{T\in\mathcal{T}}\int_{\partial T}v\boldsymbol{\gamma}\cdot\boldsymbol{n}_T, \tag{49}$$

for all $\boldsymbol{\phi}\in\boldsymbol{\Theta}_h$ and $v\in W_h$. Following the proof of Theorem 7.1, we obtain

$$\|\boldsymbol{\theta}^I-\boldsymbol{\theta}_h\|_1^2+t^2\|\boldsymbol{\gamma}^I-\boldsymbol{\gamma}_h\|_0^2\le C\Big(\|\boldsymbol{\theta}-\boldsymbol{\theta}^I\|_1^2+t^2\|\boldsymbol{\gamma}-\boldsymbol{\gamma}^I\|_0^2 \\ +h^2\|\boldsymbol{\gamma}-\boldsymbol{\Pi}^0\boldsymbol{\gamma}\|_0^2+\Big|\sum_{T\in\mathcal{T}}\int_{\partial T}(w^I-w_h)\boldsymbol{\gamma}\cdot\boldsymbol{n}_T\Big|\Big). \tag{50}$$

In this case, $\boldsymbol{\Pi}^0$ is $\boldsymbol{L}^2$ projection into piecewise constants, so we can use the trivial estimate $\|\boldsymbol{\gamma}-\boldsymbol{\Pi}^0\boldsymbol{\gamma}\|_0\le\|\boldsymbol{\gamma}\|_0$. As in Theorem 7.3, we need to define $\boldsymbol{\theta}^I$ and w^I and hence $\boldsymbol{\gamma}^I$ to satisfy (30) and (31). The choice of $\boldsymbol{\theta}^I$ is the same as that used for the MINI element for the Stokes problem. This satisfies (31) with $n=1$ (and the 1-norm replaced by the discrete 1-norm) and also the condition $\boldsymbol{\Pi}^{\boldsymbol{\Gamma}}\boldsymbol{\theta}^I=\boldsymbol{\Pi}^{\boldsymbol{\Gamma}}\boldsymbol{\theta}$. Hence, to satisfy (30), we need only to find w^I such that

$$\mathbf{grad}_h\,w^I=\boldsymbol{\Pi}^{\boldsymbol{\Gamma}}\,\mathbf{grad}\,w. \tag{51}$$

This is easily accomplished by choosing w^I to satisfy $\int_e w^I=\int_e w$ on each edge e. Then for all $\boldsymbol{\eta}\in\boldsymbol{\Gamma}_h$

$$\int_T\mathbf{grad}\,w\cdot\boldsymbol{\eta}\,dx=\int_{\partial T}w\boldsymbol{\eta}\cdot\boldsymbol{n}_T\,ds=\int_{\partial T}w^I\boldsymbol{\eta}\cdot\boldsymbol{n}_T\,ds=\int_T\mathbf{grad}\,w^I\cdot\boldsymbol{\eta}\,dx,$$

which implies (51). Then (32) is satisfied with $n=1$.

It only remains to estimate the term arising from the nonconforming approximation. Unfortunately, we cannot estimate this term by applying Lemma 8.9 directly, since the result would then contain the term $\|\boldsymbol{\gamma}\|_1$ which is not bounded independent of the thickness t. Instead, we use the Helmholtz decomposition to write $\boldsymbol{\gamma}=\mathbf{grad}\,r+\mathbf{curl}\,p$ with $r\in\mathring{H}^1(\Omega)$ and $p\in H^1(\Omega)$. Recalling that

$$\mathbf{grad}_h(w^I-w_h)=\lambda^{-1}t^2(\boldsymbol{\gamma}^I-\boldsymbol{\gamma}_h)+\boldsymbol{\Pi}^{\boldsymbol{\Gamma}}(\boldsymbol{\theta}^I-\boldsymbol{\theta}_h),$$

we first use Lemma 8.9 to get

$$\Big|\sum_{T\in\mathcal{T}}\int_{\partial T}(w^I-w_h)\,\mathbf{grad}\,r\cdot\boldsymbol{n}_T\,ds\Big|\le Ch\|r\|_2\|\mathbf{grad}_h(w^I-w_h)\|_0 \\ \le Ch\|r\|_2\left(t^2\|\boldsymbol{\gamma}^I-\boldsymbol{\gamma}_h\|_0+\|\boldsymbol{\Pi}^{\boldsymbol{\Gamma}}(\boldsymbol{\theta}^I-\boldsymbol{\theta}_h)\|_0\right) \\ \le Ch\|r\|_2\left(t^2\|\boldsymbol{\gamma}^I-\boldsymbol{\gamma}_h\|_0+\|\boldsymbol{\theta}^I-\boldsymbol{\theta}_h\|_0\right).$$

Now for all $p^I \in M_1^0$,

$$\sum_{T\in\mathcal{T}} \int_{\partial T} (w^I - w_h)\, \mathbf{curl}\, p \cdot \boldsymbol{n}_T \,\mathrm{d}s = \sum_{T\in\mathcal{T}} \int_T \mathbf{grad}(w^I - w_h) \cdot \mathbf{curl}\, p \,\mathrm{d}x$$
$$= \sum_{T\in\mathcal{T}} \int_T \mathbf{grad}(w^I - w_h) \cdot \mathbf{curl}(p - p^I) \,\mathrm{d}x$$
$$= \lambda^{-1} t^2 (\boldsymbol{\gamma}^I - \boldsymbol{\gamma}_h, \mathbf{curl}[p - p^I]) + (\boldsymbol{\Pi}^{\boldsymbol{\Gamma}}[\boldsymbol{\theta}^I - \boldsymbol{\theta}_h], \mathbf{curl}[p - p^I])$$
$$= \lambda^{-1} t^2 (\boldsymbol{\gamma}^I - \boldsymbol{\gamma}_h, \mathbf{curl}[p - p^I]) + ([\boldsymbol{\Pi}^{\boldsymbol{\Gamma}} - I](\boldsymbol{\theta}^I - \boldsymbol{\theta}_h), \mathbf{curl}[p - p^I])$$
$$+ (\mathrm{rot}[\boldsymbol{\theta}^I - \boldsymbol{\theta}_h], p - p^I).$$

Choosing p^I to satisfy

$$\|p - p^I\|_0 + h\|p - p^I\|_1 \le Ch^s \|p^I\|_s, \quad s = 1, 2,$$

(e.g., the Clement interpolant), we have by standard estimates that

$$\Big|\sum_{T\in\mathcal{T}} \int_{\partial T} (w^I - w_h)\, \mathbf{curl}\, p \cdot \boldsymbol{n}_T \,\mathrm{d}s\Big|$$
$$\le C \left(t^2 \|\boldsymbol{\gamma}^I - \boldsymbol{\gamma}_h\|_0 h \|p\|_2 + h \|\boldsymbol{\theta}^I - \boldsymbol{\theta}_h\|_1 \|p\|_1 \right).$$

Combining these results, we obtain

$$\Big|\sum_{T\in\mathcal{T}} \int_{\partial T} (w^I - w_h) \boldsymbol{\gamma} \cdot \boldsymbol{n}_T \,\mathrm{d}x\Big|$$
$$\le Ch \left(t\|\boldsymbol{\gamma}^I - \boldsymbol{\gamma}_h\|_0 + \|\boldsymbol{\theta}^I - \boldsymbol{\theta}_h\|_1 \right) (\|r\|_2 + \|p\|_1 + t\|p\|_2).$$

The first two estimates of the theorem now follow by combining all these results and using the a priori estimate (21). To obtain an error estimate on the transverse displacement, we need a nonconforming version of Theorem 7.5.

Choosing $\eta = \mathbf{grad}_h v_h$, $v_h \in W_h$, we get for all $w_I \in W_h$,

$$(\mathbf{grad}_h[w_I - w_h], \mathbf{grad}_h v_h) = (\mathbf{grad}_h[w_I - w], \mathbf{grad}\, v_h)$$
$$+ (\boldsymbol{\theta} - \boldsymbol{\Pi}^{\boldsymbol{\Gamma}} \boldsymbol{\theta}_h, \mathbf{grad}_h v_h) + \sum_T \int_{\partial T} v_h \frac{\partial w}{\partial n} \,\mathrm{d}s. \quad (52)$$

Then choosing $v_h = w_h - w_I$, it easily follows using Lemma 8.9 that

$$\|\mathbf{grad}_h[w_I - w_h]\|_0 \le \|\mathbf{grad}_h[w_I - w]\|_0 + \|\boldsymbol{\theta} - \boldsymbol{\Pi}^{\boldsymbol{\Gamma}} \boldsymbol{\theta}_h\|_0 + Ch\|w\|_2$$
$$\le \|\mathbf{grad}_h[w_I - w]\|_0 + \|[\boldsymbol{I} - \boldsymbol{\Pi}^{\boldsymbol{\Gamma}}]\boldsymbol{\theta}\|_0 + \|[\boldsymbol{I} - \boldsymbol{\Pi}^{\boldsymbol{\Gamma}}][\boldsymbol{\theta}_h - \boldsymbol{\theta}]\|_0 + \|\boldsymbol{\theta} - \boldsymbol{\theta}_h\|_0 + Ch\|w\|_2$$
$$\le \|\mathbf{grad}_h[w_I - w]\|_0 + \|[\boldsymbol{I} - \boldsymbol{\Pi}^{\boldsymbol{\Gamma}}]\boldsymbol{\theta}\|_0 + Ch\|\boldsymbol{\theta} - \boldsymbol{\theta}_h\|_1 + \|\boldsymbol{\theta} - \boldsymbol{\theta}_h\|_1 + Ch\|w\|_2.$$

The desired result now follows from the triangle inequality and standard estimates.

Using a nonconforming version of Theorem 7.4, we can also establish the following L^2 error estimate.

$$\|\boldsymbol{\theta} - \boldsymbol{\theta}_h\|_0 + \|w - w_h\|_0 \leq Ch^2(\|\boldsymbol{f}\|_0 + \|g\|_0).$$

See also [36] and [30] for a modification of this element, and [2] for a relationship between these two approaches.

9 Some Rectangular Reissner–Mindlin Elements

Now let $\mathcal{T}_h$ denote a rectangular mesh of Ω and R an element of $\mathcal{T}_h$. We denote by Q_{k_1,k_2} the set of polynomials of separate degree $\leq k_1$ in x and $\leq k_2$ in y and set $Q_k = Q_{k,k}$. We also define the serendipity polynomials $Q_k^s = P_k \oplus x^k y \oplus xy^k$. Finally, we will also use the rotated versions of the rectangular Raviart–Thomas, Brezzi–Douglas–Marini, and Brezzi–Douglas–Fortin–Marini spaces, which we define locally for $k \geq 1$ as follows.

$$\begin{aligned}
\boldsymbol{RT}_{k-1}^{\perp}(R) &= \{\boldsymbol{\eta} : \boldsymbol{\eta} = (Q_{k-1,k}(R), Q_{k,k-1}(R))\},\\
\boldsymbol{BDM}_k^{\perp}(R) &= \{\boldsymbol{\eta} : \boldsymbol{\eta} \in \boldsymbol{P}_k(R) \oplus \nabla(xy^{k+1}) \oplus \nabla(x^{k+1}y)\},\\
\boldsymbol{BDFM}_k^{\perp}(R) &= \{\boldsymbol{\eta} : \boldsymbol{\eta} = (P_k(R) \setminus \{x^k\}, P_k(R) \setminus \{y^k\})\}.
\end{aligned}$$

9.1 Rectangular MITC Elements and Generalizations [20, 17, 23, 48]

In the original MITC family, we choose for $k \geq 1$,

$$\boldsymbol{\Theta}_h = \{\boldsymbol{\phi} \in \mathring{\boldsymbol{H}}^1(\Omega) : \boldsymbol{\phi}|_R \in \boldsymbol{Q}_k(R)\}, \qquad W_h = \{v \in \mathring{H}^1(\Omega) : v|_R \in Q_k^s(R)\},$$
$$\boldsymbol{\Gamma}_h = \{\boldsymbol{\eta} \in \boldsymbol{L}^2(\Omega) : \boldsymbol{\eta}|_R \in \boldsymbol{BDFM}_k^{\perp}(R)\}.$$

The auxiliary pressure space

$$Q_h = \{q \in L_0^2(\Omega) : q|_R \in P_{k-1}\}$$

and the reduction operator $\boldsymbol{\Pi}^{\boldsymbol{\Gamma}}$ is defined by

$$\int_e (\boldsymbol{\Pi}^{\boldsymbol{\Gamma}}\gamma - \gamma) \cdot \boldsymbol{s}\, p_{k-1}(s)\, \mathrm{d}s = 0, \quad \forall e, \quad \forall p_{k-1} \in P_{k-1}(e),$$
$$\int_R (\boldsymbol{\Pi}^{\boldsymbol{\Gamma}}\gamma - \gamma) \cdot \boldsymbol{p}_{k-2}\, \mathrm{d}x\, \mathrm{d}y = 0, \quad \forall R, \quad \forall \boldsymbol{p}_{k-2} \in \boldsymbol{P}_{k-2}(R).$$

The lowest order element ($k = 1$) is called MITC4. In this case, the space $\boldsymbol{BDFM}_1^{\perp}(R)$ has the form $(a + by, c + dx)$ and coincides with the lowest order rotated rectangular Raviart–Thomas element $\boldsymbol{RT}_0^{\perp}(R)$. The space $Q_1^s(R) = Q_1(R)$. The MITC4 element was proposed in [20] and analyzed in [17], [18], [33], and most recently in [31], where the proof is extended to more general quadrilateral meshes using a macro-element technique and the results obtained under less regularity than previously required. For rectangular meshes, this method coincides with the T1 method of Hughes and Tezduyar [38]. The $k = 2$ method is known as MITC9 and has been analyzed in [23, 33].

For $k \geq 3$, it is shown in [48] and [45] that it is possible to reduce the number of degrees of freedom in the rotation space $\boldsymbol{\Theta}_h$ without affecting the locking-free convergence. In particular, one can choose

$$\boldsymbol{\Theta}_h = \{\boldsymbol{\phi} \in \mathring{\boldsymbol{H}}^1(\Omega) : \boldsymbol{\phi}|_R \in [\boldsymbol{Q}_k(R) \cap \boldsymbol{P}_{k+2}(R)]\}.$$

Another possibility (cf. [48]) is to choose for $k \geq 2$

$$\begin{aligned}
\boldsymbol{\Theta}_h &= \{\boldsymbol{\phi} \in \mathring{\boldsymbol{H}}^1(\Omega) : \boldsymbol{\phi}|_R \in [\boldsymbol{Q}_k(R) \cap \boldsymbol{P}_{k+2}(R)]\}, \\
W_h &= \{v \in \mathring{H}^1(\Omega) : v|_R \in Q^s_{k+1}(R)\}, \\
\boldsymbol{\Gamma}_h &= \{\boldsymbol{\eta} \in \boldsymbol{L}^2(\Omega) : \boldsymbol{\eta}|_R \in \boldsymbol{BDM}^{\perp}_k(R)\}.
\end{aligned}$$

The auxiliary pressure space is again $Q_h = \{q \in L^2_0(\Omega) : q|_R \in P_{k-1}\}$ and the reduction operator $\boldsymbol{\Pi}^{\boldsymbol{\Gamma}}$ is defined by

$$\int_e (\boldsymbol{\Pi}^{\boldsymbol{\Gamma}}\gamma - \gamma) \cdot \boldsymbol{s}\, p_k(s)\, \mathrm{d}s = 0, \quad \forall e, \quad \forall p_k \in P_k(e),$$

$$\int_R (\boldsymbol{\Pi}^{\boldsymbol{\Gamma}}\gamma - \gamma) \cdot \boldsymbol{p}_{k-2}\, \mathrm{d}x\, \mathrm{d}y = 0, \quad \forall R, \ \forall \boldsymbol{p}_{k-2} \in \boldsymbol{P}_{k-2}(R).$$

A fourth possibility discussed in [48] is to choose for $k \geq 2$

$$\begin{aligned}
\boldsymbol{\Theta}_h &= \{\boldsymbol{\phi} \in \mathring{\boldsymbol{H}}^1(\Omega) : \boldsymbol{\phi}|_R \in [\boldsymbol{Q}_{k+1}(R), \boldsymbol{\phi}|_e \in \boldsymbol{P}_k(e)]\}, \\
W_h &= \{v \in \mathring{H}^1(\Omega) : v|_R \in Q^s_k(R)\}, \\
\boldsymbol{\Gamma}_h &= \{\boldsymbol{\eta} \in \boldsymbol{L}^2(\Omega) : \boldsymbol{\eta}|_R \in \boldsymbol{RT}^{\perp}_{k-1}(R)\}.
\end{aligned}$$

In this case, the auxiliary pressure space is now $Q_h = \{q \in L^2_0(\Omega) : q|_R \in Q_{k-1}\}$ and the reduction operator $\boldsymbol{\Pi}^{\boldsymbol{\Gamma}}$ is defined by

$$\int_e (\boldsymbol{\Pi}^{\boldsymbol{\Gamma}}\gamma - \gamma) \cdot \boldsymbol{s}\, p_{k-1}(s)\, \mathrm{d}s = 0, \quad \forall e, \quad \forall p_{k-1} \in P_{k-1}(e),$$

$$\int_R (\boldsymbol{\Pi}^{\boldsymbol{\Gamma}}\gamma - \gamma) \cdot \boldsymbol{r}_{k-2}\, \mathrm{d}x\, \mathrm{d}y = 0, \quad \forall R, \forall \boldsymbol{r}_{k-2} \in Q_{k-1,k-2}(R) \times Q_{k-2,k-1}(R).$$

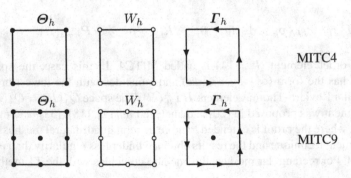

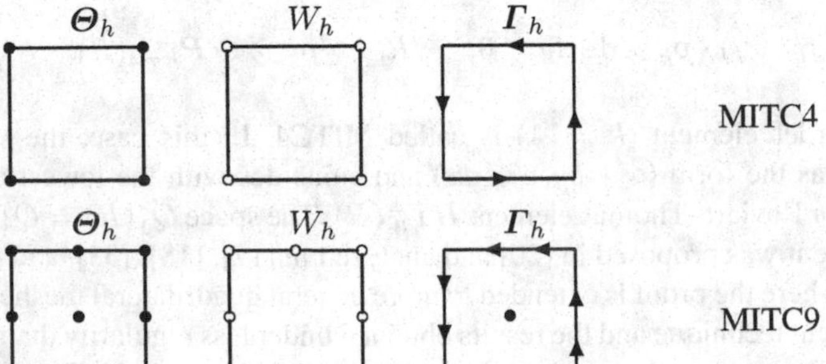

One can also consider a low order element, associated with the choice

$$W_h = \{v \in \mathring{H}^1(\Omega) : v|_R \in Q_2^s(R)\},\ \boldsymbol{\Gamma}_h = \{\boldsymbol{\eta} \in \boldsymbol{L}^2(\Omega) : \boldsymbol{\eta}|_R \in \boldsymbol{BDM}_1^{\perp}(R)\},$$

where we choose $\boldsymbol{\Theta}_h = \{\boldsymbol{\phi} \in \mathring{\boldsymbol{H}}^1(\Omega) : \boldsymbol{\phi}|_R \in Q_2^s(R)\}$. This element, MITC8 (cf. [21]), is depicted below.

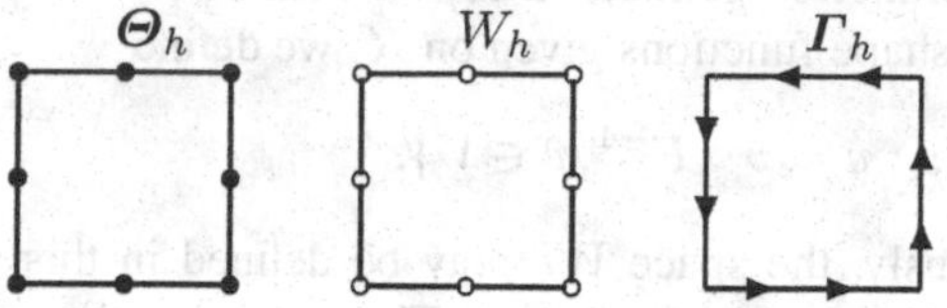

MITC8

9.2 DL4 Method [31]

The DL4 method is the extension to rectangles of the Durán–Liberman triangular element defined previously. The spaces W_h and $\boldsymbol{\Gamma}_h$ are the same as those chosen for the MITC4 method, while the space of rotations is now chosen to be:

$$\boldsymbol{\Theta}_h = \{\boldsymbol{\phi} \in \mathring{\boldsymbol{H}}^1(\Omega) : \boldsymbol{\phi}|_K \in \boldsymbol{Q}_1(K) \oplus \langle \boldsymbol{b}_1, , \boldsymbol{b}_2, \boldsymbol{b}_3, \boldsymbol{b}_4 \rangle, \forall K \in \mathcal{T}_h\},$$

where $\boldsymbol{b}_i = b_i \boldsymbol{s}_i$, with $\boldsymbol{s}_i$ the counterclockwise unit tangent vector to the edge e_i of K and $b_i \in Q_2(K)$ vanishes on the edges e_j, $j \neq i$.

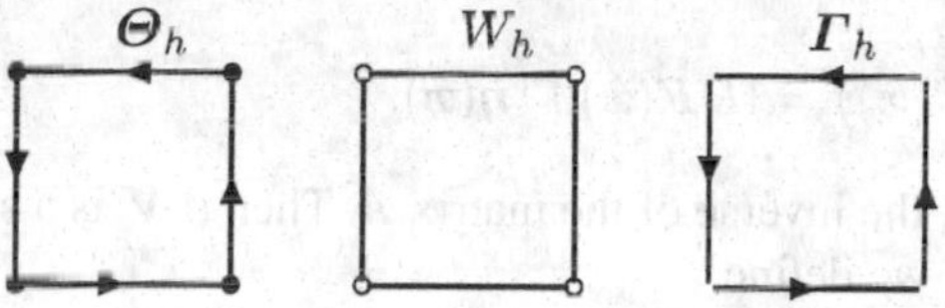

DL4

9.3 Ye's Method

Ye's method is the extension to rectangles of the Arnold–Falk element. This is not completely straightforward, since the values at the midpoints of the edges of a rectangle are not a unisolvent set of degrees of freedom for a bilinear function (consider $(x-1/2)(y-1/2)$ on the unit square). Hence, the nonconforming space W_h must be chosen differently.

$$\boldsymbol{\Theta}_h = \{\boldsymbol{\phi} \in \mathring{\boldsymbol{H}}^1(\Omega) : \boldsymbol{\phi}|_R \in \boldsymbol{Q}_2(R)\},$$
$$\boldsymbol{\Gamma}_h = \{\boldsymbol{\eta} \in \boldsymbol{L}^2(\Omega) : \boldsymbol{\eta}|_R = (b + dx, c - dy) \equiv S\}.$$
$$W_h = \{v \in \mathring{H}^1(\mathcal{T}_h) : v|_R = a + bx + cy + d(x^2 - y^2)/2\},$$

and $\boldsymbol{\Pi}^{\Gamma}$ is the $\boldsymbol{L}^2$ projection.

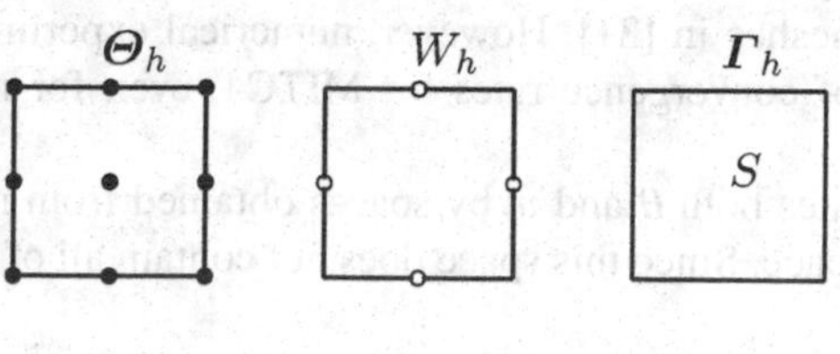

Ye

10 Extension to Quadrilaterals

Meshes of rectangular elements are very restrictive, so one would like to extend the elements defined above to quadrilaterals. To do so, we let $\boldsymbol{F}$ be an invertible bilinear mapping from the reference element $\hat{K} = [0,1] \times [0,1]$ to a convex quadrilateral K. For scalar functions, if $\hat{v}(\hat{\boldsymbol{x}})$ is function defined on $\hat{K}$, we define $v(\boldsymbol{x})$ on K by $v = \hat{v} \circ \boldsymbol{F}^{-1}$. Then, for $\hat{V}$ a set of shape functions given on $\hat{K}$, we define

$$V_F(K) = \{v : v = \hat{v} \circ \boldsymbol{F}^{-1}, \hat{v} \in \hat{V}\}.$$

For all the examples given previously, the space W_h may be defined in this way, beginning with the shape functions denoted in the figures. This preserves the appropriate interelement continuity when the usual degrees of freedom are chosen. The same mapping, applied to each component, can be used with minor exceptions to define the space $\boldsymbol{\Theta}_h$. One exception occurs for the Durán–Liberman element, where one now defines the edge bubbles $\boldsymbol{b}_i = (\hat{b}_i \circ \boldsymbol{F}^{-1})\boldsymbol{s}_i$ where $\boldsymbol{s}_i$ denotes the unit tangent on the ith edge of K. There is also the possibility of using a different mapping to define the interior degrees of freedom for the space $\boldsymbol{\Theta}_h$, since this will not affect the interelement continuity.

To define the space $\boldsymbol{\Gamma}_h$, we use a rotated version of the Piola transform. Letting $D\boldsymbol{F}$ denote the Jacobian matrix of the transformation $\boldsymbol{F}$, if $\hat{\boldsymbol{\eta}}$ is a vector function defined on $\hat{K}$, we define $\boldsymbol{\eta}$ on K by

$$\boldsymbol{\eta}(\boldsymbol{x}) = \boldsymbol{\eta}(\boldsymbol{F}(\hat{\boldsymbol{x}})) = [D\boldsymbol{F}(\hat{\boldsymbol{x}})]^{-t}\hat{\boldsymbol{\eta}}(\hat{\boldsymbol{x}}),$$

where A^{-t} denotes the transpose of the inverse of the matrix A. Then if $\hat{\boldsymbol{V}}$ is a set of vector shape functions given on $\hat{K}$, we define

$$\boldsymbol{V}_F(K) = \{\boldsymbol{\eta} : \boldsymbol{\eta} = [D\boldsymbol{F}]^{-t}\hat{\boldsymbol{\eta}} \circ \boldsymbol{F}^{-1}, \hat{\boldsymbol{\eta}} \in \hat{\boldsymbol{V}}\}.$$

For $w \in W_h$, $\mathbf{grad}\, w = D\boldsymbol{F}^{-t}\widehat{\mathbf{grad}}\,\hat{w}$. Hence, if on the reference square $\widehat{\mathbf{grad}}\,\hat{w} \subseteq \hat{\boldsymbol{V}}$, we will also have $\mathbf{grad}\, w \subseteq \boldsymbol{\Gamma}_h$, a key condition in our analysis.

Although the extensions to quadrilaterals are in most cases straightforward to define, the question is whether the method retains the same order of approximation as in the rectangular case. The problem, as discussed in [3, 6, 5, 4], is that the approximation properties of some of the elements can deteriorate, depending on the way that the mesh is refined. Thus, much of the existing analysis for quadrilateral elements is restricted to the case of parallelograms (e.g., [48]), where the mapping $\boldsymbol{F}$ is affine, or to elements that are $O(h^2)$ perturbations of parallelograms. Another possibility is to restrict the refinement strategy to produce asymptotically affine meshes, so that the deterioration in approximation is also avoided. Error estimates are obtained for the DL4 method for shape-regular quadrilateral meshes and for the MITC4 method for asymptotically parallelogram meshes in [31]. However, numerical experiments do not indicate any deterioration of convergence rates for MITC4, even for more general shape regular meshes.

The MITC8 element approximates both $\boldsymbol{\theta}$ and w by spaces obtained from mappings of the quadratic serendipity space. Since this space does not contain all of Q_2, (i.e, it is missing the basis function x^2y^2), we expect to see only $O(h)$ convergence. The space $\boldsymbol{\Gamma}_h$ is obtained by mapping the $\boldsymbol{BDM}_1^{\perp}$ space, which also degrades in convergence after a bilinear mapping. The MITC9 element uses the full Q_2 approximation for $\boldsymbol{\theta}$, but the use of the Q_2 serendipity space to approximate w and the $\boldsymbol{BDFM}_2^{\perp}$ space to approximate γ will cause degradation in the convergence rate on general quadrilateral meshes.

11 Other Approaches

So far, all the finite element methods discussed have basically followed the common approach of modifying the original variational formulation only by the introduction of the reduction operator $\boldsymbol{\Pi}^{\Gamma}$. However, there are a number of other approaches that produce locking-free approximation schemes by modifying the variational formulation in other ways. Although we will not analyze these methods in detail, the main ideas are presented for a sampling of such methods in the following subsections.

11.1 Expanded Mixed Formulations

One of the first approaches to developing locking-free finite elements for the Reissner–Mindlin plate problem was the method proposed by Brezzi and Fortin [24], based on the expanded mixed formulation (17)–(20). There are now four variables to approximate and piecewise linear functions are used to approximate r, p, and w, while piecewise linears plus cubic bubble functions are used to approximate $\boldsymbol{\theta}$. The key idea was that equations (18) are (19) are perturbations of the stationary Stokes equations, and so a stable conforming approximation is obtained by Stokes elements with continuous pressures (note that (19) requires $p \in H^1(\Omega)$). The choice made for these two variables was the mini element. In fact, the Arnold–Falk method presented earlier was developed as a modification of this method that had the added feature that the finite element method was also equivalent to a method using only the primitive variables $\boldsymbol{\theta}$ and w. The new idea in [11] was to use a discrete Helmholtz decomposition of piecewise constant functions as the element-wise gradient of nonconforming piecewise linear functions plus the curl of continuous piecewise linear functions to reduce the discrete expanded mixed formulation back to a discrete formulation using only the primitive variables.

11.2 Simple Modification of the Reissner–Mindlin Energy

In this method by Arnold and Brezzi [7], the definition of the variable γ is modified to be

$$\gamma = \lambda(t^{-2} - 1)(\boldsymbol{\theta} - \mathbf{grad}\, w)$$

and a new bilinear form is defined:

$$a(\boldsymbol{\theta}, w; \boldsymbol{\phi}, v) = (\mathfrak{C}\varepsilon(\boldsymbol{\theta}), \varepsilon(\boldsymbol{\phi})) + \lambda(\boldsymbol{\theta} - \mathbf{grad}\, w, \boldsymbol{\psi} - \mathbf{grad}\, v).$$

Then a modified weak formulation of the Reissner–Mindlin equations is to find $(\boldsymbol{\theta}, w, \boldsymbol{\gamma}) \in \mathring{\boldsymbol{H}}^1(\Omega) \times \mathring{H}^1(\Omega) \times \boldsymbol{L}^2(\Omega)$ such that

$$a(\boldsymbol{\theta}, w; \boldsymbol{\phi}, v) + \lambda^{-1} t^2 (\boldsymbol{\gamma}, \boldsymbol{\phi} - \mathbf{grad}\, v) = (g, v) - (\boldsymbol{f}, \boldsymbol{\phi}),$$
$$\boldsymbol{\phi} \in \mathring{\boldsymbol{H}}^1(\Omega), v \in \mathring{H}^1(\Omega),$$
$$(\mathbf{grad}\, w - \boldsymbol{\theta}, \boldsymbol{\eta}) - \frac{t^2}{\lambda(1-t^2)}(\boldsymbol{\gamma}, \boldsymbol{\eta}) = 0, \quad \boldsymbol{\eta} \in \boldsymbol{L}^2(\Omega).$$

When this formulation is discretized by finite elements, we no longer need the condition that $\mathbf{grad}\, W_h \subset \boldsymbol{\Gamma}_h$, since the form $a(\boldsymbol{\theta}, w; \boldsymbol{\phi}, v)$ is coercive over $\mathring{\boldsymbol{H}}^1(\Omega) \times \mathring{H}^1(\Omega)$. Hence, greater flexibility is allowed in the design of stable elements. Using this formulation, the choice

$$\boldsymbol{\Theta}_h = \mathring{\boldsymbol{M}}_1^0 + \boldsymbol{B}_3, \quad W_h = \mathring{M}_2^0, \quad \boldsymbol{\Gamma}_h = \boldsymbol{M}_0.$$

gives a stable discretization and the error estimate

$$\|\boldsymbol{\theta} - \boldsymbol{\theta}_h\|_1 + t\|\boldsymbol{\gamma} - \boldsymbol{\gamma}_h\|_0 + \|w - w_h\|_1 \le Ch(\|\boldsymbol{f}\|_0 + \|g\|_0).$$

11.3 Least-Squares Stabilization Schemes

In this approach by Hughes–Franca [37] and Stenberg [47], the bilinear forms defining the method are modified by adding least-squares type stabilization terms. The approach of Stenberg is simpler and we present that here. A weak formulation of the Reissner–Mindlin equations without the introduction of the shear stress is to find $(\boldsymbol{\theta}, w) \in \mathring{\boldsymbol{H}}^1(\Omega) \times \mathring{H}^1(\Omega)$ such that

$$B(\boldsymbol{\theta}, w; \boldsymbol{\phi}, v) = (g, v) - (\boldsymbol{f}, \boldsymbol{\phi}), \quad \boldsymbol{\psi} \in \mathring{\boldsymbol{H}}^1(\Omega), v \in \mathring{H}^1(\Omega), \tag{53}$$

where

$$B(\boldsymbol{\theta}, w; \boldsymbol{\phi}, v) = a(\boldsymbol{\theta}, \boldsymbol{\phi}) + \lambda t^{-2}(\boldsymbol{\theta} - \mathbf{grad}\, w, \boldsymbol{\phi} - \mathbf{grad}\, v).$$

In the stabilized scheme, we define

$$B_h(\boldsymbol{\theta}, w; \boldsymbol{\phi}, v) = a(\boldsymbol{\theta}, \boldsymbol{\phi}) - \alpha \sum_{T \in \mathcal{T}_h} h_T^2 (\boldsymbol{L}\boldsymbol{\theta}, \boldsymbol{L}\boldsymbol{\psi})_T$$
$$+ \sum_{T \in \mathcal{T}_h} (\lambda^{-1} t^2 + \alpha h_T^2)^{-1} (\boldsymbol{\theta} - \mathbf{grad}\, w + \alpha h_T^2 \boldsymbol{L}\boldsymbol{\theta}, \boldsymbol{\phi} - \mathbf{grad}\, v + \alpha h_T^2 \boldsymbol{L}\boldsymbol{\phi})_T,$$

where $\boldsymbol{L}\boldsymbol{\theta} = \operatorname{div} \mathfrak{C}\varepsilon(\boldsymbol{\theta})$, and then seek an approximate solution $(\boldsymbol{\theta}_h, w_h) \in \boldsymbol{\Theta}_h \times W_h$ such that

$$B_h(\boldsymbol{\theta}_h, w_h; \boldsymbol{\phi}, v) = (g, v) - (\boldsymbol{f}, \boldsymbol{\phi}), \quad \boldsymbol{\psi} \in \boldsymbol{\Theta}_h, v \in W_h,$$

The new bilinear form B_h is constructed so that the new formulation is both consistent and stable independent of the choice of finite element spaces. Dictated by approximation theory estimates with respect to the norms used, the choices $\boldsymbol{\Theta}_h = \boldsymbol{M}_k^0$,

$W_h = M^0_{k+1}$ are considered for $k \geq 1$. In the lowest order case $k = 1$, $L\phi|_T = 0$ for all $T \in \mathcal{T}_h$ and all $\phi \in \Theta_h$ and hence the bilinear form reduces to:

$$B_h(\theta_h, w_h; \phi, v) = a(\theta, \phi) + \sum_{T \in \mathcal{T}_h} (\lambda^{-1}t^2 + \alpha h_T^2)^{-1}(\theta - \mathbf{grad}\, w, \phi - \mathbf{grad}\, v)_T,$$

a method proposed in Pitkäranta [46]. Under the hypothesis $0 < \alpha < C_I$ (for an appropriately chosen constant C_I), it is shown that

$$\|\theta - \theta_h\|_1 + \|w - w_h\|_1 \leq Ch^k(\|w\|_{k+2} + \|\theta\|_{k+1}),$$

Estimates in other norms and for additional quantities are also obtained.

A modification of this method is also considered in [25]. In the modified method, $\Theta_h = \mathring{M}^0_1$, $W_h = \mathring{M}^0_1$, and the term $(\theta - \mathbf{grad}\, w, \phi - \mathbf{grad}\, v)$ is modified to $(\Pi^\Gamma\theta - \mathbf{grad}\, w, \Pi^\Gamma\phi - \mathbf{grad}\, v)$ by adding the interpolation operator Π^Γ into the space $RT_0^\perp$. Thus, the method uses only linear elements. We also note that a stabilized version of the MITC4 element is proposed and analyzed in [42].

In Lyly [41], it is shown that the linked interpolation method discussed previously has close connections (and in some cases is equivalent) to the stabilized method of [25] and also to a stabilized linked method proposed by Tessler and Hughes [50]. The connection to the method of [25] is established by proving that for $\phi \in \mathring{M}^1_0$, $\phi - \mathbf{grad}\, L\phi = \Pi^\Gamma\phi$, where Π^Γ denotes the usual interpolant in $RT^0_\perp$. Connections to the stabilized methods are then established by using static condensation to eliminate the cubic bubble functions.

11.4 Discontinuous Galerkin Methods [9], [8]

In this approach, the bilinear forms are modified to include terms that allow the use of totally discontinuous elements. We use the notation $H^s(\mathcal{T}_h)$ to denote functions whose restrictions to T belong to $H^s(T)$ for all $T \in \mathcal{T}_h$. To define the modified forms, we first define the jump and average of a function in $H^1(\mathcal{T}_h)$ as functions on the union of the edges of the triangulation. Let e be an internal edge of $\mathcal{T}_h$, shared by two elements T^+ and T^-, and let n^+ and n^- denote the unit normals to e, pointing outward from T^+ and T^-, respectively. For a scalar function $\varphi \in H^1(\mathcal{T}_h)$, its average and jump on on e are defined respectively, by

$$\{\varphi\} = \frac{\varphi^+ + \varphi^-}{2}, \qquad [\![\varphi]\!] = \varphi^+ n^+ + \varphi^- n^-.$$

Note that the jump is a vector normal to e. The jump of a vector $\phi \in H^1(\mathcal{T}_h)$ is the symmetric matrix-valued function given on e by:

$$[\![\phi]\!] = \phi^+ \odot n^+ + \phi^- \odot n^-,$$

where $\phi \odot n = (\phi \otimes n + n \otimes \phi)/2$ is the symmetric part of the tensor product of ϕ and n. On a boundary edge, the average $\{\varphi\}$ is defined simply as the trace of φ, while for a scalar-valued function, we define $[\![\varphi]\!]$ to be $\varphi\boldsymbol{n}$ (with $\boldsymbol{n}$ the outward unit normal), and for a vector-valued function we define $[\![\boldsymbol{\phi}]\!] = \boldsymbol{\phi} \odot \boldsymbol{n}$.

To obtain a DG discretization, we have to choose finite-dimensional subspaces $\boldsymbol{\Theta}_h \subset \boldsymbol{H}^2(\mathcal{T}_h)$, $W_h \subset H^1(\mathcal{T}_h)$, and $\boldsymbol{\Gamma}_h \subset \boldsymbol{H}^1(\mathcal{T}_h)$. The method then takes the form:

Find $(\boldsymbol{\theta}_h, w_h) \in \boldsymbol{\Theta}_h \times W_h$ *and* $\gamma_h \in \boldsymbol{\Gamma}_h$ *such that*

$$\begin{aligned}
&(\mathfrak{C}\,\mathcal{E}_h(\boldsymbol{\theta}_h), \mathcal{E}_h(\boldsymbol{\phi})) - \langle\{\mathfrak{C}\,\mathcal{E}_h(\boldsymbol{\theta}_h)\}, [\![\boldsymbol{\phi}]\!]\rangle - \langle[\![\boldsymbol{\theta}_h]\!], \{\mathfrak{C}\,\mathcal{E}_h(\boldsymbol{\phi})\}\rangle\\
&\quad + (\gamma_h, \mathbf{grad}_h\, v - \boldsymbol{\phi}) - \langle\{\gamma_h\}, [\![v]\!]\rangle\\
&\quad + p_\Theta(\boldsymbol{\theta}_h, \boldsymbol{\phi}) + p_W(w_h, v) = (g, v) - (\boldsymbol{f}, \boldsymbol{\phi}), \quad (\boldsymbol{\phi}, v) \in \boldsymbol{\Theta}_h \times W_h,\\
&(\mathbf{grad}_h\, w_h - \boldsymbol{\theta}_h, \boldsymbol{\eta}) - \langle[\![w_h]\!], \{\boldsymbol{\eta}\}\rangle - t^2(\gamma_h, \boldsymbol{\eta}) = 0, \quad \boldsymbol{\eta} \in \boldsymbol{\Gamma}_h.
\end{aligned}$$

We make a standard choice for the interior penalty terms p_Θ and p_W:

$$p_\Theta(\boldsymbol{\theta}, \boldsymbol{\phi}) = \sum_{e\in\mathcal{E}_h} \frac{\kappa^\Theta}{|e|}\int_e [\![\boldsymbol{\theta}]\!] : [\![\boldsymbol{\phi}]\!]\, ds, \quad p_W(w, v) = \sum_{e\in\mathcal{E}_h} \frac{\kappa^W}{|e|}\int_e [\![w]\!] \cdot [\![v]\!]\, ds,$$

so that $p_\Theta(\boldsymbol{\phi}, \boldsymbol{\phi})$, $(p_W(v, v)$, resp.) can be viewed as a measure of the deviation of $\boldsymbol{\phi}$ (v, resp.) from being continuous. The parameters κ^Θ and κ^W are positive constants to be chosen; they must be sufficiently large to ensure stability. In the case when W_h consists of continuous elements, the penalty term p_W will not be needed.

In the simplest of such methods, one chooses for $k \geq 1$, $W_h = \mathring{M}^0_{k+1}$, i.e., continuous piecewise polynomials of degree $\leq k+1$. We then choose $w^I = \Pi^W w$, where Π^W is defined as for the MITC elements. Since the space $\boldsymbol{\Theta}_h$ need not be continuous, we can now choose $\boldsymbol{\Theta}_h$ so that condition (30) is satisfied without the need for a reduction operator $\boldsymbol{\Pi}^\Gamma$. The simplest choice is $\boldsymbol{\Theta}_h = \boldsymbol{BDM}^\perp_{k-1}$. We note that $\mathbf{grad}\, W_h \subset \boldsymbol{\Theta}_h$. We next define $\boldsymbol{\theta}^I = \boldsymbol{\Pi}^\Theta\boldsymbol{\theta}$, where $\boldsymbol{\Pi}^\Theta : \boldsymbol{H}^1(\Omega) \mapsto \boldsymbol{\Theta}_h$ is defined by the conditions:

$$\begin{aligned}
&\int_e (\boldsymbol{\phi} - \boldsymbol{\Pi}^\Theta\boldsymbol{\phi}) \cdot \boldsymbol{s}\, q\, \mathrm{d}s = 0, \quad q \in P_{k-1}(e),\\
&\int_T (\boldsymbol{\phi} - \boldsymbol{\Pi}^\Theta\boldsymbol{\phi}) \cdot \boldsymbol{q}\, \mathrm{d}x = 0, \quad \boldsymbol{q} \in \boldsymbol{RT}_{k-3}(T),
\end{aligned}$$

where $\boldsymbol{RT}_{k-3}$ is the usual (unrotated) Raviart–Thomas space of index $k-3$. We note that the interior degrees of freedom are not the original degrees of freedom defined for these spaces. However, the natural interpolant defined by these modified degrees of freedom satisfies the additional and key property that

$$\boldsymbol{\Pi}^\Theta\, \mathbf{grad}\, w = \mathbf{grad}\, \Pi^W w.$$

From this condition, we get

$$\begin{aligned}
\gamma^I = \lambda t^{-2}(\mathbf{grad}\, w^I - \boldsymbol{\theta}^I) &= \lambda t^{-2}(\mathbf{grad}\, \Pi^W w - \boldsymbol{\Pi}^\Theta\boldsymbol{\theta})\\
&= \lambda t^{-2}\boldsymbol{\Pi}^\Theta(\mathbf{grad}\, w - \boldsymbol{\theta}) = \boldsymbol{\Pi}^\Theta\gamma.
\end{aligned}$$

11.5 Methods Using Nonconforming Finite Elements

In the nonconforming element of Oñate, Zarate, and Flores [43], one chooses

$$\boldsymbol{\Theta}_h = \mathring{\boldsymbol{M}}_1^*, \quad W_h = \mathring{M}_1^0, \quad \boldsymbol{\Gamma}_h = \boldsymbol{RT}_0^{\perp}.$$

In this case, $\boldsymbol{\Theta}_h$ is not contained in $\mathring{\boldsymbol{H}}^1(\Omega)$, and so $\mathcal{E}$ must be replaced by $\mathcal{E}_h$. The main problem with this method is that $\| \mathcal{E}_h(\boldsymbol{\Theta}_h)\|_0^2$ is not a norm on $\boldsymbol{\Theta}_h$ because Korn's inequality fails for nonconforming piecewise linear functions. To partially compensate for this fact, one can use the following result, established in [14]. Define

$$\boldsymbol{Z}_h = \left\{ (\boldsymbol{\psi}, \boldsymbol{\eta}) \in \mathring{\boldsymbol{M}}_1^* \times \boldsymbol{\Gamma}_h : \lambda^{-1} t^2 \operatorname{rot} \boldsymbol{\eta} = \operatorname{rot}_h \boldsymbol{\psi} \right\}. \tag{54}$$

Lemma 11.1 *There exists a constant c independent of h and t such that*

$$a_h(\boldsymbol{\psi}, \boldsymbol{\psi}) + \lambda^{-1} t^2 (\boldsymbol{\eta}, \boldsymbol{\eta}) \geq c[\min(1, h^2/t^2) \|\boldsymbol{\psi}\|_{1,h}^2 + \| \mathcal{E}_h \boldsymbol{\psi}\|_0^2$$
$$+ t^2 \|\boldsymbol{\eta}\|_0^2 + h^2 t^2 \| \operatorname{rot} \boldsymbol{\eta}\|_0^2] \quad \text{forall } (\boldsymbol{\psi}, \boldsymbol{\eta}) \in \boldsymbol{Z}_h.$$

Note that the bilinear form is not uniformly coercive. It is then possible to establish the following error estimates (cf. [14]).

Theorem 11.2 *There exists a constant C independent of h and t such that*

$$\|\boldsymbol{\theta} - \boldsymbol{\theta}_h\|_{1,h} + t^2 \| \operatorname{rot}(\boldsymbol{\gamma} - \boldsymbol{\gamma}_h)\|_0^2 \leq Ch \max(1, t^2/h^2) \|g\|_0,$$
$$\| \mathcal{E}(\boldsymbol{\theta} - \boldsymbol{\theta}_h)\|_0^2 + t\|\boldsymbol{\gamma} - \boldsymbol{\gamma}_h\|_0 \leq Ch \max(1, t/h) \|g\|_0.$$
$$\|\boldsymbol{\theta} - \boldsymbol{\theta}_h\|_0 + \|w - w_h\|_0 \leq C \max(h^2, t^2) \|g\|_0.$$

Note that this theorem does not imply convergence of the method. If $h \sim t$, however, the error will be small.

In the method proposed by Lovadina [40],

$$\boldsymbol{\Theta}_h = \mathring{\boldsymbol{M}}_1^*, \quad W_h = \mathring{M}_1^*, \quad \boldsymbol{\Gamma}_h = \mathring{\boldsymbol{M}}_0,$$

so two of the spaces are nonconforming. Hence, both $\mathcal{E}$ and **grad** are replaced by their element-wise counterparts. In addition, the bilinear form $a(\boldsymbol{\theta}, \boldsymbol{\phi})$ is replaced by

$$a_h(\boldsymbol{\theta}, \boldsymbol{\phi}) = \sum_{T \in \mathcal{T}_h} a_T(\boldsymbol{\theta}, \boldsymbol{\phi}) + p_\Theta(\boldsymbol{\theta}, \boldsymbol{\phi}), \quad a_T(\boldsymbol{\theta}, \boldsymbol{\phi}) = \int_T (\mathfrak{C} \mathcal{E}\, \boldsymbol{\theta}) : \mathcal{E}(\boldsymbol{\phi})\, dx,$$

where p_Θ has the same definition as in the discontinuous Galerkin method. By adding the term p_Θ, one is able to establish a discrete Korn's inequality.

This method is a simplified version of a method proposed earlier by Brezzi–Marini [26]. Using a similar formulation, they made the choices

$$\boldsymbol{\Theta}_h = \mathring{\boldsymbol{M}}_1^* + \boldsymbol{B}_2^*, \quad W_h = \mathring{M}_1^* + B_2^*, \quad \boldsymbol{\Gamma}_h = \boldsymbol{M}_0 + \mathbf{grad}_h\, B_2^*,$$

where B_2^* denotes the nonconforming quadratic bubble function that vanishes at the two Gauss points of each edge of a triangle. See also [27] for L^2 estimates for the method of [40].

11.6 A Negative-Norm Least Squares Method

This method, proposed by Bramble–Sun [22], begins with the expanded mixed formulation used by Brezzi–Fortin. The problem is then reformulated as a least squares method using a special minus one norm developed previously by Bramble, Lazarov, and Pasciak. Only continuous finite elements are needed to approximate all the variables, and piecewise linears can be used. Optimal order error estimates are established uniformly in the thickness t. The stability result also gives a natural block diagonal preconditioner, using only standard preconditioners for second order elliptic problems, for the solution of the resulting least squares system.

12 Summary

We have treated in these notes only a selection of the finite element methods that have been developed for the approximation of the Reissner–Mindlin plate problem, concentrating on those for which there is a mathematical analysis. There are many other methods available in the engineering literature, and the list is too long to give proper citations.

References

1. S. M. Alessandrini, D. N. Arnold, R. S. Falk, and A. L. Madureira. Derivation and justification of plate models by variational methods. In *Plates and Shells (Quebec 1996)*, pp. 1–20. American Mathematical Society, Providence, RI, 1999.
2. D. N. Arnold. Innovative finite element methods for plates. *Math. Appl. Comput.*, 10(2):77–88, 1991.
3. D. N. Arnold, D. Boffi, and R. S. Falk. Approximation by quadrilateral finite elements. *Math. Comput.*, 71(239):909–922 (electronic), 2002.
4. D. N. Arnold, D. Boffi, and R. S. Falk. Remarks on quadrilateral Reissner-Mindlin plate elements. In *Proceedings of the Fifth World Congress on Computational Mechanics (WCCM V)*. Vienna University of Technology, Austria, 2002.
5. D. N. Arnold, D. Boffi, and R. S. Falk. Quadrilateral $H(\mathrm{div})$ finite elements. *SIAM J. Numer. Anal.*, 42(6):2429–2451 (electronic), 2005.
6. D. N. Arnold, D. Boffi, R. S. Falk, and L. Gastaldi. Finite element approximation on quadrilateral meshes. *Commun. Numer. Methods Eng.*, 17(11):805–812, 2001.
7. D. N. Arnold and F. Brezzi. Some new elements for the Reissner–Mindlin plate model. In *Boundary Value Problems for Partial Differential Equations and Applications*, *RMA Res. Notes Appl. Math.*, vol. 29, pp. 287–292. Masson, Paris, 1993.
8. D. N. Arnold, F. Brezzi, R. S. Falk, and L. D. Marini. Locking-free Reissner–Mindlin elements without reduced integration. *Comput. Methods Appl. Mech. Eng.*, 196(37–40): 3660–3671, 2007.
9. D. N. Arnold, F. Brezzi, and L. D. Marini. A family of discontinuous Galerkin finite elements for the Reissner–Mindlin plate. *J. Sci. Comput.*, 22/23:25–45, 2005.
10. D. N. Arnold and R. S. Falk. Edge effects in the Reissner–Mindlin plate theory. In *Analytic and Computational Models of Shells*, pp. 71–90. A.S.M.E., New York, 1989.

11. D. N. Arnold and R. S. Falk. A uniformly accurate finite element method for the Reissner–Mindlin plate. *SIAM J. Numer. Anal.*, 26:1276–1290, 1989.
12. D. N. Arnold and R. S. Falk. The boundary layer for the Reissner–Mindlin plate model. *SIAM J. Math. Anal.*, 21(2):281–312, 1990.
13. D. N. Arnold and R. S. Falk. Asymptotic analysis of the boundary layer for the Reissner–Mindlin plate model. *SIAM J. Math. Anal.*, 27(2):486–514, 1996.
14. D. N. Arnold and R. S. Falk. Analysis of a linear–linear finite element for the Reissner–Mindlin plate model. *Math. Models Methods Appl. Sci.*, 7(2):217–238, 1997.
15. F. Auricchio and C. Lovadina. Analysis of kinematic linked interpolation methods for Reissner–Mindlin plate problems. *Comput. Methods Appl. Mech. Eng.*, 190(18–19):2465–2482, 2001.
16. F. Auricchio and R. L. Taylor. A triangular thick plate element with an exact thin limit. *Finite Elem. Anal. Des.*, 19:57–68, 1995.
17. K.-J. Bathe and F. Brezzi. On the convergence of a four-node plate bending element based on Mindlin–Reissner plate theory and a mixed interpolation. In *The Mathematics of Finite Elements and Applications, V (Uxbridge, 1984)*, pp. 491–503. Academic Press, London, 1985.
18. K.-J. Bathe and F. Brezzi. A simplified analysis of two plate bending elements – the MITC4 and MITC9 elements. In *Numerical Techniques for Engineering Analysis and Design*, vol. 1. Martinus Nijhoff, Amsterdam, 1987.
19. K.-J. Bathe and F. Brezzi and S. W. Cho. The MITC7 and MITC9 plate bending elements. In *Comput. Struct*, 32:797–841, 1989.
20. K.-J. Bathe and E. N. Dvorkin. A four-node plate bending element based on Mindlin/Reissner plate theory and mixed interpolation. *Int. J. Numer. Methods Eng.*, 21:367–383, 1985.
21. K.-J. Bathe and E. N. Dvorkin. A formulation of general shell elements—the use of mixed interpolation of tensorial components. *Int. J. Numer. Methods Eng.*, 22:697–722, 1986.
22. J. H. Bramble and T. Sun. A negative-norm least squares method for Reissner–Mindlin plates. *Math. Comp.*, 67(223):901–916, 1998.
23. F. Brezzi, K.-J. Bathe, and M. Fortin. Mixed-interpolated elements for Reissner–Mindlin plates. *Int. J. Numer. Methods Eng.*, 28(8):1787–1801, 1989.
24. F. Brezzi and M. Fortin. Numerical approximation of Mindlin–Reissner plates. *Math. Comput.*, 47(175):151–158, 1986.
25. F. Brezzi, M. Fortin, and R. Stenberg. Error analysis of mixed-interpolated elements for Reissner–Mindlin plates. *Math. Models Methods Appl. Sci.*, 1(2):125–151, 1991.
26. F. Brezzi and L. D. Marini. A nonconforming element for the Reissner–Mindlin plate. *Comput. & Struct.*, 81(8–11):515–522, 2003. In honour of Klaus-Jürgen Bathe.
27. C. Chinosi, C. Lovadina, and L. D. Marini. Nonconforming finite elements for Reissner-Mindlin plates. In *Applied and Industrial Mathematics in Italy*, *Ser. Adv. Math. Appl. Sci.*, vol. 69, pp. 213–224. World Sci. Publ., Hackensack, NJ, 2005.
28. M. Crouzeix and P.-A. Raviart. Conforming and nonconforming finite element methods for solving the stationary Stokes equations. I. *Rev. Française Automat. Informat. Recherche Opérationnelle Sér. Rouge*, 7(R-3):33–75, 1973.
29. R. Durán. The inf–sup condition and error estimates for the Arnold–Falk plate bending element. *Numer. Math.*, 59(8):769–778, 1991.
30. R. Durán, A. Ghioldi, and N. Wolanski. A finite element method for the Mindlin–Reissner plate model. *SIAM J. Numer. Anal.*, 28(4):1004–1014, 1991.
31. R. Durán, E. Hernández, L. Hervella-Nieto, E. Liberman, and R. Rodríguez. Error estimates for low-order isoparametric quadrilateral finite elements for plates. *SIAM J. Numer. Anal.*, 41(5):1751–1772 (electronic), 2003.

32. R. Durán, L. Hervella-Nieto, E. Liberman, R. Rodríguez, and J. Solomin. Approximation of the vibration modes of a plate by Reissner–Mindlin equations. *Math. Comp.*, 68(228):1447–1463, 1999.
33. R. Durán and E. Liberman. On mixed finite element methods for the Reissner–Mindlin plate model. *Math. Comp.*, 58(198):561–573, 1992.
34. R. Durán and E. Liberman. On the convergence of a triangular mixed finite element method for Reissner–Mindlin plates. *Math. Models Methods Appl. Sci.*, 6(3):339–352, 1996.
35. R. S. Falk and T. Tu. Locking-free finite elements for the Reissner–Mindlin plate. *Math. Comp.*, 69(231):911–928, 2000.
36. L. P. Franca and R. Stenberg. A modification of a low-order Reissner–Mindlin plate bending element. In *The Mathematics of Finite Elements and Applications, VII (Uxbridge, 1990)*, pp. 425–436. Academic Press, London, 1991.
37. T. J. R. Hughes and L. P. Franca. A mixed finite element formulation for Reissner–Mindlin plate theory: uniform convergence of all higher-order spaces. *Comput. Methods Appl. Mech. Eng.*, 67(2):223–240, 1988.
38. T. J. R. Hughes and T. E. Tezuyar. Finite elements based upon Mindlin plate theory with particular reference to the four node blinear isoparamtric element. *J. Appl. Mech. Eng.*, 48:587–598, 1981.
39. C. Lovadina. Analysis of a mixed finite element method for the Reissner–Mindlin plate problems. *Comput. Methods Appl. Mech. Eng.*, 163(1–4):71–85, 1998.
40. C. Lovadina. A low-order nonconforming finite element for Reissner–Mindlin plates. *SIAM J. Numer. Anal.*, 42(6):2688–2705 (electronic), 2005.
41. M. Lyly. On the connection between some linear triangular Reissner–Mindlin plate bending elements. *Numer. Math.*, 85(1):77–107, 2000.
42. M. Lyly, R. Stenberg, and T. Vihinen. A stable bilinear element for the Reissner–Mindlin plate model. *Comput. Methods Appl. Mech. Eng.*, 110(3–4):343–357, 1993.
43. E. Oñate, F. Zarate, and F. Flores. A simple triangular element for thick and thin plate and shell analysis. *Int. J. Numer. Methods Eng.*, 37:2569–2582, 1994.
44. P. Peisker and D. Braess. Uniform convergence of mixed interpolated elements for Reissner–Mindlin plates. *RAIRO Modél. Math. Anal. Numér.*, 26(5):557–574, 1992.
45. I. Perugia and T. Scapolla. Optimal rectangular MITC finite elements for Reissner–Mindlin plates. *Numer. Methods Partial Differen. Equations*, 13(5):575–585, 1997.
46. J. Pitkäranta. Analysis of some low-order finite element schemes for Mindlin–Reissner and Kirchhoff plates. *Numer. Math.*, 53(1–2):237–254, 1988.
47. R. Stenberg. A new finite element formulation for the plate bending problem. In *Asymptotic Methods for Elastic Structures (Lisbon, 1993)*, pp. 209–221. de Gruyter, Berlin, 1995.
48. R. Stenberg and M. Suri. An *hp* error analysis of MITC plate elements. *SIAM J. Numer. Anal.*, 34(2):544–568, 1997.
49. R. L. Taylor and F. Auricchio. Linked interpolation for Reissner–Mindlin plate elements: Part II – a simple triangle. *Int. J. Num. Meths. Engrg.*, 50:71–101, 1985.
50. A. Tessler and T. J. R. Hughes. A three-node mindlin plate element with improved transverse shear. *Comput. Methods Appl. Mech. Eng.*, 50:71–101, 1985.
51. Z. Xu. A thick-thin triangular plate element. *Int. J. Numer. Methods Eng.*, 33:963–973, 1992.
52. O. C. Zienkiewicz and D. Lefebvre. A robust triangular plate bending element of the Reissner–Mindlin plate. *Int. J. Numer. Methods Eng.*, 26:1169–1184, 1998.
53. O. C. Zienkiewicz, R. L. Taylor, P. Papadopoulos, and E. Oñate. Plate bending elements with discrete constraints: New triangular elements. *Comput. & Struct.*, 35:505–522, 1990.

List of Participants

1. Antonietti Paola Francesca
   University of Pavia, Italy
   paola.antonietti@unipv.it
2. Ayuso Blanca
   IMATI-CNR, Italy
   blanca@imati.cnr.it
3. Boal Natalia
   University of Zaragoza, Spain
   nboal@unizar.es
4. Boffi Daniele
   University of Pavia, Italy
   daniele.boffi@unipv.it
   editor, lecturer
5. Brezzi Franco
   IMATI-CNR, Italy
   brezzi@imati.cnr.it
   lecturer
6. Busa Jan
   Ghent University, Belgium
   Jan.Busa@ugent.be
7. Cangiani Andrea
   University of Pavia, Italy
   cangiani@unipv.it
8. Colli Franzone Piero
   IMATI-CNR, Italy
   colli@imati.cnr.it
9. Demkowicz Leszek
   University of Texas at Austin, USA
   lesrek@ices.utexas.edu
   lecturer
10. Denis Ivanov
    Saint-Petersburg State University, Russia
    denisiv@rol.ru
11. Dolores Gomez
    University of Santiago de Compostela, Spain
    malola@usc.es
12. Drelichman Irene
    University of Buenos Aires, Argentina
    irene@drelichman.com
13. Durán Ricardo G.
    University of Buenos Aires, Argentina
    rduran@dm.uba.ar
    lecturer
14. Euler Timo
    Technical University of Darmstadt, Germany
    euler@temf.tu-darmstadt.de
15. Falk Richard S.
    Rutgers University, USA
    falk@math.rutgers.edu
    lecturer

16. Ferrandi Paolo Giacomo
Politecnico of Milano, Italy
ferrandi@mate.polimi.it

17. Fortin Michel
University of Laval, Quebec, Canada
mfortin@mat.ulaval.ca
lecturer

18. Fransos Davide
Politecnico of Torino, Italy
davide.fransos@polito.it

19. Garceta Giovanni
University of Calabria, Italy
giovanni.garceta@unical.it

20. Gardini Francesca
University of Ulm, Germany
francesca.gardini@uni-ulm.de

21. Gastaldi Lucia
University of Brescia, Italy
gastaldi@ing.unibs.it
editor

22. Gatto Paolo
University of Pavia, Italy
paologatto81@gmail.cam

23. Hamelinck Wouter
Ghent University, Belgium
wh@cage.ugent.be

24. Heltai Luca
University of Pavia, Italy
luca.heltai@unipv.it

25. Janikova Edita
Ghent University, Belgium
edita.janikova@ugent.be

26. Karakatsani Fotini
University of Crete, Greece
fotini@math.uoc.gr

27. Karamzin Dmitry
Computing Centre of RAS, Russia
dmitry_karamzin@mail.ru

28. Khattri Sanjay
University of Bergen, Belgium
sanjaykhatri1976@yahoo.com

29. Klausen Runhild
University of Oslo, Norway
runhildk@ifi.uio.no

30. Koch Stephan
Technical University of Darmstadt, German
koch@temf.tu-darmstadt.de

31. Kondratyuk Yaroslav
Utrecht University, The Netherlands
Kondratyuk@math.uu.nl

32. Kuteeva Galina
Saint-Petersburg State University, Russia
gkut@rambler.ru

33. Layal Lizaik
University of Pau, France
layal.lizaik@total.com

34. Leykekhman Dmitriy
Rice University, USA
dmitriy@caam.rice.edu

35. Lopez Salvatore
University of Calabria, Italy
salvatore.lopez@strutture.unical.it

36. Manzini Gianmarco
IMATI-CNR, Italy
marco.manzini@imati.cnr.it

37. Marazzina Daniele
University of Pavia, Italy
daniele.marazzina@unipv.it

38. Marini Donatella
IMATI-CNR, Italy
marini@imati.cnr.it

39. Muniz Wagner
University of Karlsruhe, Germany
muniz@math.uni-karlsruhe.de

40. Naumova Natalia
Saint-Petersburg State University, Russia
nat_n75@mail.ru

41. Olech Michal
University of Wroclaw, Poland
olech@math.uni.wroc.pl

42. Osorio Eduardo
Rutgers University, USA
eduardos@math.rutgers.edu

43. Perugia Ilaria
University of Pavia, Italy
ilaria.perugia@unipv.it

44. Phillips Joel
McGill University, Canada
phillips@math.mcgill.ca

45. Rodrguez Salgado
University of Santiago de Compostela, Spain
mpilar@usc.es

46. Rognes Marie Elisabeth
University of Oslo, Norway
meg@math.uio.no

47. Russo Katia
University of Calabria, Italy
katia.russo@labmec.unical.it

48. Saedpanah Fardin
Chalmers University of Technology, Sweden
fardin@math.chalmers.se

49. Salas Oscar
University of Pavia, Italy
oscar.salas@unipv.it

50. Sanhueza Frank Emilio
Concepcion University, Chile
fsanhuez@ing-mat.udec.cl

51. Sacchi Simone
University of Pavia, Italy
simone.scacchi@unipv.it

52. Turco Alessandro
Sissa Trieste, Italy
turco@sissa.it

53. Valli Alberto
University of Trento, Italy
valli@science.unitn.it

54. Zemanova Viera
Ghent University, Belgium
viera@cage.ugent.be

55. Zikatanov Ludmil
Penn State University, USA
ludmil02@gmail.com

LIST OF C.I.M.E. SEMINARS

Published by C.I.M.E

1954
1. Analisi funzionale
2. Quadratura delle superficie e questioni connesse
3. Equazioni differenziali non lineari

1955
4. Teorema di Riemann-Roch e questioni connesse
5. Teoria dei numerl
6. Topologia
7. Teorie non linearizzate in elasticità, idrodinamica, aerodinamic
8. Geometria proiettivo-differenziale

1956
9. Equazioni alle derivate parziali a caratteristiche reali
10. Propagazione delle onde elettromagnetiche automorfe
11. Teoria della funzioni di più variabili complesse e delle funzioni

1957
12. Geometria aritmetica e algebrica (2 vol.)
13. Integrali singolari e questioni connesse
14. Teoria della turbolenza (2 vol.)

1958
15. Vedute e problemi attuali in relatività generale
16. Problemi di geometria differenziale in grande
17. Il principio di minimo e le sue applicazioni alle equazioni funzionali

1959
18. Induzione e statistica
19. Teoria algebrica dei meccanismi automatici (2 vol.)
20. Gruppi, anelli di Lie e teoria della coomologia

1960
21. Sistemi dinamici e teoremi ergodici
22. Forme differenziali e loro integrali

1961
23. Geometria del calcolo delle variazioni (2 vol.)
24. Teoria delle distribuzioni
25. Onde superficiali

1962
26. Topologia differenziale
27. Autovalori e autosoluzioni
28. Magnetofluidodinamica

1963
29. Equazioni differenziali astratte
30. Funzioni e varietà complesse
31. Proprietà di media e teoremi di confronto in Fisica Matematica

1964
32. Relatività generale
33. Dinamica dei gas rarefatti
34. Alcune questioni di analisi numerica
35. Equazioni differenziali non lineari

1965
36. Non-linear continuum theories
37. Some aspects of ring theory
38. Mathematical optimization in economics

Published by Ed. Cremonese, Firenze

1966 39. Calculus of variations
40. Economia matematica
41. Classi caratteristiche e questioni connesse
42. Some aspects of diffusion theory

1967 43. Modern questions of celestial mechanics
44. Numerical analysis of partial differential equations
45. Geometry of homogeneous bounded domains

1968 46. Controllability and observability
47. Pseudo-differential operators
48. Aspects of mathematical logic

1969 49. Potential theory
50. Non-linear continuum theories in mechanics and physics and their applications
51. Questions of algebraic varieties

1970 52. Relativistic fluid dynamics
53. Theory of group representations and Fourier analysis
54. Functional equations and inequalities
55. Problems in non-linear analysis

1971 56. Stereodynamics
57. Constructive aspects of functional analysis (2 vol.)
58. Categories and commutative algebra

1972 59. Non-linear mechanics
60. Finite geometric structures and their applications
61. Geometric measure theory and minimal surfaces

1973 62. Complex analysis
63. New variational techniques in mathematical physics
64. Spectral analysis

1974 65. Stability problems
66. Singularities of analytic spaces
67. Eigenvalues of non linear problems

1975 68. Theoretical computer sciences
69. Model theory and applications
70. Differential operators and manifolds

Published by Ed. Liguori, Napoli

1976 71. Statistical Mechanics
72. Hyperbolicity
73. Differential topology

1977 74. Materials with memory
75. Pseudodifferential operators with applications
76. Algebraic surfaces

Published by Ed. Liguori, Napoli & Birkhäuser

1978 77. Stochastic differential equations
78. Dynamical systems

1979 79. Recursion theory and computational complexity
80. Mathematics of biology

1980 81. Wave propagation
82. Harmonic analysis and group representations
83. Matroid theory and its applications

Published by Springer-Verlag

1981	84. Kinetic Theories and the Boltzmann Equation	(LNM 1048)
	85. Algebraic Threefolds	(LNM 947)
	86. Nonlinear Filtering and Stochastic Control	(LNM 972)
1982	87. Invariant Theory	(LNM 996)
	88. Thermodynamics and Constitutive Equations	(LNP 228)
	89. Fluid Dynamics	(LNM 1047)
1983	90. Complete Intersections	(LNM 1092)
	91. Bifurcation Theory and Applications	(LNM 1057)
	92. Numerical Methods in Fluid Dynamics	(LNM 1127)
1984	93. Harmonic Mappings and Minimal Immersions	(LNM 1161)
	94. Schrödinger Operators	(LNM 1159)
	95. Buildings and the Geometry of Diagrams	(LNM 1181)
1985	96. Probability and Analysis	(LNM 1206)
	97. Some Problems in Nonlinear Diffusion	(LNM 1224)
	98. Theory of Moduli	(LNM 1337)
1986	99. Inverse Problems	(LNM 1225)
	100. Mathematical Economics	(LNM 1330)
	101. Combinatorial Optimization	(LNM 1403)
1987	102. Relativistic Fluid Dynamics	(LNM 1385)
	103. Topics in Calculus of Variations	(LNM 1305)
1988	104. Logic and Computer Science	(LNM 1429)
	105. Global Geometry and Mathematical Physics	(LNM 1451)
1989	106. Methods of nonconvex analysis	(LNM 1446)
	107. Microlocal Analysis and Applications	(LNM 1495)
1990	108. Geometric Topology: Recent Developments	(LNM 1504)
	109. H_∞ Control Theory	(LNM 1496)
	110. Mathematical Modelling of Industrial Processes	(LNM 1521)
1991	111. Topological Methods for Ordinary Differential Equations	(LNM 1537)
	112. Arithmetic Algebraic Geometry	(LNM 1553)
	113. Transition to Chaos in Classical and Quantum Mechanics	(LNM 1589)
1992	114. Dirichlet Forms	(LNM 1563)
	115. D-Modules, Representation Theory, and Quantum Groups	(LNM 1565)
	116. Nonequilibrium Problems in Many-Particle Systems	(LNM 1551)
1993	117. Integrable Systems and Quantum Groups	(LNM 1620)
	118. Algebraic Cycles and Hodge Theory	(LNM 1594)
	119. Phase Transitions and Hysteresis	(LNM 1584)
1994	120. Recent Mathematical Methods in Nonlinear Wave Propagation	(LNM 1640)
	121. Dynamical Systems	(LNM 1609)
	122. Transcendental Methods in Algebraic Geometry	(LNM 1646)
1995	123. Probabilistic Models for Nonlinear PDE's	(LNM 1627)
	124. Viscosity Solutions and Applications	(LNM 1660)
	125. Vector Bundles on Curves. New Directions	(LNM 1649)
1996	126. Integral Geometry, Radon Transforms and Complex Analysis	(LNM 1684)
	127. Calculus of Variations and Geometric Evolution Problems	(LNM 1713)
	128. Financial Mathematics	(LNM 1656)

1997	129. Mathematics Inspired by Biology	(LNM 1714)
	130. Advanced Numerical Approximation of Nonlinear Hyperbolic Equations	(LNM 1697)
	131. Arithmetic Theory of Elliptic Curves	(LNM 1716)
	132. Quantum Cohomology	(LNM 1776)
1998	133. Optimal Shape Design	(LNM 1740)
	134. Dynamical Systems and Small Divisors	(LNM 1784)
	135. Mathematical Problems in Semiconductor Physics	(LNM 1823)
	136. Stochastic PDE's and Kolmogorov Equations in Infinite Dimension	(LNM 1715)
	137. Filtration in Porous Media and Industrial Applications	(LNM 1734)
1999	138. Computational Mathematics driven by Industrial Applications	(LNM 1739)
	139. Iwahori-Hecke Algebras and Representation Theory	(LNM 1804)
	140. Hamiltonian Dynamics - Theory and Applications	(LNM 1861)
	141. Global Theory of Minimal Surfaces in Flat Spaces	(LNM 1775)
	142. Direct and Inverse Methods in Solving Nonlinear Evolution Equations	(LNP 632)
2000	143. Dynamical Systems	(LNM 1822)
	144. Diophantine Approximation	(LNM 1819)
	145. Mathematical Aspects of Evolving Interfaces	(LNM 1812)
	146. Mathematical Methods for Protein Structure	(LNCS 2666)
	147. Noncommutative Geometry	(LNM 1831)
2001	148. Topological Fluid Mechanics	to appear
	149. Spatial Stochastic Processes	(LNM 1802)
	150. Optimal Transportation and Applications	(LNM 1813)
	151. Multiscale Problems and Methods in Numerical Simulations	(LNM 1825)
2002	152. Real Methods in Complex and CR Geometry	(LNM 1848)
	153. Analytic Number Theory	(LNM 1891)
	154. Inverse Problems and Imaging	(LNM 1943)
2003	155. Stochastic Methods in Finance	(LNM 1856)
	156. Hyperbolic Systems of Balance Laws	(LNM 1911)
	157. Symplectic 4-Manifolds and Algebraic Surfaces	(LNM 1938)
	158. Mathematical Foundation of Turbulent Viscous Flows	(LNM 1871)
2004	159. Representation Theory and Complex Analysis	(LNM 1931)
	160. Nonlinear and Optimal Control Theory	(LNM 1932)
	161. Stochastic Geometry	(LNM 1892)
2005	162. Enumerative Invariants in Algebraic Geometry and String Theory	to appear
	163. Calculus of Variations and Non-linear Partial Differential Equations	(LNM 1927)
	164. SPDE in Hydrodynamic. Recent Progress and Prospects	(LNM 1942)
2006	165. Pseudo-Differential Operators, Quantization and Signals	to appear
	166. Mixed Finite Elements, Compatibility Conditions, and Applications	(LNM 1939)
	167. Multiscale Problems in the Life Sciences. From Microscopic to Macroscopic	(LNM 1940)
	168. Quantum Transport: Modelling, Analysis and Asymptotics	to appear
2007	169. Geometric Analysis and Partial Differential Equations	to appear
	170. Nonlinear Optimization	to appear
	171. Arithmetic Geometry	to appear
2008	172. Nonlinear Partial Differential Equations and Applications	announced
	173. Holomorphic Dynamical Systems	announced
	174. Level Set and PDE based Reconstruction Methods: Applications to Inverse Problems and Image Processing	announced
	175. Mathematical models in the manufacturing of glass, polymers and textiles	announced

Lecture Notes in Mathematics

For information about earlier volumes please contact your bookseller or Springer LNM Online archive: springerlink.com

Vol. 1755: J. Azéma, M. Émery, M. Ledoux, M. Yor (Eds.), Séminaire de Probabilités XXXV (2001)
Vol. 1756: P. E. Zhidkov, Korteweg de Vries and Nonlinear Schrödinger Equations: Qualitative Theory (2001)
Vol. 1757: R. R. Phelps, Lectures on Choquet's Theorem (2001)
Vol. 1758: N. Monod, Continuous Bounded Cohomology of Locally Compact Groups (2001)
Vol. 1759: Y. Abe, K. Kopfermann, Toroidal Groups (2001)
Vol. 1760: D. Filipović, Consistency Problems for Heath-Jarrow-Morton Interest Rate Models (2001)
Vol. 1761: C. Adelmann, The Decomposition of Primes in Torsion Point Fields (2001)
Vol. 1762: S. Cerrai, Second Order PDE's in Finite and Infinite Dimension (2001)
Vol. 1763: J.-L. Loday, A. Frabetti, F. Chapoton, F. Goichot, Dialgebras and Related Operads (2001)
Vol. 1764: A. Cannas da Silva, Lectures on Symplectic Geometry (2001)
Vol. 1765: T. Kerler, V. V. Lyubashenko, Non-Semisimple Topological Quantum Field Theories for 3-Manifolds with Corners (2001)
Vol. 1766: H. Hennion, L. Hervé, Limit Theorems for Markov Chains and Stochastic Properties of Dynamical Systems by Quasi-Compactness (2001)
Vol. 1767: J. Xiao, Holomorphic Q Classes (2001)
Vol. 1768: M. J. Pflaum, Analytic and Geometric Study of Stratified Spaces (2001)
Vol. 1769: M. Alberich-Carramiñana, Geometry of the Plane Cremona Maps (2002)
Vol. 1770: H. Gluesing-Luerssen, Linear Delay-Differential Systems with Commensurate Delays: An Algebraic Approach (2002)
Vol. 1771: M. Émery, M. Yor (Eds.), Séminaire de Probabilités 1967-1980. A Selection in Martingale Theory (2002)
Vol. 1772: F. Burstall, D. Ferus, K. Leschke, F. Pedit, U. Pinkall, Conformal Geometry of Surfaces in S^4 (2002)
Vol. 1773: Z. Arad, M. Muzychuk, Standard Integral Table Algebras Generated by a Non-real Element of Small Degree (2002)
Vol. 1774: V. Runde, Lectures on Amenability (2002)
Vol. 1775: W. H. Meeks, A. Ros, H. Rosenberg, The Global Theory of Minimal Surfaces in Flat Spaces. Martina Franca 1999. Editor: G. P. Pirola (2002)
Vol. 1776: K. Behrend, C. Gomez, V. Tarasov, G. Tian, Quantum Comohology. Cetraro 1997. Editors: P. de Bartolomeis, B. Dubrovin, C. Reina (2002)
Vol. 1777: E. García-Río, D. N. Kupeli, R. Vázquez-Lorenzo, Osserman Manifolds in Semi-Riemannian Geometry (2002)
Vol. 1778: H. Kiechle, Theory of K-Loops (2002)
Vol. 1779: I. Chueshov, Monotone Random Systems (2002)
Vol. 1780: J. H. Bruinier, Borcherds Products on O(2,1) and Chern Classes of Heegner Divisors (2002)
Vol. 1781: E. Bolthausen, E. Perkins, A. van der Vaart, Lectures on Probability Theory and Statistics. Ecole d' Eté de Probabilités de Saint-Flour XXIX-1999. Editor: P. Bernard (2002)
Vol. 1782: C.-H. Chu, A. T.-M. Lau, Harmonic Functions on Groups and Fourier Algebras (2002)
Vol. 1783: L. Grüne, Asymptotic Behavior of Dynamical and Control Systems under Perturbation and Discretization (2002)
Vol. 1784: L. H. Eliasson, S. B. Kuksin, S. Marmi, J.-C. Yoccoz, Dynamical Systems and Small Divisors. Cetraro, Italy 1998. Editors: S. Marmi, J.-C. Yoccoz (2002)
Vol. 1785: J. Arias de Reyna, Pointwise Convergence of Fourier Series (2002)
Vol. 1786: S. D. Cutkosky, Monomialization of Morphisms from 3-Folds to Surfaces (2002)
Vol. 1787: S. Caenepeel, G. Militaru, S. Zhu, Frobenius and Separable Functors for Generalized Module Categories and Nonlinear Equations (2002)
Vol. 1788: A. Vasil'ev, Moduli of Families of Curves for Conformal and Quasiconformal Mappings (2002)
Vol. 1789: Y. Sommerhäuser, Yetter-Drinfel'd Hopf algebras over groups of prime order (2002)
Vol. 1790: X. Zhan, Matrix Inequalities (2002)
Vol. 1791: M. Knebusch, D. Zhang, Manis Valuations and Prüfer Extensions I: A new Chapter in Commutative Algebra (2002)
Vol. 1792: D. D. Ang, R. Gorenflo, V. K. Le, D. D. Trong, Moment Theory and Some Inverse Problems in Potential Theory and Heat Conduction (2002)
Vol. 1793: J. Cortés Monforte, Geometric, Control and Numerical Aspects of Nonholonomic Systems (2002)
Vol. 1794: N. Pytheas Fogg, Substitution in Dynamics, Arithmetics and Combinatorics. Editors: V. Berthé, S. Ferenczi, C. Mauduit, A. Siegel (2002)
Vol. 1795: H. Li, Filtered-Graded Transfer in Using Non-commutative Gröbner Bases (2002)
Vol. 1796: J.M. Melenk, hp-Finite Element Methods for Singular Perturbations (2002)
Vol. 1797: B. Schmidt, Characters and Cyclotomic Fields in Finite Geometry (2002)
Vol. 1798: W.M. Oliva, Geometric Mechanics (2002)
Vol. 1799: H. Pajot, Analytic Capacity, Rectifiability, Menger Curvature and the Cauchy Integral (2002)
Vol. 1800: O. Gabber, L. Ramero, Almost Ring Theory (2003)
Vol. 1801: J. Azéma, M. Émery, M. Ledoux, M. Yor (Eds.), Séminaire de Probabilités XXXVI (2003)
Vol. 1802: V. Capasso, E. Merzbach, B. G. Ivanoff, M. Dozzi, R. Dalang, T. Mountford, Topics in Spatial Stochastic Processes. Martina Franca, Italy 2001. Editor: E. Merzbach (2003)

Vol. 1803: G. Dolzmann, Variational Methods for Crystalline Microstructure – Analysis and Computation (2003)
Vol. 1804: I. Cherednik, Ya. Markov, R. Howe, G. Lusztig, Iwahori-Hecke Algebras and their Representation Theory. Martina Franca, Italy 1999. Editors: V. Baldoni, D. Barbasch (2003)
Vol. 1805: F. Cao, Geometric Curve Evolution and Image Processing (2003)
Vol. 1806: H. Broer, I. Hoveijn. G. Lunther, G. Vegter, Bifurcations in Hamiltonian Systems. Computing Singularities by Gröbner Bases (2003)
Vol. 1807: V. D. Milman, G. Schechtman (Eds.), Geometric Aspects of Functional Analysis. Israel Seminar 2000-2002 (2003)
Vol. 1808: W. Schindler, Measures with Symmetry Properties (2003)
Vol. 1809: O. Steinbach, Stability Estimates for Hybrid Coupled Domain Decomposition Methods (2003)
Vol. 1810: J. Wengenroth, Derived Functors in Functional Analysis (2003)
Vol. 1811: J. Stevens, Deformations of Singularities (2003)
Vol. 1812: L. Ambrosio, K. Deckelnick, G. Dziuk, M. Mimura, V. A. Solonnikov, H. M. Soner, Mathematical Aspects of Evolving Interfaces. Madeira, Funchal, Portugal 2000. Editors: P. Colli, J. F. Rodrigues (2003)
Vol. 1813: L. Ambrosio, L. A. Caffarelli, Y. Brenier, G. Buttazzo, C. Villani, Optimal Transportation and its Applications. Martina Franca, Italy 2001. Editors: L. A. Caffarelli, S. Salsa (2003)
Vol. 1814: P. Bank, F. Baudoin, H. Föllmer, L.C.G. Rogers, M. Soner, N. Touzi, Paris-Princeton Lectures on Mathematical Finance 2002 (2003)
Vol. 1815: A. M. Vershik (Ed.), Asymptotic Combinatorics with Applications to Mathematical Physics. St. Petersburg, Russia 2001 (2003)
Vol. 1816: S. Albeverio, W. Schachermayer, M. Talagrand, Lectures on Probability Theory and Statistics. Ecole d'Eté de Probabilités de Saint-Flour XXX-2000. Editor: P. Bernard (2003)
Vol. 1817: E. Koelink, W. Van Assche (Eds.), Orthogonal Polynomials and Special Functions. Leuven 2002 (2003)
Vol. 1818: M. Bildhauer, Convex Variational Problems with Linear, nearly Linear and/or Anisotropic Growth Conditions (2003)
Vol. 1819: D. Masser, Yu. V. Nesterenko, H. P. Schlickewei, W. M. Schmidt, M. Waldschmidt, Diophantine Approximation. Cetraro, Italy 2000. Editors: F. Amoroso, U. Zannier (2003)
Vol. 1820: F. Hiai, H. Kosaki, Means of Hilbert Space Operators (2003)
Vol. 1821: S. Teufel, Adiabatic Perturbation Theory in Quantum Dynamics (2003)
Vol. 1822: S.-N. Chow, R. Conti, R. Johnson, J. Mallet-Paret, R. Nussbaum, Dynamical Systems. Cetraro, Italy 2000. Editors: J. W. Macki, P. Zecca (2003)
Vol. 1823: A. M. Anile, W. Allegretto, C. Ringhofer, Mathematical Problems in Semiconductor Physics. Cetraro, Italy 1998. Editor: A. M. Anile (2003)
Vol. 1824: J. A. Navarro González, J. B. Sancho de Salas, $\mathscr{C}^\infty$ – Differentiable Spaces (2003)
Vol. 1825: J. H. Bramble, A. Cohen, W. Dahmen, Multiscale Problems and Methods in Numerical Simulations, Martina Franca, Italy 2001. Editor: C. Canuto (2003)
Vol. 1826: K. Dohmen, Improved Bonferroni Inequalities via Abstract Tubes. Inequalities and Identities of Inclusion-Exclusion Type. VIII, 113 p, 2003.
Vol. 1827: K. M. Pilgrim, Combinations of Complex Dynamical Systems. IX, 118 p, 2003.
Vol. 1828: D. J. Green, Gröbner Bases and the Computation of Group Cohomology. XII, 138 p, 2003.
Vol. 1829: E. Altman, B. Gaujal, A. Hordijk, Discrete-Event Control of Stochastic Networks: Multimodularity and Regularity. XIV, 313 p, 2003.
Vol. 1830: M. I. Gil', Operator Functions and Localization of Spectra. XIV, 256 p, 2003.
Vol. 1831: A. Connes, J. Cuntz, E. Guentner, N. Higson, J. E. Kaminker, Noncommutative Geometry, Martina Franca, Italy 2002. Editors: S. Doplicher, L. Longo (2004)
Vol. 1832: J. Azéma, M. Émery, M. Ledoux, M. Yor (Eds.), Séminaire de Probabilités XXXVII (2003)
Vol. 1833: D.-Q. Jiang, M. Qian, M.-P. Qian, Mathematical Theory of Nonequilibrium Steady States. On the Frontier of Probability and Dynamical Systems. IX, 280 p, 2004.
Vol. 1834: Yo. Yomdin, G. Comte, Tame Geometry with Application in Smooth Analysis. VIII, 186 p, 2004.
Vol. 1835: O.T. Izhboldin, B. Kahn, N.A. Karpenko, A. Vishik, Geometric Methods in the Algebraic Theory of Quadratic Forms. Summer School, Lens, 2000. Editor: J.-P. Tignol (2004)
Vol. 1836: C. Năstăsescu, F. Van Oystaeyen, Methods of Graded Rings. XIII, 304 p, 2004.
Vol. 1837: S. Tavaré, O. Zeitouni, Lectures on Probability Theory and Statistics. Ecole d'Eté de Probabilités de Saint-Flour XXXI-2001. Editor: J. Picard (2004)
Vol. 1838: A.J. Ganesh, N.W. O'Connell, D.J. Wischik, Big Queues. XII, 254 p, 2004.
Vol. 1839: R. Gohm, Noncommutative Stationary Processes. VIII, 170 p, 2004.
Vol. 1840: B. Tsirelson, W. Werner, Lectures on Probability Theory and Statistics. Ecole d'Eté de Probabilités de Saint-Flour XXXII-2002. Editor: J. Picard (2004)
Vol. 1841: W. Reichel, Uniqueness Theorems for Variational Problems by the Method of Transformation Groups (2004)
Vol. 1842: T. Johnsen, A. L. Knutsen, K_3 Projective Models in Scrolls (2004)
Vol. 1843: B. Jefferies, Spectral Properties of Noncommuting Operators (2004)
Vol. 1844: K.F. Siburg, The Principle of Least Action in Geometry and Dynamics (2004)
Vol. 1845: Min Ho Lee, Mixed Automorphic Forms, Torus Bundles, and Jacobi Forms (2004)
Vol. 1846: H. Ammari, H. Kang, Reconstruction of Small Inhomogeneities from Boundary Measurements (2004)
Vol. 1847: T.R. Bielecki, T. Björk, M. Jeanblanc, M. Rutkowski, J.A. Scheinkman, W. Xiong, Paris-Princeton Lectures on Mathematical Finance 2003 (2004)
Vol. 1848: M. Abate, J. E. Fornaess, X. Huang, J. P. Rosay, A. Tumanov, Real Methods in Complex and CR Geometry, Martina Franca, Italy 2002. Editors: D. Zaitsev, G. Zampieri (2004)
Vol. 1849: Martin L. Brown, Heegner Modules and Elliptic Curves (2004)
Vol. 1850: V. D. Milman, G. Schechtman (Eds.), Geometric Aspects of Functional Analysis. Israel Seminar 2002-2003 (2004)
Vol. 1851: O. Catoni, Statistical Learning Theory and Stochastic Optimization (2004)
Vol. 1852: A.S. Kechris, B.D. Miller, Topics in Orbit Equivalence (2004)
Vol. 1853: Ch. Favre, M. Jonsson, The Valuative Tree (2004)

Vol. 1854: O. Saeki, Topology of Singular Fibers of Differential Maps (2004)
Vol. 1855: G. Da Prato, P.C. Kunstmann, I. Lasiecka, A. Lunardi, R. Schnaubelt, L. Weis, Functional Analytic Methods for Evolution Equations. Editors: M. Iannelli, R. Nagel, S. Piazzera (2004)
Vol. 1856: K. Back, T.R. Bielecki, C. Hipp, S. Peng, W. Schachermayer, Stochastic Methods in Finance, Bressanone/Brixen, Italy, 2003. Editors: M. Fritelli, W. Runggaldier (2004)
Vol. 1857: M. Émery, M. Ledoux, M. Yor (Eds.), Séminaire de Probabilités XXXVIII (2005)
Vol. 1858: A.S. Cherny, H.-J. Engelbert, Singular Stochastic Differential Equations (2005)
Vol. 1859: E. Letellier, Fourier Transforms of Invariant Functions on Finite Reductive Lie Algebras (2005)
Vol. 1860: A. Borisyuk, G.B. Ermentrout, A. Friedman, D. Terman, Tutorials in Mathematical Biosciences I. Mathematical Neurosciences (2005)
Vol. 1861: G. Benettin, J. Henrard, S. Kuksin, Hamiltonian Dynamics – Theory and Applications, Cetraro, Italy, 1999. Editor: A. Giorgilli (2005)
Vol. 1862: B. Helffer, F. Nier, Hypoelliptic Estimates and Spectral Theory for Fokker-Planck Operators and Witten Laplacians (2005)
Vol. 1863: H. Führ, Abstract Harmonic Analysis of Continuous Wavelet Transforms (2005)
Vol. 1864: K. Efstathiou, Metamorphoses of Hamiltonian Systems with Symmetries (2005)
Vol. 1865: D. Applebaum, B.V. R. Bhat, J. Kustermans, J. M. Lindsay, Quantum Independent Increment Processes I. From Classical Probability to Quantum Stochastic Calculus. Editors: M. Schürmann, U. Franz (2005)
Vol. 1866: O.E. Barndorff-Nielsen, U. Franz, R. Gohm, B. Kümmerer, S. Thorbjønsen, Quantum Independent Increment Processes II. Structure of Quantum Lévy Processes, Classical Probability, and Physics. Editors: M. Schürmann, U. Franz, (2005)
Vol. 1867: J. Sneyd (Ed.), Tutorials in Mathematical Biosciences II. Mathematical Modeling of Calcium Dynamics and Signal Transduction. (2005)
Vol. 1868: J. Jorgenson, S. Lang, $Pos_n(R)$ and Eisenstein Series. (2005)
Vol. 1869: A. Dembo, T. Funaki, Lectures on Probability Theory and Statistics. Ecole d'Eté de Probabilités de Saint-Flour XXXIII-2003. Editor: J. Picard (2005)
Vol. 1870: V.I. Gurariy, W. Lusky, Geometry of Müntz Spaces and Related Questions. (2005)
Vol. 1871: P. Constantin, G. Gallavotti, A.V. Kazhikhov, Y. Meyer, S. Ukai, Mathematical Foundation of Turbulent Viscous Flows, Martina Franca, Italy, 2003. Editors: M. Cannone, T. Miyakawa (2006)
Vol. 1872: A. Friedman (Ed.), Tutorials in Mathematical Biosciences III. Cell Cycle, Proliferation, and Cancer (2006)
Vol. 1873: R. Mansuy, M. Yor, Random Times and Enlargements of Filtrations in a Brownian Setting (2006)
Vol. 1874: M. Yor, M. Émery (Eds.), In Memoriam Paul-André Meyer - Séminaire de Probabilités XXXIX (2006)
Vol. 1875: J. Pitman, Combinatorial Stochastic Processes. Ecole d'Eté de Probabilités de Saint-Flour XXXII-2002. Editor: J. Picard (2006)
Vol. 1876: H. Herrlich, Axiom of Choice (2006)
Vol. 1877: J. Steuding, Value Distributions of *L*-Functions (2007)
Vol. 1878: R. Cerf, The Wulff Crystal in Ising and Percolation Models, Ecole d'Eté de Probabilités de Saint-Flour XXXIV-2004. Editor: Jean Picard (2006)
Vol. 1879: G. Slade, The Lace Expansion and its Applications, Ecole d'Eté de Probabilités de Saint-Flour XXXIV-2004. Editor: Jean Picard (2006)
Vol. 1880: S. Attal, A. Joye, C.-A. Pillet, Open Quantum Systems I, The Hamiltonian Approach (2006)
Vol. 1881: S. Attal, A. Joye, C.-A. Pillet, Open Quantum Systems II, The Markovian Approach (2006)
Vol. 1882: S. Attal, A. Joye, C.-A. Pillet, Open Quantum Systems III, Recent Developments (2006)
Vol. 1883: W. Van Assche, F. Marcellàn (Eds.), Orthogonal Polynomials and Special Functions, Computation and Application (2006)
Vol. 1884: N. Hayashi, E.I. Kaikina, P.I. Naumkin, I.A. Shishmarev, Asymptotics for Dissipative Nonlinear Equations (2006)
Vol. 1885: A. Telcs, The Art of Random Walks (2006)
Vol. 1886: S. Takamura, Splitting Deformations of Degenerations of Complex Curves (2006)
Vol. 1887: K. Habermann, L. Habermann, Introduction to Symplectic Dirac Operators (2006)
Vol. 1888: J. van der Hoeven, Transseries and Real Differential Algebra (2006)
Vol. 1889: G. Osipenko, Dynamical Systems, Graphs, and Algorithms (2006)
Vol. 1890: M. Bunge, J. Funk, Singular Coverings of Toposes (2006)
Vol. 1891: J.B. Friedlander, D.R. Heath-Brown, H. Iwaniec, J. Kaczorowski, Analytic Number Theory, Cetraro, Italy, 2002. Editors: A. Perelli, C. Viola (2006)
Vol. 1892: A. Baddeley, I. Bárány, R. Schneider, W. Weil, Stochastic Geometry, Martina Franca, Italy, 2004. Editor: W. Weil (2007)
Vol. 1893: H. Hanßmann, Local and Semi-Local Bifurcations in Hamiltonian Dynamical Systems, Results and Examples (2007)
Vol. 1894: C.W. Groetsch, Stable Approximate Evaluation of Unbounded Operators (2007)
Vol. 1895: L. Molnár, Selected Preserver Problems on Algebraic Structures of Linear Operators and on Function Spaces (2007)
Vol. 1896: P. Massart, Concentration Inequalities and Model Selection, Ecole d'Été de Probabilités de Saint-Flour XXXIII-2003. Editor: J. Picard (2007)
Vol. 1897: R. Doney, Fluctuation Theory for Lévy Processes, Ecole d'Été de Probabilités de Saint-Flour XXXV-2005. Editor: J. Picard (2007)
Vol. 1898: H.R. Beyer, Beyond Partial Differential Equations, On linear and Quasi-Linear Abstract Hyperbolic Evolution Equations (2007)
Vol. 1899: Séminaire de Probabilités XL. Editors: C. Donati-Martin, M. Émery, A. Rouault, C. Stricker (2007)
Vol. 1900: E. Bolthausen, A. Bovier (Eds.), Spin Glasses (2007)
Vol. 1901: O. Wittenberg, Intersections de deux quadriques et pinceaux de courbes de genre 1, Intersections of Two Quadrics and Pencils of Curves of Genus 1 (2007)
Vol. 1902: A. Isaev, Lectures on the Automorphism Groups of Kobayashi-Hyperbolic Manifolds (2007)
Vol. 1903: G. Kresin, V. Maz'ya, Sharp Real-Part Theorems (2007)
Vol. 1904: P. Giesl, Construction of Global Lyapunov Functions Using Radial Basis Functions (2007)

Vol. 1905: C. Prévôt, M. Röckner, A Concise Course on Stochastic Partial Differential Equations (2007)
Vol. 1906: T. Schuster, The Method of Approximate Inverse: Theory and Applications (2007)
Vol. 1907: M. Rasmussen, Attractivity and Bifurcation for Nonautonomous Dynamical Systems (2007)
Vol. 1908: T.J. Lyons, M. Caruana, T. Lévy, Differential Equations Driven by Rough Paths, Ecole d'Été de Probabilités de Saint-Flour XXXIV-2004 (2007)
Vol. 1909: H. Akiyoshi, M. Sakuma, M. Wada, Y. Yamashita, Punctured Torus Groups and 2-Bridge Knot Groups (I) (2007)
Vol. 1910: V.D. Milman, G. Schechtman (Eds.), Geometric Aspects of Functional Analysis. Israel Seminar 2004-2005 (2007)
Vol. 1911: A. Bressan, D. Serre, M. Williams, K. Zumbrun, Hyperbolic Systems of Balance Laws. Cetraro, Italy 2003. Editor: P. Marcati (2007)
Vol. 1912: V. Berinde, Iterative Approximation of Fixed Points (2007)
Vol. 1913: J.E. Marsden, G. Misiołek, J.-P. Ortega, M. Perlmutter, T.S. Ratiu, Hamiltonian Reduction by Stages (2007)
Vol. 1914: G. Kutyniok, Affine Density in Wavelet Analysis (2007)
Vol. 1915: T. Bıyıkoğlu, J. Leydold, P.F. Stadler, Laplacian Eigenvectors of Graphs. Perron-Frobenius and Faber-Krahn Type Theorems (2007)
Vol. 1916: C. Villani, F. Rezakhanlou, Entropy Methods for the Boltzmann Equation. Editors: F. Golse, S. Olla (2008)
Vol. 1917: I. Veselić, Existence and Regularity Properties of the Integrated Density of States of Random Schrödinger (2008)
Vol. 1918: B. Roberts, R. Schmidt, Local Newforms for GSp(4) (2007)
Vol. 1919: R.A. Carmona, I. Ekeland, A. Kohatsu-Higa, J.-M. Lasry, P.-L. Lions, H. Pham, E. Taflin, Paris-Princeton Lectures on Mathematical Finance 2004. Editors: R.A. Carmona, E. Çinlar, I. Ekeland, E. Jouini, J.A. Scheinkman, N. Touzi (2007)
Vol. 1920: S.N. Evans, Probability and Real Trees. Ecole d'Été de Probabilités de Saint-Flour XXXV-2005 (2008)
Vol. 1921: J.P. Tian, Evolution Algebras and their Applications (2008)
Vol. 1922: A. Friedman (Ed.), Tutorials in Mathematical BioSciences IV. Evolution and Ecology (2008)
Vol. 1923: J.P.N. Bishwal, Parameter Estimation in Stochastic Differential Equations (2008)
Vol. 1924: M. Wilson, Littlewood-Paley Theory and Exponential-Square Integrability (2008)
Vol. 1925: M. du Sautoy, L. Woodward, Zeta Functions of Groups and Rings (2008)
Vol. 1926: L. Barreira, V. Claudia, Stability of Nonautonomous Differential Equations (2008)
Vol. 1927: L. Ambrosio, L. Caffarelli, M.G. Crandall, L.C. Evans, N. Fusco, Calculus of Variations and Non-Linear Partial Differential Equations. Cetraro, Italy 2005. Editors: B. Dacorogna, P. Marcellini (2008)
Vol. 1928: J. Jonsson, Simplicial Complexes of Graphs (2008)
Vol. 1929: Y. Mishura, Stochastic Calculus for Fractional Brownian Motion and Related Processes (2008)
Vol. 1930: J.M. Urbano, The Method of Intrinsic Scaling. A Systematic Approach to Regularity for Degenerate and Singular PDEs (2008)
Vol. 1931: M. Cowling, E. Frenkel, M. Kashiwara, A. Valette, D.A. Vogan, Jr., N.R. Wallach, Representation Theory and Complex Analysis. Venice, Italy 2004. Editors: E.C. Tarabusi, A. D'Agnolo, M. Picardello (2008)
Vol. 1932: A.A. Agrachev, A.S. Morse, E.D. Sontag, H.J. Sussmann, V.I. Utkin, Nonlinear and Optimal Control Theory. Cetraro, Italy 2004. Editors: P. Nistri, G. Stefani (2008)
Vol. 1933: M. Petkovic, Point Estimation of Root Finding Methods (2008)
Vol. 1934: C. Donati-Martin, M. Émery, A. Rouault, C. Stricker (Eds.), Séminaire de Probabilités XLI (2008)
Vol. 1935: A. Unterberger, Alternative Pseudodifferential Analysis (2008)
Vol. 1936: P. Magal, S. Ruan (Eds.), Structured Population Models in Biology and Epidemiology (2008)
Vol. 1937: G. Capriz, P. Giovine, P.M. Mariano (Eds.), Mathematical Models of Granular Matter (2008)
Vol. 1938: D. Auroux, F. Catanese, M. Manetti, P. Seidel, B. Siebert, I. Smith, G. Tian, Symplectic 4-Manifolds and Algebraic Surfaces. Cetraro, Italy 2003. Editors: F. Catanese, G. Tian (2008)
Vol. 1939: D. Boffi, F. Brezzi, L. Demkowicz, R.G. Durán, R.S. Falk, M. Fortin, Mixed Finite Elements, Compatibility Conditions, and Applications. Cetraro, Italy 2006. Editors: D. Boffi, L. Gastaldi (2008)
Vol. 1940: J. Banasiak, V. Capasso, M.A.J. Chaplain, M. Lachowicz, J. Miękisz, Multiscale Problems in the Life Sciences. From Microscopic to Macroscopic. Będlewo, Poland 2006. Editors: V. Capasso, M. Lachowicz (2008)
Vol. 1941: S.M.J. Haran, Arithmetical Investigations. Representation Theory, Orthogonal Polynomials, and Quantum Interpolations (2008)
Vol. 1942: S. Albeverio, F. Flandoli, Y.G. Sinai, SPDE in Hydrodynamic. Recent Progress and Prospects. Cetraro, Italy 2005. Editors: G. Da Prato, M. Röckner (2008)
Vol. 1943: L.L. Bonilla (Ed.), Inverse Problems and Imaging. Martina Franca, Italy 2002 (2008)
Vol. 1944: A. Di Bartolo, G. Falcone, P. Plaumann, K. Strambach, Algebraic Groups and Lie Groups with Few Factors (2008)

Recent Reprints and New Editions

Vol. 1702: J. Ma, J. Yong, Forward-Backward Stochastic Differential Equations and their Applications. 1999 – Corr. 3rd printing (2007)
Vol. 830: J.A. Green, Polynomial Representations of GL_n, with an Appendix on Schensted Correspondence and Littelmann Paths by K. Erdmann, J.A. Green and M. Schoker 1980 – 2nd corr. and augmented edition (2007)
Vol. 1693: S. Simons, From Hahn-Banach to Monotonicity (Minimax and Monotonicity 1998) – 2nd exp. edition (2008)
Vol. 470: R.E. Bowen, Equilibrium States and the Ergodic Theory of Anosov Diffeomorphisms. With a preface by D. Ruelle. Edited by J.-R. Chazottes. 1975 – 2nd rev. edition (2008)
Vol. 523: S.A. Albeverio, R.J. Høegh-Krohn, S. Mazzucchi, Mathematical Theory of Feynman Path Integral. 1976 – 2nd corr. and enlarged edition (2008)